Frontiers in Clinical Drug Research-Alzheimer Disorders

(Volume 8)

Edited by

Atta-ur-Rahman, *FRS*
Honorary Life Fellow, Kings College, University of Cambridge, Cambridge, UK

Frontiers in Clinical Drug Research Alzheimer Disorders

Volume # 8

Editor: Atta-ur-Rahman

ISSN (Online): 2214-5168

ISSN (Print): 2451-8743

ISBN (Online): 978-981-14-0189-3

ISBN (Print): 978-981-14-0190-9

©2019, Bentham eBooks imprint.

Published by Bentham Science Publishers Pte. Ltd. Singapore. All Rights Reserved.

First published in 2019.

need for a court order if at any point you breach any terms of this License Agreement. In no event will any delay or failure by Bentham Science Publishers in enforcing your compliance with this License Agreement constitute a waiver of any of its rights.

3. You acknowledge that you have read this License Agreement, and agree to be bound by its terms and conditions. To the extent that any other terms and conditions presented on any website of Bentham Science Publishers conflict with, or are inconsistent with, the terms and conditions set out in this License Agreement, you acknowledge that the terms and conditions set out in this License Agreement shall prevail.

Bentham Science Publishers Pte Ltd.
80 Robinson Road #02-00
Singapore 068898
Singapore
Email: subscriptions@benthamscience.net

BENTHAM SCIENCE

CONTENTS

PREFACE

The book series *Frontiers in Clinical Drug Research-Alzheimer Disorders* presents important recent developments in the form of cutting edge reviews written by the authorities in the field. The chapters in the 8th volume are focused on different therapeutic agents being employed or under development for treating Alzheimer's disease.

Salama *et al.*, in Chapter 1 discuss the structure, functions and interactions of tau protein with reference to the potential therapeutic targets for the treatment of tauopathies. Piemontese *et al.*, in Chapter 2 present the new generation of drugs as potential therapy for Alzheimer's disease. Chapter 3 by Burgos-Ramos *et al.*, reviews the neuroprotective role of some antibiotics as therapeutic means in Alzheimer's disease.

Chapter 4 by Xu *et al.*, describes the use of antipsychotics in patients with Alzheimer's disease. ÖZDEMİR *et al.*, in chapter 5, discuss the role of cholinergic hypothesis and cholinesterase inhibitors for the treatment of Alzheimer's disease. The last chapter by Stovicek *et al.*, discusses the potential biological mechanisms with prophylactic action in rapid cognitive impairment in late-onset Alzheimer's disease.

I am grateful to all the eminent scientists for their excellent contributions. The efforts of Ms. Fariya Zulfiqar (Manager Publications) & Mr. Shehzad Naqvi (Editorial Manager Publications) and the leadership of Mr. Mahmood Alam (Director Publications) are greatly appreciated.

Prof. Atta-ur-Rahman, *FRS*
Kings College,
University of Cambridge,
Cambridge,
UK

List of Contributors

Azime-Berna ÖZÇELİK Department of Pharmaceutical Chemistry, Gazi University, Faculty of Pharmacy, Ankara, Turkey

Cristina Pintado-Losa Biochemistry Area, Faculty of Environmental Sciences and Biochemistry, University of Castilla-La Mancha, Toledo, Spain

Dragoş Marinescu Member of the Romanian Academy of Medical Sciences, University of Medicine and Pharmacy of Craiova, Doctoral School, Interim President of the Romanian Society for Biological Psychiatry and Psychopharmacology, Romania

Emma Burgos-Ramos Biochemistry Area, Faculty of Environmental Sciences and Biochemistry, University of Castilla-La Mancha, Toledo, Spain

Fulvio Loiodice Dipartimento di Farmacia–Scienze del Farmaco, Università degli Studi di Bari "Aldo Moro", Via E. Orabona 4, 70125 Bari, Italy

Haiyun Xu The Mental Health Center, Shantou University Medical College, Shantou, PR China

Ileana Marinescu Department of Psychiatry, University of Medicine and Pharmacy of Craiova, Craiova, Romania

Jue He First Affiliated Hospital, Institute of Neurological Disease, Henan University, Henan, PR China; Xiamen Xian Yue Hospital, Xiamen, PR China

Laurenţiu Mogoantă Department of Morphopathology, University of Medicine and Pharmacy of Craiova, Craiova, Romania

Luca Piemontese Dipartimento di Farmacia–Scienze del Farmaco, Università degli Studi di Bari "Aldo Moro", Via E. Orabona 4, 70125 Bari, Italy

Maria Amélia Santos Centro de Química Estrutural, Instituto Superior Técnico, Universidade de Lisboa, Av. Rovisco Pais 1, 1049-001 Lisboa, Portugal

María Rodríguez-Pérez Biochemistry Area, Faculty of Environmental Sciences and Biochemistry, University of Castilla-La Mancha, Toledo, Spain

Mehtap UYSAL Department of Pharmaceutical Chemistry, Gazi University, Faculty of Pharmacy, Ankara, Turkey

Mohamed El-Gamal Toxicology Department & Medical Experimental Research Center (MERC), Faculty of Medicine, Mansoura University, Mansoura, Egypt; IUF – Leibniz Research Institute for Environmental Medicine, Düsseldorf, Germany

Mohamed Salama Toxicology Department & Medical Experimental Research Center (MERC), Faculty of Medicine, Mansoura University, Mansoura, Egypt

Oscar Gómez-Torres Biochemistry Area, Faculty of Environmental Sciences and Biochemistry, University of Castilla-La Mancha, Toledo, Spain

Puiu Olivian Stovicek Department of Pharmacology, "Titu Maiorescu" University of Bucharest, Faculty of Nursing, Târgu Jiu, Romania,

Sílvia Chaves Centro de Química Estrutural, Instituto Superior Técnico, Universidade de Lisboa, Av. Rovisco Pais 1, 1049-001 Lisboa, Portugal

Xiaoyin Zhuang The Mental Health Center, Shantou University Medical College, Shantou, PR China

Yasmeen M. Taalab Toxicology Department, Faculty of Medicine, Mansoura University, Mansoura, Mansoura;
German Institute of Disaster Medicine and Emergency Medicine, Tubingen, Germany

Yuan Shao Shenzhen Mental Health Center, Shenzhen Kangning Hospital, Shenzhen Shi, PR China

Zeynep ÖZDEMİR Department of Pharmaceutical Chemistry, İnönü University, Faculty of Pharmacy, Malatya, Turkey

CHAPTER 1

Novel Molecular Targets of Tauopathy; Therapeutic and Diagnostic Applications

Yasmeen M. Taalab[1,2]**, Mohamed El-Gamal**[1,3,4] **and Mohamed Salama**[1,4,5,*]

[1] *Toxicology Department, Faculty of Medicine, Mansoura University, Mansoura, Egypt*

[2] *German Institute of Disaster Medicine and Emergency Medicine, Tubingen, Germany*

[3] *IUF – Leibniz Research Institute for Environmental Medicine, Düsseldorf, Germany*

[4] *Medical Experimental Research Center (MERC), Faculty of Medicine, Mansoura University, Mansoura, Egypt*

[5] *Global Brain Health Institute (GBHI), Trinity college Dublin (TCD), Dublin, Ireland*

Abstract: Neurodegenerative diseases (NDDs) are heterogeneous group of disorders that lately become among the most life-threatening disorders affecting the elderly people. The neurodegenerative disorders that are collectively grouped under the term of tauopathies are featured by the presence of abundant neurofibrillary lesions made by accumulation of abnormal hyperphosphorylated microtubule associated protein tau inside the neurons and/or glial cells. Undoubtedly, tau protein plays a fundamental role in axonal microtubule network stabilization however, the flexible unfolded structure of tau enables modification of tau by several intracellular enzymes which in turn extends tau function and interaction spectrum. The distinctive characteristics of tau protein alongside the essential role of tau interaction partners in the development and progression of neuronal neurodegeneration suggest tau and its binding partners as potential drug targets for the treatment of neurodegenerative diseases. This chapter aims to discuss interaction between mitochondria and tau, and the key molecular players that interfere with tau proteins in physiological and pathological conditions. We outline the putative molecular targets and address the mitochondrial critical role based on research efforts that previously identify their influence on diseases models. Taken together, no solitary player would trigger the whole pathogenic pathway, we attempted to give a detailed description of structure, functions and interactions of tau protein in order to provide insight into potential therapeutic targets for treatment of tauopathies.

Keywords: Mitochondrial Complexes, Molecules, mTOR, PERK, Tauopathy, Targets.

---

[*] **Corresponding author Mohamed Salama:** Medical Experimental Research Center (MERC), Faculty of Medicine, Mansoura University, Mansoura, Egypt & Global Brain Health Institute (GBHI), Trinity college Dublin (TCD), Dublin, Ireland; Tel: +201060556633, Fax:+20(50)239733; E-mail: toxicsalama@hotmail.com

Atta-ur-Rahman (Ed.)

The microtubule associated protein tau (MAPT) plays a fundamental role in the physiological functions of the nerve cell. As a cytoskeletal protein, it is not only involved in maintenance of cellular structure but in signaling pathway as well. MAPT has been extensively studied for regulation of microtubules (MTs) assembly and stabilization under normal physiological conditions, however, in pathological circumstances resulting in neurodegeneration.

Tau as a stigma of neurodegenerative diseases (NDDs), was initially discussed in Alzheimer's disease (AD). However, with more research efforts in this topic, other disorders appear to share tau protein abnormalities, like atypical Parkinson's disease (PD+). Classically, α-Synucleinopathies staining of Lewy bodies present the hallmark inclusion bodies in diagnosis of PD, however, multiple tau antibodies have been additionally reported which denote the involvement of tau in PD pathologies.

Tauopathy is a terminology that has been emerged to describe a pathological condition comprises aggregation of hyperphosphorylated tau protein. Tauopathies include, but are not limited to, AD, argyrophilic grain disease, corticobasal degeneration, dementia pugilistica, Down's syndrome, frontotemporal dementia with parkinsonism related to chromosome 17 (FTDP-17), Parkinson's disease complex of Guam, Pick's disease, postencephalitic parkinsonism and progressive supranuclear palsy (PSP) [1].

In tauopathy, the soluble tau proteins become hyperphosphorylated and detach from microtubules losing their functions and consequently aggregate intracellularly into abnormal toxic filaments so-called neurofibrillary tangles (NFTs) [2]. Alongside the abnormal aggregations of tau, the pathological form of tau proteins, tend to follow a somatodendritic pattern of distribution. On the contrary, under physiological condition tau proteins show axonal localization.

Among all human tauopathies, age is a shared event which triggers tau accumulation even in the presence of obvious genetic component, therefore, it is believed that mTOR principally controls normal physiological development and growth process whilst, during adulthood, where there is relatively little growth, mTOR controls aging and nutrient-related physiology [3]. This dual role was supported by the results obtained from transgenic mice model, triggering tau expression during development or even shortly after birth, however, tau accumulation and the linked phenotype develops only as the age advances [4]. This has led to the assumption that molecular events that accompanied aging may contribute and/or promote tau aggregation [5].

Nevertheless, the trigger initiating transformation of functional tau into pathological one is yet unraveled despite several theories have been suggested to advocate different molecular targets as possible key players in tauopathy. In this context, we will discuss various pathways that might be involved in tauopathy including; mTOR pathway with defective autophagy, protein kinase RNA-activated-like ER kinase (PERK) with its role in unfolded protein response, neuronal autoantibodies formation and autoimmunity, and mitochondrial dysfunction with impairment of the energy production. We believe that better understanding of these molecular targets will offer new therapeutic and diagnostic windows for tauopathies and related NDDs.

IMPLICATION OF MTOR IN NEURODEGENERATION AND TAUOPATHIES

With increasing number of population aging above 60 years together with recorded age-related diseases and the resultant cognitive deficit, understanding the molecular mechanisms underlying the NDDs has become a matter of global concern. Promoting healthy aging process and decreasing the incidence of age-related debilitating diseases that affect even individuals who are otherwise healthy would have a lifelong beneficial impact on the society [6 - 8].

In review of evidences coupling mTOR to tau deposition, mTOR activation pathway is thought to be the key element in tauopathies pathogenesis [9]. Since its discovery in 1994, as the direct target of the rapamycin-FKBP12 complex in mammals, the mechanistic (formerly "mammalian") target of rapamycin (mTOR) has become a hotspot molecular target that attracts dozens of studies to reveal its role in cell physiology [10 - 13].

mTOR is a conserved 289-kDa Serine/Threonine protein kinase belonging to the phosphoinositide 3-kinase (PI3K)-related kinase family [14]. It is an extensive eukaryotic signaling network that gains enormous value due to its involvement in various critical cell processes; protein homeostasis and autophagy in response to various environmental cues. Such pathway generates and/or utilizes huge amount of energy and nutrients in organizing cell growth, proliferation and metabolism. Therefore, disturbed mTOR signaling is implicated in disease progression including type 2 diabetes, cancer and NDDs particularly the pathophysiological changes occurred with the aging process rendering it a plausible target aiming at modulating its activity to cure diseases [3, 13, 14].

Structurally, mTOR is the common catalytic subunit which comprises the two well-identified large protein complexes; mTORC1 and mTORC2 [8, 13, 14]. Both complexes share mammalian lethal with sec-13 protein 8 (mLST8, also known as GbL) [15, 16], DEP domain containing mTOR-interacting protein (DEPTOR) [17], and the Tti1/Tel2 complex [18]. Further, mTORC1 is defined by regulatory-associated protein of mammalian target of rapamycin (raptor) [19, 20], which enables substrate enrollment to mTORC1 *via* binding to the TOR signaling (TOS) modifies which are located on many canonical mTORC1 substrates and are crucial for the proper subcellular localization of mTORC1 [13] and proline-rich Akt substrate 40 kDa (PRAS40) which is specific to mTORC1 [21 - 23]. Instead of raptor, mTORC2 has a rapamycin-insensitive companion of mTOR (rictor) which serves as an analogous function [16]. mTORC2 also involves mammalian stress-activated MAP kinase-interacting protein 1 (mSin1) [24, 25], and protein observed with rictor 1 and 2 (protor1/2) [26]. By activating mTORC1, it activates p70S6 kinase (S6K), and mTORC2 as an essential component of the PI3K pathway/phosphoinositide-dependent kinase 2 phosphorylates Akt eventually to stimulate cell survival and protein homeostasis [14].

mTOR is the essential regulator of protein synthesis, longevity and degradation, and cytoskeletal formation [6, 8, 27]. It interacts with several proteins resulting in identification of two distinct complexes; mTOR complex 1 (mTORC1) that facilitates mRNA translation, controls the balance between protein synthesis and degradation and consequently promotes cell growth and proliferation, lipid biogenesis, controls mitochondrial metabolism and modulates autophagy [28 - 30]. The way by which mTORC1 regulates protein translation is thought to be mainly *via* controlling the activity of ribosomal protein S6 kinase-1 (S6K1) and eukaryotic initiation factor 4E-binding protein 1 (4EBP1), which consequently control the activity of several initiation factors in mRNA translation process [31 - 33]. Oppositely, mTOR complex 2 (mTORC2) endorses cell survival, normal cellular shape and size by modulating actin function primarily in cytoskeleton assembly [14, 24, 34, 35]. These two complexes vary in their sensitivity to rapamycin and have different upstream inputs and downstream outputs [14].

Several factors including growth factors, amino acids, glucose nutrients, insulin and oxidative stress together with various signaling pathways converge on mTOR and may differentially activate/inhibit protein synthesis [31, 36]. Stress induces mTOR complex 1 through phosphoinositide 3-kinase (PI3K)/v-Akt murine thymoma viral oncogene homolog-1 (Akt) and Ras/extracellular signal regulated kinase 1 and 2 (ERK1/2) pathways and is suppressed by deficient energy through 5-adenosine monophosphate-activated protein kinase and glycogen synthase kinase-3β (GSK-3β). cAMP-dependent protein kinase (PKA) induces up-regulation of mTORC1 by stimulating ERK1/2 [37], whereas cyclin-dependent protein kinase 5 (Cdk5) is a downstream substrate of PI3K-mTorC1 [9]. In contrast, mTOR complex 2 appears to be strictly controlled by growth factor [8, 24, 35].

The importance of mTOR arises from its essential implication as a regulator of neural development. Both mTORC1 and mTORC2 signaling is essential for proper brain development in which deletion of either Raptor or Rictor in neurons results in a reduction in the neuronal size, and early neuronal death. However, mTORC1 signaling hyperactivation has been observed in several neurological disorders, including autism, epilepsy, and benign brain tumors [13].

In the synaptic vicinity, mTORC1 controls a fundamental step in neuronal circuit formation where it promotes activity-dependent mRNA translation. The finding of mTORC1 signaling activation by the NMDA receptor antagonist ketamine in mouse neurons with a subsequent increase in synaptic proteins translation strongly support this function [38]. Further, the role of mTORC1 in regulating autophagy is likely a strong contributor in the pathogenesis of NDDs, including AD and PD. Inhibiting mTOR signaling induces beneficial effects on mouse models of AD [39].

Generally, autophagy is the process responsible for getting rid of the misfolded/unfolded aberrant proteins and/or the degraded organelles with their constituents recycled in lysosomes aiming at maintaining proper cell function [29, 33, 40, 41]. Three types of autophagy are well-identified; microautophagy, macroautophagy and chaperone-mediated autophagy. Micro- and macroautophagy take over the degradation process in the regions of the cytosol whilst chaperone-mediated autophagy is known to be a more selective pathway and involves degradation of only proteins with a lysosomal targeting sequence [42 - 44].

mTOR is a negative autophagic regulator by which when inhibited results in autophagy stimulation. This can be achieved by regulating Atg1 in yeast and its Atg1 kinase complex mammalian homolog that is so-called serine/threonine protein kinase ULK1 (UNC-51-like kinase) [45, 46]. Atg1 and its regulatory subunits Atg13 and Atg17 together with the well-conserved serine/threonine kinase mammalian counterparts; ULK1/2, mATG13 and FIP200 are the most upstream components of the core autophagy machinery. It was demonstrated that AMPK regulation of ULK1 is necessary for autophagy [45]. Both mTORC1 and AMPK retain cellular homeostasis through regulation of nutrition and cellular energy signals that control autophagy. The disturbance in ATP/ADP ratio could result in the activation of AMPK, which then suppresses the mTOR activity resulting in autophagy [46]. Indeed, suppression of mTORC1 reduces the phosphorylation of ULK1, which sequentially activates several autophagy-related proteins, ended by the formation of the autophagosome (AV) [47 - 50]. Therefore, the AMP activated protein kinase/ mammalian target of rapamycin/p70ribosomal S6 Kinase (AMPK/mTOR/S6K) is a major signaling pathway that regulates autophagy [51], in which autophagy is stimulated by AMPK, and the mTOR/S6K signaling pathway is the basic inhibitory for autophagy [45, 52 - 54].

In NDDs, autophagy and the ubiquitin-proteasome system are considered the key players [14]. Consequently, mTORC1 signaling as the main regulator of autophagy, has been extensively studied over the last decade. The observation made by deletion of the essential autophagy genes; Atg 5 or 7 in the central nervous system of mice resulting in accumulation of polyubiquitinated proteins and neurodegeneration, despite the absence of any disease-associated mutant proteins, has supported the idea that autophagy is crucial for the survival of neural cells and its impairment may play an important role in the pathogenesis of NDDs [55, 56]. Further, inhibition of mTORC1 with rapamycin diminishes the severity of neurodegeneration in several *in vivo* models and facilitates the autophagic degradation of aggregate-prone proteins *in vitro* [14]. Interestingly, other downstream effectors of mTORC1 signaling are thought to be implicated in the development of neurodegenerative pathologies, the effect which was suggested by rapamycin reducing the aggregation of misfolded proteins by slowing the rate of the protein synthesis [57].

With the fact that mTOR kinase inhibitors are considered more efficient than the first generation of rapalogs in stimulating autophagy and blocking protein synthesis [58, 59], it is quite logical to suppose that these molecules could be even more superior in treating diseases that are associated with the formation and accumulation of protein aggregates. Nevertheless, the prolonged use of mTOR kinase inhibitors could result in impairment of metabolism and damage of the tissue. The development of small molecules that can selectively modulate the activity of proteins controlling autophagy downstream of mTORC1 could represent a possible hope to make this process more specific [14].

Since the work conducted in S. cerevisiae investigating the longevity and stress resistance, the doubling of the chronological lifespan as a result of deletion of the gene encoding the yeast orthologue of S6K1 was the first evidence of mTOR's involvement in aging [60]. Afterward, an extended lifespan was demonstrated in C. elegans following inhibition of raptor and S6K1 [61, 62]. Ultimately, the role of mTOR in aging has been extensively discussed elsewhere and aging has become the main risk factor for the development of NDDs even with well-documented genetic background therefore, understanding the molecular mechanisms regulating aging process might help to delay the age-related pathologies and could eventually extend human healthy lifespan [8].

The relation that links rapamycin, an mTOR inhibitor, to lifespan has been studied in which remarkable results revealed that rapamycin, was administrated to genetically heterogeneous mice, prolonged their lifespan [63]. The involvement of mTOR in controlling lifespan in mammals has also been investigated using two independent genetic approaches. The first one revealed that the deletion of S6K1, a downstream target of mTOR, prolongs lifespan and health span in both female and male mice by 9% [64]. The second study reported that mice having two hypomorphic alleles, which reduce mTOR expression by 25%, compared to wild-type mice had a 20% rise in the median lifespan [65]. Surprisingly, complete suppression of TOR signaling during development results in premature lethality [66 - 68], indicating that TOR signaling is a crucial and evolutionarily conserved regulator of longevity, which can only function within a narrow range for maintaining the homeostasis and health [8].

Additionally, among all human tauopathies, age is a shared event which trigger tau accumulation even in presence of obvious genetic component, therefore, it is believed that mTOR principally controls normal physiological development and growth process whilst, during adulthood, where there is relatively little growth, mTOR controls aging and nutrient-related physiology [3]. This dual role was supported by the results obtained from transgenic mice model which trigger tau expression during development or shortly after birth, however, tau accumulation and the associated phenotype develops only as the age advances [4]. This has led to assume that molecular events that accompanied aging may contribute or facilitate tau aggregation [5].

AD is the most prevalent NDD [69]. When AD occur without any apparent association with known genetic mutation, it is described as sporadic form of the disease in which several different pathophysiological mechanisms are involved [70] and various risk factors have been implicated including aging, neuroinflammation, head trauma, and diabetes [9].

Generally, it is characterized by the presence of the filamentous lesions, which are composed of the 39-43-amino acid, ß-amyloid peptide and hyperphosphorylated tau [71], the later has been strongly linked to disease pathogenesis [72, 73]. Strong lines of evidence obtained from cell-based and transgenic animal models overexpressing tau emphasize the neurotoxicity of aggregated and hyperphosphorylated tau [74 - 78]. Importantly, when tau hyperphosphorylates, its association with microtubules reduces and its propensity for aggregation increases displaying the hallmark in AD brains. Though, the severe form of the cortical neurodegeneration and synapse loss of AD was not observed among the genetically modified mouse lines with cerebral amyloid deposition [79]. Initially, the incriminating role of the phosphorylated and/or aggregated intracellular tau protein in induction of neuronal degeneration and death was postulated. Both forms of abnormally hyperphosphorylated tau, the soluble and insoluble one, exist in AD brains without interaction with tubulin [80, 81]. Moreover, when the soluble form of abnormally hyperphosphorylated tau exists, it sequesters normal tau and microtubule-associated proteins 1 and 2 [82], resulting in accelerating disruption of the microtubule network [9]. However, the presence of non-phosphorylated and non-aggregated tau in the brain extracellular space has been linked to neurotoxicity [69].

In transgenic mouse brains, before formation of NFTs and occurrence of the neuronal loss, the abnormal hyperphosphorylated tau is aggregated [74]. *In vitro* expression of tau pseudophosphorylated at Thr-212, Thr-231, and Ser-262 induces apoptosis [83], which is associated by tau aggregation and breakdown of the microtubule network [83, 84]. On the other hand, *in vivo* expression of wild type tau results in synaptic loss, while deletion of tau exerts a protective effect against ß-amyloid peptide- induced toxicity at the synapse [85]. Conclusively, the formation of NFTs is thought to be the major contributor to AD pathogenesis and correlates with the duration and progression of neuronal degeneration of AD [86]. It is yet so far to wholly understand the upstream intracellular effectors underlying the molecular events involving in the process of tau deposition that leads to the changes of neuronal function and cognitive decline. However, mammalian target of rapamycin (mTOR), has been robustly proposed with its multiple signal pathway [87 - 90].

Regarding AD treatment approaches, several tau-based therapeutic strategies are supposed aiming to stabilize microtubules, block tau aggregation, inhibit protein kinases involved in tau hyperphosphorylation, and immunologically remove extracellular tau. Another promising strategy is enhancement of the degradation of tau multimers through stimulation of macroautophagy; a cellular pathway responsible for getting rid of protein aggregates and dysfunctional organelles. Autophagy can be stimulated by either pharmacologic or molecular genetic suppression of the protein kinase mechanistic target of rapamycin (mTOR) has been shown to reduce tau pathology [5, 91, 92], exerts a neuroprotective effect against neuronal loss and reduce behavioral impairment in tau-overexpressing transgenic mice [93].

Interestingly, a tau-directed therapeutic modality targeting the preservation of both the function and the structure of the perforant pathway, this pathway is projected from layer II of the entorhinal cortex to the hippocampal dentate gyrus and plays an essential role in long-term memory formation, and it is preferentially susceptible to develop a degenerative tauopathy early in AD which may spread later trans-synaptically, has been studied and concluded that the mTOR inhibitor and autophagy stimulator rapamycin can mitigate pathological tau-induced loss of perforant pathway neurons, axons, and synapses, and antagonizes tau-induced neuroinflammation. The data indicate the potential of slowing the progression of the disease pharmacologically mainly at its earliest neuropathological phase. Molecular understanding how rapamycin exerts the neuroprotective effect in the perforant pathway and mechanistic understanding how the neuron-specific signaling regulates the degradation of pathological protein aggregates would be pivotal in the development of novel therapeutic strategies for attenuating or even hindering the tau-mediated pathology and reducing the progressive brain atrophy in AD [94].

Conclusion

The complex role of mTOR is at the crossroad between age-related cognitive decline and pathological tau accumulation in which mTOR regulates tau phosphorylation and degradation *via* autophagy and directly increases overall tau levels by regulating translation of its mRNA. Hence, environmental and genetic factors that increase mTOR may contribute to the development of tau pathology. Beside what was extensively and previously described for the key role of mTOR in aging and AD pathogenesis, it is very attractive avenue that modulating mTOR activity could serve as a novel therapy to enhance healthy brain, as well as attenuating age-related disease progression and treating age-related diseases not only limited to AD but further those involve tau accumulation. Logically, suppression of mTOR molecule would enhance brain healthspan and halt cognitive aging effect.

IMPLICATIONS OF PERK IN NEURODEGENERATION AND TAUOPATHIES

Three tauopathy models were used in a study conducted by Bruch and colleagues using a cultured human neuronal cell line overexpressing the 4-repeat wild-type tau, treated with annonacin, an environmental neurotoxin, and P301S tau transgenic mice. The results indicated that treatment with CCT020312, a pharmacological protein kinase RNA-like endoplasmic reticulum kinase (PERK) activator, reduced tau phosphorylation and tau isoforms, and improve the cell viability *in vitro*. Furthermore, *in vivo*, the PERK activator reduced pathological findings linked to tau aggregates and alleviated the impaired memory and locomotor function. In P301S tau mice, PERK overexpression was protective against dendritic spine and motoneuron loss. EIF2A, the PERK substrate, was observed to be down-regulated in PSP brains and tauopathy models, exploring PERK-NRF2 as another pathway responsible for beneficial effects in the tauopathies. Eventually, they came to the conclusion that PERK stimulation could be a promising novel intervention for tauopathies [95].

PERK is the major signaling pathways of the endoplasmic reticulum (ER) stress and unfolded protein response (UPR) which perfectly reduces the misfolded protein loads in the ER by inhibiting general protein synthesis and consequently reducing their entrance into the ER lumen [96]. The process of PERK activation comprises autophosphorylation and dimerization resulting in the formation of large clusters [97]. Activated PERK, a member of eIF2α kinase family, phosphorylates the α-subunit of eukaryotic translation-initiation factor 2α (eIF2α). In the brain, eIF2α is essential for regulating learning and memory function and for preserving neuronal integrity in health and disease [98]. The phosphorylation of eIF2α is carried out by a family of four protein kinases, PERK, protein kinase double-stranded RNA-dependent (PKR), general control non-derepressible-2 (GCN2), and heme-regulated inhibitor (HRI) [99 - 101]. Among four eIF2α kinases, PERK is considered as a key regulator responsible for neurodegeneration and memory impairments in AD [98].

Upon severe sustained cellular stress which is robustly accused for NDDs, PERK activation is sustained with subsequent abnormal prolonged hyperphosphorylated eIF2α leading to a long-lasting suppression of mRNA translation; such pathway which causes harmful effects on the cognitive function as new protein synthesis is crucial for long-term neuronal plasticity and memory formation [102, 103]. This has led to a considerable discussion on whether or not PERK activation during cellular stress is harmful rather than protective? And hence, PERK inhibition could serve as a promising therapeutic modality for the cognitive and memory deficits in NDDs.

Overactivation of PERK downstream pathway of UPR is observed in several mouse models of NDDs, including AD [96, 104, 105], prion [106], tauopathy [104, 107] and amyotrophic lateral sclerosis (ALS) [108, 109]. In a studied mice model of prion-disease, misfolded prion protein (PrP) aggregate promotes persistently higher levels of phosphorylated PERK and eIF2α, and subsequent sustained translational failure that results in a devastating decrease in levels of key synaptic proteins, resulting in synaptic failure and finally neuronal loss [106]. Both genetic interventions that lower eIF2α-P levels [106] and pharmacological inhibition of PERK by the inhibitors like; GSK2606414 [110], or with the small molecule inhibitor, ISRIB acting a downstream of eIF2α-P, resulting in restoration of the vital protein synthesis rates and antagonizing the neurodegeneration in prion-infected mice [106, 111]. On the other hand, pharmacological treatment preventing the reduction of eIF2α-P levels worse the disease and aggravated toxicity in prion-diseased mice [106]. PERK branch of UPR has found to be dysregulated in a wide range of NDDs hence, it is of a great importance to understand its role in facilitating neuronal loss in disorders associated with protein-misfolding more broadly, mainly in the AD and other tauopathies [112].

Moreover, with relevance to AD and other tauopathies, PERK branch activation has significant contribution to the phosphorylation of tau; which is linked to tauopathies at both *in vitro* and *in vivo* [113 - 115]. *In vitro*, this pathology has been revealed to be attributed to PERK-mediated induction of glycogen synthase kinase-3 (GSK3β), a serine/threonine kinase that induces phosphorylation of tau at disease-relevant epitopes [116], an effect that can be reversed by suppression of PERK [115]. Therefore, UPR activation could have a dual pathological role in tauopathies, influencing on two processes essential to induce neurodegeneration in these disorders.

The effect of PERK pathway activation in the rTg4510 mouse model of FTD induced by the human tau P301L mutation has been examined. Mice carrying the mutant tau transgene (tau+P301L) express high levels of P301L tau. They show age-related tau pathology, progressive memory decline and, significantly, massive forebrain neurodegeneration from 5.5 months of age. Tau+P301L rTg4510 mice demonstrate abnormal tau aggregation from 2.5 months, from which hyperphosphorylated tau was manifested by 4 months of age, while the mature neurofibrillary tangles were apparent with marked neurodegeneration in the hippocampus from 5.5 months. Extensive neuronal loss has been demonstrated throughout the forebrain after the age of 7 months, resulted in massive forebrain atrophy associated with clinical signs of inferior grooming performance and motor impairment. In contrast, the transgene-negative (tau−P301L) rTg4510 mice did not suffer from a behavioral impairment. Their brains show normal morphology without phosphorylation of tau [78, 117, 118]. High levels of PERK-P in the brain of these mice have been detected during advanced disease stages (from 9 months), while the role of PERK-P signaling in mediating neurodegeneration in these animals has not previously been reported [104, 112].

A highly selective PERK inhibitor molecule so-called GSK2606414 that easily cross blood–brain barrier, was administrated to prion-infected mice and resulted in successful management from the pathological and behavioral prospectives [119, 120]. Additionally, the genetic knockdown of PERK in the forebrain of AD model mice was shown to improve AD- associated manifestation regarding synaptic plasticity and memory impairments [121]. Nevertheless, the putative role of the PERK inhibitors as an effective target therapy in different NDDs models should not be taken for granted cause massive suppression of PERK, as this could induce disruption of the normal protein regulation and consequently neuronal response to cellular stress leading to synaptic dysfunction [122]. Thus, it is mandatory to adjust both the dose and duration of treatment with PERK inhibitor for achieving the desired beneficial effect with maintenance of cellular capability of restoring the protein synthesis in response to stress [123].

Taken together, genetic and pharmacological manipulations of PERK-eIF2α signaling pathway have demonstrated that its excessive activation is not a mere consequence of the neurodegenerative process but play pivotal roles in AD pathogenesis and the development of memory deficits and thus the expectation of the therapeutic perspectives targeting the PERK-eIF2α pathway are providing multiple beneficial outcomes in AD, including; neuroprotection, antagonizing the memory impairment, and disease modification [98].

Conclusion

PERK branch of UPR and its downstream signaling pathway offers extremely rich subject for research in the context of NDDS and targeting therapeutic modalities. Simply, this is due to the contradictory and sometimes, unpredicted effects on disease pathophysiology and disease progression. It is not unforeseen that PERK contribution to NDDs is not an easy task owing to the crosstalk with other important stress responses involved in neuroprotection, such as autophagy. Hence, it becomes an extremely convincing to carefully and systematically assess PERK signaling and highlight the complex interaction to distinct NDDs.

IMPLICATIONS OF MITOCHONDRIAL DYSFUNCTION IN NEURODEGENERATION AND TAUOPATHIES

The powerhouses of the cell. This how mitochondria are known in the scientific community. Mitochondria are intracellular organelles, which are responsible for the production of the majority of the energy needed for our bodies. The 1960s was one of the critical times showing great progress in understanding how the mitochondria perform their function. This occurred when Peter Mitchell proposed the chemiosmotic theory for linking the respiration with the energy production in mitochondria. In spite of controversy in his work, he won the Nobel Prize and his theory was fully accepted at the 1970s [124 - 126].

With the advance in science, we could understand more deeply how is energy is obtained from the mitochondria. This can be achieved through oxidative phosphorylation of food by electron transport chain (ETC). Mitochondria are surrounded by double membranes creating two spaces; matrix and intermembrane space. The inner membrane is folded to form cristae for increasing its surface area. One of the unique features of the mitochondria is that they have their own DNA. Mitochondrial DNA (mtDNA) is present in the matrix and ETC complexes are embedded in the inner membrane. The oxidative phosphorylation process started by entering of high energy electron mainly through the mitochondrial complex I (NADH dehydrogenase) and to less extent from the complex II (succinate-ubiquinone oxidoreductase) from NADH or succinate, respectively. These highly energetic electrons are carried through a mobile carrier called ubiquinone to complex III (cytochrome bc1 complex). Cytochrome c transfers these electrons to the complex IV (cytochrome c oxidase). Finally, the electrons are carried to oxygen forming water. This creates transmembrane proton gradients due to the movement of the protons to the intermembrane space of the mitochondria, which acts as a driving force stimulating the mitochondrial complex V (ATP synthase) for ATP production [126 - 129].

Mitochondrial complex I is the largest mitochondrial complex composed of 45 subunits and represents the main site for entrance of the electrons in ECT. 14 subunits of the complex I named the core subunits, which are mostly similar between the different creatures and are responsible for basic function of the mitochondrial complex I. They are categorized into three modules; N module receives the electrons from NADH, Q module which is responsible for transferring these electrons to ubiquinone through the iron-sulfur clusters and P one translocates the protons into the intermembrane space. Seven of the core subunits are encoded by mtDNA, while the other subunits are controlled by nuclear DNA. This means that some of the subunits should be imported first into the mitochondria, and then get assembled by assembling factors [130 - 133]. The oxidative phosphorylation is associated with the production of the reactive oxygen species (ROS) mainly superoxide, however, when function of the mitochondrial complex I is preserved, it is responsible for getting rid of ROS. Dysfunction of the mitochondrial complex I can result from the mutation of nuclear DNA and/ or mtDNA encoding for its subunits, defect in assembling factors and/ or exposure to environmental exposure to mitochondrial complex I toxins. The dysfunction mitochondrial complex I results in impairment of the energy production and oxidative stress [131, 134 - 136].

The mitochondria are not only the powerhouses of the cells, but also the central regulator of the programmed cell death [137, 138]. Mitochondrial complex I dysfunction [139] and exposure of the environmental contaminants such as rotenone [140, 141] and OPs pesticides; chlorpyrifos [142 - 144], malathion [145], and dichlrovos [146, 147] resulted in mitochondrial apoptosis through translocation of Bax to mitochondria, efflux of the cytochrome *c* from the mitochondria to the cytoplasm and activation of the caspases.

Furthermore, mitochondria have essential metabolic functions and their number and size of mitochondria are variable from cell to another cell depending on the metabolic activities. Mitochondria play a pivotal role in calcium homeostasis and in the metabolism of carbohydrate, fat, and proteins. The Mitochondria are capable of performing specific functions depending on the organ. In the liver, mitochondria are responsible for the detoxification of ammonia [148 - 152].

The mitochondrial dysfunction was detected earlier in the myopathies than the NDDs. The mitochondrial muscle disease was reported in 1959 with a defect in the oxidative phosphorylation process [153, 154]. This was associated with a defect in the functions of the mitochondrial complex I and III, and mitochondrial respiration in the muscle biopsy of patient diagnosed with myopathies [155 - 157]. Deletion of the mitochondrial DNA was detected among 36% of muscles of the mitochondrial myopathies' patients [158]. Patients suffered from Kearns-Sayre syndrome have deletion of the mitochondrial DNA ranging from 45% to 75% with partial reduction of the mitochondrial complex IV activity [159].

A young man developed Parkinsonism like manifestation after injecting himself with a drug of abuse, meperidine like structure and also called synthetic heroin or MPTP, after his trial of synthesis. Years later, another four drugs abuser developed Parkinsonism. Subsequently, MPTP was used for modeling PD in monkey and mice. Due to the fact that MPTP is one of the potent mitochondrial complex I, this opened the door for investigating the role of mitochondrial dysfunction in PD [160].

At the end of the 1980s and beginning of the 1990s, a deficit of the mitochondrial complex I was firstly detected in the substantia nigra (SN) (39-42%) [161, 162] and in platelets (16-55%) of PD patients [163, 164]. Nearly at the same period, Parker *et al.*, 1990 reported that the activity of mitochondrial complex IV was reduced to 50.09% of the healthy control individuals in the platelet of AD patients [165]. In the frontal and temporal cortex of AD patients, the activity of mitochondrial complex IV was 74% and 83% values of the healthy individuals, in that order [166]. The activity of the mitochondrial ECT was suppressed among AD patients' brain, as the activities of the mitochondrial complex I, II, III and IV were 60.58%, 64.45%, 57.48% and 46.64% of the healthy persons, respectively [167]. Furthermore, the mitochondrial dysfunction was also found in another NDS. The activity of mitochondrial complex I in platelets of Huntington patients is 28,03% of the value of the healthy individuals [168], while in the caudate nucleus of Huntington patients, the activities of mitochondrial complexes I, II, III and IV were 85.71%, 47.44%, 40.91% and 61,83%, respectively [169].

Combinations of affection of the skeletal and nervous systems were detected among four members of an English family in the form retinitis pigmentosa, developmental impairment, dementia, ataxia, seizures, sensory neuropathy and neurogenic muscle weakness. This might be attributed to the point mutation of nucleotide 8993 resulting in changing of leucine to arginine in the subunit 6 of the mitochondrial complex V. The clinical severity was partially related to the degree of the mutation [170]. This came in agreement with the finding among the members of an Italian family suffering from of similar clinical presentations in form of retinitis pigmentosa, psychomotor retardation, memory impairment, muscle weakness and wasting, hypotonia, ataxia, and dysarthria. The affected members suffered from the same mutation and the severity of the clinical manifestations was related to the amount of this point mutation [171].

Mitochondria dysfunction is observed among the idiopathic and familial AD due to mutation of APP, PS1 and PS2, and both idiopathic and genetic PD due to DJ1, PARK2 and PINK1 mutations. Additionally, these mutations increase the susceptibility of the neurons when they exposure to environmental contaminates [172 - 177].

Growing scientific evidence supports the incriminating role of the pathological form of tau in aggravating the mitochondrial dysfunction in NDDs. Stereotaxic injection of tau oligomers in the hippocampus in C57BL/6 mice resulted in synaptic dysfunction and memory impairment. Tau oligomers induced inhibition of the mitochondrial complex I. Mitochondrial apoptotic pathway was stimulated through activation of caspase 9 [178]. *In vitro* application of prefibrillar amyloid aggregates, alpha-synuclein, and tau 441 induced permeabilization of the mitochondrial membranes, mainly through the affection of the cardiolipin, which facilitated the efflux of cytochrome *c* [179]. Pathological tau induced somatodendritic distribution of the mitochondria. These findings were observed in the temporal cortex of AD patients and primary somatosensory of rTg4510 transgenic mice, overexpressing the mutant tau P301L. This can be explained by impairment of the mitochondrial trafficking [180].

Another animal model of FTDP-17T was established by knocking in mutant P301 tau show an age-dependent alteration of axonal trafficking of mitochondria. In the young mice, the antegrade mitochondrial movement increased in the young mice by 22.5%, however, in the older ones, the antegrade movement reduced by 27.5%. No significant effects were observed regarding the retrograde movement, average nor the maximal velocity of the mitochondrial movement [181]. Expression of P301L tau in mouse cortical neurons resulted in a reduction of the phosphorylation of the tau and impairment of binding of the tau to the microtubules without affecting its membrane localization. Reductions of the number of the mitochondria in axons, increasing in the size of the moving mitochondria without affection of the velocity or the direction of mitochondrial movement were apparent tau effects [182].

In vitro, NH2-tau fragments induced synaptic dysfunction and impaired of the mitochondrial dynamic by which the mitochondria are redistributed mainly to the cell body. NH2-tau fragments led to a reduction of the number and size of the mitochondria, disruption of mitochondrial morphology, enhancement of selective autophagic clearance of the mitochondria, inhibition of the mitochondrial complex IV and stimulation of the oxidative stress [183]. Phosphorylated tau was detected among the cerebral cortex among the AD patients and three animal model of AD; single (mutant APP), double (mutant APP and PS1), and triple (mutant APP, PS1 and tau P301L) transgenic mice. The interaction between the phosphorylated tau and a mitochondrial fission protein called Drp1 was obvious among the AD patients and all animal AD models. The activity of one the enzyme responsible for the mitochondrial fragment (GTPase) was elevated among the AD patients and the AD animal models. A positive correlation existed between the phosphorylated tau and the GTPase activity [184].

The pathological forms of the tau induce mitochondrial dysfunction in different ways, as previously described, but can the mitochondrial dysfunction trigger the formation of abnormal tau? The deficit in mitochondrial complex I due to impairment of the mitochondrial assembly factors like; NDUFA 12L is associated with taupathies [131, 185].

Rotenone is a potent mitochondrial complex I inhibitor. It one of the most widely used animal models of PD. This is due to its capability to mimic the chronic and progressive nature of the PD with the formation of alpha-synuclein in dopaminergic neurons. Additionally, the epidemiological studies show that exposure to rotenone increase the risk of PD [186 - 190]. Rotenone not only induced degeneration of the dopaminergic neurons in SN and fibers in corpus striatum (CS) and locomotor impairment of Lewis rats but also it provoked hyperphosphorylated tau in neuronal fibers, astrocytes, and oligodendrocytes in CS [191]. Rotenone promoted destabilization of the microtubules and phosphorylation of tau, inhibited tau binding to the microtubules through activation of the glycogen synthase 3β. Taxol was used for preventing displacement of the tau from the microtubule resulting in significant attenuation of the rotenone-induced cell death in SH-SY5Y cells [192].

High prevalence of PD+ which shares some feature of the progressive supranuclear palsy was observed in Guadeloupe in the French West Indies. Tau accumulations are obvious in the neuronal cell body in the brain of the patients. There was a close association between consuming annonaceous plants containing annonacin, one of the potent mitochondrial complex I inhibitor, and this form of the neurological impairment [193, 194]. Annonacin induced dose-dependent cell death, reduction of ATP level and somatodendritic redistribution of the tau and retrograde transportation of the mitochondria in primary culture of fetal striatum Wister rat. The tau was attached to the outer mitochondrial membrane and taxol hindered the redistribution of both mitochondria and tau to the cell body. Other mitochondrial complex I inhibitors; used for modeling PD (MPP+) and Huntington disease (3-nitropropionic acid) induced dose-dependent cell death and redistribution of tau to the cell bodies [195]. Administration of the annonacin to the transgenic mice with R406W-tau mutation led to an increase in tau level by inhibition of tau degradation, and stimulation of both phosphorylation of tau and redistribution of tau to the cell body in frontal and parietal cortex and hippocampus [196].

Fenazaquin (64 nM) is another mitochondrial complex I induced somatodendritic redistribution of the tau in among 37% of Lund human mesencephalic cells (LUHMES) [197]. Twenty-four mitochondrial environmental complex I inhibitors induced neuronal cell death in the primary culture of fetal striatum Wister rat and induced the redistribution of tau to the cell bodies. A positive correlation existed between the degree of the inhibition of the mitochondrial complex I and tau redistribution [198].

Another possibility is that both pathological tau accumulation and mitochondrial dysfunction synergically interact together facilitating the neurodegeneration. Both mitochondrial affection and tau aggregation was observed among AD and animal model with APP and PS1 mutations [184]. Among the familial PD patients, PINK1 and LRRK2 mutations induce both pathological forms. Mitochondrial dysfunction was observed in the SH-SY5Y cells with PINK1 (R492X) mutation in the form of inhibition of the mitochondrial complex I, impairment of the mitochondrial membrane potential and activation of the mitochondrial apoptosis pathway through the release of the cytochrome c to the cytoplasm. This mitochondrial impairment was aggravated by MPP+ exposure [199]. Furthermore, PINK1 (G309D) mutation induced the phosphorylation of tau in N2a neuroblastoma cells [200]. Additionally, skin biopsies from PD patients having LRRK2 (G2019S) mutation show disruption of the mitochondrial morphology, inhibition of mitochondrial complex I, II and IV, and impairment of the mitochondrial membrane potential with a significant reduction of ATP level [201]. Hyperphosphorylated tau was observed in the brainstem of PD patients having LRRK2 (I2020T) mutation. This form of the mutation induced also *in vitro* hyperphosphorylation of tau [202].

Conclusion

The epidemiological and experimental studies emphasize the incriminating role of the mitochondrial dysfunction in NDDs. Strong scientific evidence supports the hazardous consequences of the pathological form of tau in worsening the mitochondrial dysfunction, however, others reveal the role of mitochondrial dysfunction in aggravation of tau pathology. This still unresolved issue whether the pathological tau or the mitochondrial dysfunction are the initiating factors or this is a hazardous synergistic interaction between both resulting in a vicious circle ended by neurodegeneration.

ABBREVIATION

4EBP1:	Eukaryotic Initiation Factor 4E-Binding Protein 1
AD:	Alzheimer Disease
ALS	Amyotrophic Lateral Sclerosis
APP	Amyloid Precursor Protein
AMPK/mTOR/S6K	AMP Activated Protein Kinase/ Mammalian Target of Rapamycin/p70ribosomal S6 Kinase
ATP	Adenosine Triphosphate
Cdk5:	Cyclin-Dependent Protein Kinase 5
CS:	Corpus Striatum
DEPTOR:	DEP Domain Containing mTOR-Interacting Protein
ECT:	Electron Transport Chain
eIF2α	α-Subunit of Eukaryotic Translation-Initiation Factor 2α
ER:	Endoplasmic Reticulum
ERK1/2:	Extracellular Signal Regulated Kinase 1 And 2
GCN2:	General Control Non-Derepressible-2
GSK-3β:	Glycogen Synthase Kinase-3β
HRI:	Heme-Regulated Inhibitor
LRRK2:	Leucine-Rich Repeat Kinase 2
mtDNA:	mitochondrial DNA
PD:	Frontotemporal Dementia With Parkinsonism Related To Chromosome 17 (FTDP-17)
MAPT:	Microtubule Associated Protein Tau
mLST8:	Mammalian Lethal With Sec-13 Protein 8
MPP+:	1-methyl-4-phenylpyridinium
MPTP:	1-methyl-4-phenyl-1,2,3,6-tetrahydropyridine
mTOR:	Mechanistic Target of Rapamycin
mTORC1:	mTOR Complex 1
mTORC2:	mTOR Complex 2
mSin1:	Mammalian Stress-Activated MAP Kinase-Interacting Protein 1
NDDs:	Neurodegenerative Diseases
NFTs:	Neurofibrillary Tangles
PD:	Parkinson's Disease
PD+:	Atypical Parkinson's Disease
PERK:	Protein Kinase RNA-Activated-Like ER Kinase

PI3K:	phosphoinositide 3-Kinase
PINK:	PTEN-Induced Putative Kinase 1
PKA:	cAMP-Dependent Protein Kinase
PKR:	Protein Kinase Double-Stranded RNA-Dependent
PRAS40:	Proline-Rich Akt Substrate 40 kDa
protor1/2:	Protein Observed With Rictor 1 And 2
PrP:	Misfolded Prion Protein
PS1:	Presenilin1
PS2:	Presenilin2
PSP:	Progressive Supranuclear Palsy
raptor:	Regulatory-Associated Protein Of Mammalian Target Of Rapamycin
rictor:	Rapamycin-Insensitive Companion Of mTOR
ROS:	Reactive Oxygen Species
S6K:	p70S6 Kinase
S6K1:	Ribosomal Protein S6 Kinase-1
SN:	Substantia Nigra
TOS:	TOR Signaling
UNC-51-like kinase:	Serine/Threonine Protein Kinase ULK1
UPR:	Unfolded Protein Response

CONSENT FOR PUBLICATION

Not applicable.

CONFLICT OF INTEREST

The authors declare no conflict of interest, financial or otherwise.

ACKNOWLEDGEMENTS

Declared none.

REFERENCES

[1]   Williams DR. Tauopathies: classification and clinical update on neurodegenerative diseases associated with microtubule-associated protein tau. Intern Med J 2006; 36(10): 652-60.
[http://dx.doi.org/10.1111/j.1445-5994.2006.01153.x] [PMID: 16958643]

[2]   Ballatore C, Lee VM, Trojanowski JQ. Tau-mediated neurodegeneration in Alzheimer's disease and related disorders. Nat Rev Neurosci 2007; 8(9): 663-72.
[http://dx.doi.org/10.1038/nrn2194] [PMID: 17684513]

[3]   Wullschleger S, Loewith R, Hall MN. TOR signaling in growth and metabolism. Cell 2006; 124(3):

471-84.
[http://dx.doi.org/10.1016/j.cell.2006.01.016] [PMID: 16469695]

[4]   Bertram L, Tanzi R. The genetics of Alzheimer's disease Prog Mol Biol Transl Sci 107: 79-100.2012;
        [http://dx.doi.org/10.1016/B978-0-12-385883-2.00008-4]

[5]   Caccamo A, Magrì A, Medina DX, *et al.* mTOR regulates tau phosphorylation and degradation:
        implications for Alzheimer's disease and other tauopathies. Aging Cell 2013; 12(3): 370-80.
        [http://dx.doi.org/10.1111/acel.12057] [PMID: 23425014]

[6]   Vincent GK, Velkoff VA. The next four decades: The older population in the United States: 2010 to
        2050. US Department of Commerce, Economics and Statistics Administration, US Census Bureau
        2010.

[7]   Harper S. Economic and social implications of aging societies. Science 2014; 346(6209): 587-91.
        [http://dx.doi.org/10.1126/science.1254405] [PMID: 25359967]

[8]   Talboom JS, Velazquez R, Oddo S. The mammalian target of rapamycin at the crossroad between
        cognitive aging and Alzheimer's disease 2015; 1 npj Aging and Mechanisms of Disease

[9]   Tang Z, Bereczki E, Zhang H, *et al.* Mammalian target of rapamycin (mTor) mediates tau protein
        dyshomeostasis: implication for Alzheimer disease. J Biol Chem 2013; 288(22): 15556-70.
        [http://dx.doi.org/10.1074/jbc.M112.435123] [PMID: 23585566]

[10]   Brown EJ, Albers MW, Shin TB, *et al.* A mammalian protein targeted by G1-arresting rapamycin-
        receptor complex. Nature 1994; 369(6483): 756-8.
        [http://dx.doi.org/10.1038/369756a0] [PMID: 8008069]

[11]   Sabatini DM, Erdjument-Bromage H, Lui M, Tempst P, Snyder SH. RAFT1: a mammalian protein
        that binds to FKBP12 in a rapamycin-dependent fashion and is homologous to yeast TORs. Cell 1994;
        78(1): 35-43.
        [http://dx.doi.org/10.1016/0092-8674(94)90570-3] [PMID: 7518356]

[12]   Sabers CJ, Martin MM, Brunn GJ, *et al.* Isolation of a protein target of the FKBP12-rapamycin
        complex in mammalian cells. J Biol Chem 1995; 270(2): 815-22.
        [http://dx.doi.org/10.1074/jbc.270.2.815] [PMID: 7822316]

[13]   Saxton RA, Sabatini DM. mTOR signaling in growth, metabolism, and disease. Cell 2017; 168(6):
        960-76.
        [http://dx.doi.org/10.1016/j.cell.2017.02.004] [PMID: 28283069]

[14]   Laplante M, Sabatini DM. mTOR signaling in growth control and disease. Cell 2012; 149(2): 274-93.
        [http://dx.doi.org/10.1016/j.cell.2012.03.017] [PMID: 22500797]

[15]   Kim D-H, Sarbassov DD, Ali SM, *et al.* GbetaL, a positive regulator of the rapamycin-sensitive
        pathway required for the nutrient-sensitive interaction between raptor and mTOR. Mol Cell 2003;
        11(4): 895-904.
        [http://dx.doi.org/10.1016/S1097-2765(03)00114-X] [PMID: 12718876]

[16]   Jacinto E, Loewith R, Schmidt A, *et al.* Mammalian TOR complex 2 controls the actin cytoskeleton
        and is rapamycin insensitive. Nat Cell Biol 2004; 6(11): 1122-8.
        [http://dx.doi.org/10.1038/ncb1183] [PMID: 15467718]

[17]   Peterson TR, Laplante M, Thoreen CC, *et al.* DEPTOR is an mTOR inhibitor frequently
        overexpressed in multiple myeloma cells and required for their survival. Cell 2009; 137(5): 873-86.
        [http://dx.doi.org/10.1016/j.cell.2009.03.046] [PMID: 19446321]

[18]   Kaizuka T, Hara T, Oshiro N, *et al.* Tti1 and Tel2 are critical factors in mammalian target of
        rapamycin complex assembly. J Biol Chem 2010; 285(26): 20109-16.
        [http://dx.doi.org/10.1074/jbc.M110.121699] [PMID: 20427287]

[19]   Hara K, Maruki Y, Long X, *et al.* Raptor, a binding partner of target of rapamycin (TOR), mediates
        TOR action. Cell 2002; 110(2): 177-89.

[http://dx.doi.org/10.1016/S0092-8674(02)00833-4] [PMID: 12150926]

[20]   Kim D-H, Sarbassov DD, Ali SM, *et al.* mTOR interacts with raptor to form a nutrient-sensitive complex that signals to the cell growth machinery. Cell 2002; 110(2): 163-75.
[http://dx.doi.org/10.1016/S0092-8674(02)00808-5] [PMID: 12150925]

[21]   Sancak Y, Thoreen CC, Peterson TR, *et al.* PRAS40 is an insulin-regulated inhibitor of the mTORC1 protein kinase. Mol Cell 2007; 25(6): 903-15.
[http://dx.doi.org/10.1016/j.molcel.2007.03.003] [PMID: 17386266]

[22]   Vander Haar E, Lee SI, Bandhakavi S, Griffin TJ, Kim DH. Insulin signalling to mTOR mediated by the Akt/PKB substrate PRAS40. Nat Cell Biol 2007; 9(3): 316-23.
[http://dx.doi.org/10.1038/ncb1547] [PMID: 17277771]

[23]   Wang L, Harris TE, Roth RA, Lawrence JC Jr. PRAS40 regulates mTORC1 kinase activity by functioning as a direct inhibitor of substrate binding. J Biol Chem 2007; 282(27): 20036-44.
[http://dx.doi.org/10.1074/jbc.M702376200] [PMID: 17510057]

[24]   Frias MA, Thoreen CC, Jaffe JD, *et al.* mSin1 is necessary for Akt/PKB phosphorylation, and its isoforms define three distinct mTORC2s. Curr Biol 2006; 16(18): 1865-70.
[http://dx.doi.org/10.1016/j.cub.2006.08.001] [PMID: 16919458]

[25]   Jacinto E, Facchinetti V, Liu D, *et al.* SIN1/MIP1 maintains rictor-mTOR complex integrity and regulates Akt phosphorylation and substrate specificity. Cell 2006; 127(1): 125-37.
[http://dx.doi.org/10.1016/j.cell.2006.08.033] [PMID: 16962653]

[26]   Pearce LR, Huang X, Boudeau J, *et al.* Identification of Protor as a novel Rictor-binding component of mTOR complex-2. Biochem J 2007; 405(3): 513-22.
[http://dx.doi.org/10.1042/BJ20070540] [PMID: 17461779]

[27]   Johnson SC, Rabinovitch PS, Kaeberlein M. mTOR is a key modulator of ageing and age-related disease. Nature 2013; 493(7432): 338-45.
[http://dx.doi.org/10.1038/nature11861] [PMID: 23325216]

[28]   Wang X, Proud CG. The mTOR pathway in the control of protein synthesis. Physiology (Bethesda) 2006; 21(5): 362-9.
[http://dx.doi.org/10.1152/physiol.00024.2006] [PMID: 16990457]

[29]   Kim YC, Guan K-L. mTOR: a pharmacologic target for autophagy regulation. J Clin Invest 2015; 125(1): 25-32.
[http://dx.doi.org/10.1172/JCI73939] [PMID: 25654547]

[30]   Perluigi M, Di Domenico F, Butterfield DA. mTOR signaling in aging and neurodegeneration: At the crossroad between metabolism dysfunction and impairment of autophagy. Neurobiol Dis 2015; 84: 39-49.
[http://dx.doi.org/10.1016/j.nbd.2015.03.014] [PMID: 25796566]

[31]   Hay N, Sonenberg N. Upstream and downstream of mTOR. Genes Dev 2004; 18(16): 1926-45.
[http://dx.doi.org/10.1101/gad.1212704] [PMID: 15314020]

[32]   Ruvinsky I, Meyuhas O. Ribosomal protein S6 phosphorylation: from protein synthesis to cell size. Trends Biochem Sci 2006; 31(6): 342-8.
[http://dx.doi.org/10.1016/j.tibs.2006.04.003] [PMID: 16679021]

[33]   Hands SL, Proud CG, Wyttenbach A. mTOR's role in ageing: protein synthesis or autophagy? Aging (Albany NY) 2009; 1(7): 586-97.
[http://dx.doi.org/10.18632/aging.100070] [PMID: 20157541]

[34]   Bové J, Martínez-Vicente M, Vila M. Fighting neurodegeneration with rapamycin: mechanistic insights. Nat Rev Neurosci 2011; 12(8): 437-52.
[http://dx.doi.org/10.1038/nrn3068] [PMID: 21772323]

[35]   Graber TE, McCamphill PK, Sossin WS. A recollection of mTOR signaling in learning and memory.

Learn Mem 2013; 20(10): 518-30.
[http://dx.doi.org/10.1101/lm.027664.112] [PMID: 24042848]

[36]   Gingras A-C, Gygi SP, Raught B, *et al.* Regulation of 4E-BP1 phosphorylation: a novel two-step mechanism. Genes Dev 1999; 13(11): 1422-37.
[http://dx.doi.org/10.1101/gad.13.11.1422] [PMID: 10364159]

[37]   Garelick MG, Kennedy BK. TOR on the brain. Exp Gerontol 2011; 46(2-3): 155-63.
[http://dx.doi.org/10.1016/j.exger.2010.08.030] [PMID: 20849946]

[38]   Li N, Lee B, Liu RJ, *et al.* mTOR-dependent synapse formation underlies the rapid antidepressant effects of NMDA antagonists. Science 2010; 329(5994): 959-64.
[http://dx.doi.org/10.1126/science.1190287] [PMID: 20724638]

[39]   Spilman P, Podlutskaya N, Hart MJ, *et al.* Inhibition of mTOR by rapamycin abolishes cognitive deficits and reduces amyloid-β levels in a mouse model of Alzheimer's disease. PLoS One 2010; 5(4): e9979.
[http://dx.doi.org/10.1371/journal.pone.0009979] [PMID: 20376313]

[40]   Zhang C, Cuervo AM. Restoration of chaperone-mediated autophagy in aging liver improves cellular maintenance and hepatic function. Nat Med 2008; 14(9): 959-65.
[http://dx.doi.org/10.1038/nm.1851] [PMID: 18690243]

[41]   Klionsky DJ. The autophagy connection. Dev Cell 2010; 19(1): 11-2.
[http://dx.doi.org/10.1016/j.devcel.2010.07.005] [PMID: 20643346]

[42]   Mizushima N, Tsukamoto S, Kuma A. Autophagy in embryogenesis and cell differentiation. Tanpakushitsu kakusan koso Protein, nucleic acid, enzyme 2008; 53 (16 Suppl): 2170-4.

[43]   Nixon RA, Yang D-S. Autophagy failure in Alzheimer's disease--locating the primary defect. Neurobiol Dis 2011; 43(1): 38-45.
[http://dx.doi.org/10.1016/j.nbd.2011.01.021] [PMID: 21296668]

[44]   Nah J, Yuan J, Jung Y-K. Autophagy in neurodegenerative diseases: from mechanism to therapeutic approach. Mol Cells 2015; 38(5): 381-9.
[http://dx.doi.org/10.14348/molcells.2015.0034] [PMID: 25896254]

[45]   Egan D, Kim J, Shaw RJ, Guan KL. The autophagy initiating kinase ULK1 is regulated *via* opposing phosphorylation by AMPK and mTOR. Autophagy 2011; 7(6): 643-4.
[http://dx.doi.org/10.4161/auto.7.6.15123] [PMID: 21460621]

[46]   Shi WY, Xiao D, Wang L, *et al.* Therapeutic metformin/AMPK activation blocked lymphoma cell growth *via* inhibition of mTOR pathway and induction of autophagy. Cell Death Dis 2012; 3(3): e275.
[http://dx.doi.org/10.1038/cddis.2012.13] [PMID: 22378068]

[47]   He C, Klionsky DJ. Regulation mechanisms and signaling pathways of autophagy. Annu Rev Genet 2009; 43: 67-93.
[http://dx.doi.org/10.1146/annurev-genet-102808-114910] [PMID: 19653858]

[48]   Jung CH, Ro SH, Cao J, Otto NM, Kim DH. mTOR regulation of autophagy. FEBS Lett 2010; 584(7): 1287-95.
[http://dx.doi.org/10.1016/j.febslet.2010.01.017] [PMID: 20083114]

[49]   Rubinsztein DC, Mariño G, Kroemer G. Autophagy and aging. Cell 2011; 146(5): 682-95.
[http://dx.doi.org/10.1016/j.cell.2011.07.030] [PMID: 21884931]

[50]   Noda NN, Inagaki F. Mechanisms of Autophagy. Annu Rev Biophys 2015; 44: 101-22.
[http://dx.doi.org/10.1146/annurev-biophys-060414-034248] [PMID: 25747593]

[51]   Zhang L, Fang Y, Cheng X, *et al.* TRPML1 participates in the progression of alzheimer's disease by regulating the PPARγ/AMPK/mtor signalling pathway. Cell Physiol Biochem 2017; 43(6): 2446-56.
[http://dx.doi.org/10.1159/000484449] [PMID: 29131026]

[52]   Shinojima N, Yokoyama T, Kondo Y, Kondo S. Roles of the Akt/mTOR/p70S6K and ERK1/2

signaling pathways in curcumin-induced autophagy. Autophagy 2007; 3(6): 635-7.
[http://dx.doi.org/10.4161/auto.4916] [PMID: 17786026]

[53]   Kim J, Kundu M, Viollet B, Guan KL. AMPK and mTOR regulate autophagy through direct phosphorylation of Ulk1. Nat Cell Biol 2011; 13(2): 132-41.
[http://dx.doi.org/10.1038/ncb2152] [PMID: 21258367]

[54]   Alers S, Löffler AS, Wesselborg S, Stork B. Role of AMPK-mTOR-Ulk1/2 in the regulation of autophagy: cross talk, shortcuts, and feedbacks. Mol Cell Biol 2012; 32(1): 2-11.
[http://dx.doi.org/10.1128/MCB.06159-11] [PMID: 22025673]

[55]   Hara T, Nakamura K, Matsui M, *et al.* Suppression of basal autophagy in neural cells causes neurodegenerative disease in mice. Nature 2006; 441(7095): 885-9.
[http://dx.doi.org/10.1038/nature04724] [PMID: 16625204]

[56]   Komatsu M, Waguri S, Chiba T, *et al.* Loss of autophagy in the central nervous system causes neurodegeneration in mice. Nature 2006; 441(7095): 880-4.
[http://dx.doi.org/10.1038/nature04723] [PMID: 16625205]

[57]   King MA, Hands S, Hafiz F, Mizushima N, Tolkovsky AM, Wyttenbach A. Rapamycin inhibits polyglutamine aggregation independently of autophagy by reducing protein synthesis. Mol Pharmacol 2008; 73(4): 1052-63.
[http://dx.doi.org/10.1124/mol.107.043398] [PMID: 18199701]

[58]   Thoreen CC, Kang SA, Chang JW, *et al.* An ATP-competitive mammalian target of rapamycin inhibitor reveals rapamycin-resistant functions of mTORC1. J Biol Chem 2009; 284(12): 8023-32.
[http://dx.doi.org/10.1074/jbc.M900301200] [PMID: 19150980]

[59]   Yu K, Toral-Barza L, Shi C, *et al.* Biochemical, cellular, and *in vivo* activity of novel ATP-competitive and selective inhibitors of the mammalian target of rapamycin. Cancer Res 2009; 69(15): 6232-40.
[http://dx.doi.org/10.1158/0008-5472.CAN-09-0299] [PMID: 19584280]

[60]   Fabrizio P, Pozza F, Pletcher SD, Gendron CM, Longo VD. Regulation of longevity and stress resistance by Sch9 in yeast. Science 2001; 292(5515): 288-90.
[http://dx.doi.org/10.1126/science.1059497] [PMID: 11292860]

[61]   Vellai T, Takacs-Vellai K, Zhang Y, Kovacs AL, Orosz L, Müller F. Genetics: influence of TOR kinase on lifespan in C. elegans. Nature 2003; 426(6967): 620.
[http://dx.doi.org/10.1038/426620a] [PMID: 14668850]

[62]   Jia K, Chen D, Riddle DL. The TOR pathway interacts with the insulin signaling pathway to regulate C. elegans larval development, metabolism and life span. Development 2004; 131(16): 3897-906.
[http://dx.doi.org/10.1242/dev.01255] [PMID: 15253933]

[63]   Harrison DE, Strong R, Sharp ZD, *et al.* Rapamycin fed late in life extends lifespan in genetically heterogeneous mice. Nature 2009; 460(7253): 392-5.

[64]   Selman C, Tullet JM, Wieser D, *et al.* Ribosomal protein S6 kinase 1 signaling regulates mammalian life span. Science 2009; 326(5949): 140-4.
[http://dx.doi.org/10.1126/science.1177221] [PMID: 19797661]

[65]   Wu JJ, Liu J, Chen EB, *et al.* Increased mammalian lifespan and a segmental and tissue-specific slowing of aging after genetic reduction of mTOR expression. Cell Reports 2013; 4(5): 913-20.
[http://dx.doi.org/10.1016/j.celrep.2013.07.030] [PMID: 23994476]

[66]   Montagne J, Stewart MJ, Stocker H, Hafen E, Kozma SC, Thomas G. Drosophila S6 kinase: a regulator of cell size. Science 1999; 285(5436): 2126-9.
[http://dx.doi.org/10.1126/science.285.5436.2126] [PMID: 10497130]

[67]   Oldham S, Montagne J, Radimerski T, Thomas G, Hafen E. Genetic and biochemical characterization of dTOR, the Drosophila homolog of the target of rapamycin. Genes Dev 2000; 14(21): 2689-94.
[http://dx.doi.org/10.1101/gad.845700] [PMID: 11069885]

[68]   Murakami M, Ichisaka T, Maeda M, *et al.* mTOR is essential for growth and proliferation in early mouse embryos and embryonic stem cells. Mol Cell Biol 2004; 24(15): 6710-8.
[http://dx.doi.org/10.1128/MCB.24.15.6710-6718.2004] [PMID: 15254238]

[69]   Sebastián-Serrano Á, de Diego-García L, Díaz-Hernández M. The neurotoxic role of extracellular tau protein. Int J Mol Sci 2018; 19(4): 998.
[http://dx.doi.org/10.3390/ijms19040998] [PMID: 29584657]

[70]   Iqbal K, Grundke-Iqbal I. Alzheimer's disease, a multifactorial disorder seeking multitherapies. Alzheimers Dement 2010; 6(5): 420-4.
[http://dx.doi.org/10.1016/j.jalz.2010.04.006] [PMID: 20813343]

[71]   Dickson DW, Crystal HA, Bevona C, Honer W, Vincent I, Davies P. Correlations of synaptic and pathological markers with cognition of the elderly. Neurobiol Aging 1995; 16(3): 285-98.
[http://dx.doi.org/10.1016/0197-4580(95)00013-5] [PMID: 7566338]

[72]   Schneider A, Mandelkow E. Tau-based treatment strategies in neurodegenerative diseases. Neurotherapeutics 2008; 5(3): 443-57.
[http://dx.doi.org/10.1016/j.nurt.2008.05.006] [PMID: 18625456]

[73]   Lee VM, Brunden KR, Hutton M, Trojanowski JQ. Developing therapeutic approaches to tau, selected kinases, and related neuronal protein targets. Cold Spring Harb Perspect Med 2011; 1(1): a006437.
[http://dx.doi.org/10.1101/cshperspect.a006437] [PMID: 22229117]

[74]   Lewis J, Dickson DW, Lin WL, *et al.* Enhanced neurofibrillary degeneration in transgenic mice expressing mutant tau and APP. Science 2001; 293(5534): 1487-91.
[http://dx.doi.org/10.1126/science.1058189] [PMID: 11520987]

[75]   Wittmann CW, Wszolek MF, Shulman JM, *et al.* Tauopathy in Drosophila: neurodegeneration without neurofibrillary tangles. Science 2001; 293(5530): 711-4.
[http://dx.doi.org/10.1126/science.1062382] [PMID: 11408621]

[76]   Tanemura K, Murayama M, Akagi T, *et al.* Neurodegeneration with tau accumulation in a transgenic mouse expressing V337M human tau. J Neurosci 2002; 22(1): 133-41.
[http://dx.doi.org/10.1523/JNEUROSCI.22-01-00133.2002] [PMID: 11756496]

[77]   Kraemer BC, Zhang B, Leverenz JB, Thomas JH, Trojanowski JQ, Schellenberg GD. Neurodegeneration and defective neurotransmission in a Caenorhabditis elegans model of tauopathy. Proc Natl Acad Sci USA 2003; 100(17): 9980-5.
[http://dx.doi.org/10.1073/pnas.1533448100] [PMID: 12872001]

[78]   Ramsden M, Kotilinek L, Forster C, *et al.* Age-dependent neurofibrillary tangle formation, neuron loss, and memory impairment in a mouse model of human tauopathy (P301L). J Neurosci 2005; 25(46): 10637-47.
[http://dx.doi.org/10.1523/JNEUROSCI.3279-05.2005] [PMID: 16291936]

[79]   Ashe KH, Zahs KR. Probing the biology of Alzheimer's disease in mice. Neuron 2010; 66(5): 631-45.
[http://dx.doi.org/10.1016/j.neuron.2010.04.031] [PMID: 20547123]

[80]   Köpke E, Tung YC, Shaikh S, Alonso AC, Iqbal K, Grundke-Iqbal I. Microtubule-associated protein tau. Abnormal phosphorylation of a non-paired helical filament pool in Alzheimer disease. J Biol Chem 1993; 268(32): 24374-84.
[PMID: 8226987]

[81]   Iqbal K, Liu F, Gong CX, Grundke-Iqbal I. Tau in Alzheimer disease and related tauopathies. Curr Alzheimer Res 2010; 7(8): 656-64.
[http://dx.doi.org/10.2174/156720510793611592] [PMID: 20678074]

[82]   Alonso AD, Grundke-Iqbal I, Barra HS, Iqbal K. Abnormal phosphorylation of tau and the mechanism of Alzheimer neurofibrillary degeneration: sequestration of microtubule-associated proteins 1 and 2 and the disassembly of microtubules by the abnormal tau. Proc Natl Acad Sci USA 1997; 94(1): 298-303.

[http://dx.doi.org/10.1073/pnas.94.1.298] [PMID: 8990203]

[83]   Alonso AD, Di Clerico J, Li B, *et al.* Phosphorylation of tau at Thr212, Thr231, and Ser262 combined causes neurodegeneration. J Biol Chem 2010; 285(40): 30851-60.
[http://dx.doi.org/10.1074/jbc.M110.110957] [PMID: 20663882]

[84]   Brandt R, Hundelt M, Shahani N. Tau alteration and neuronal degeneration in tauopathies: mechanisms and models. Biochim Biophys Acta 2005; 1739(2-3): 331-54.
[http://dx.doi.org/10.1016/j.bbadis.2004.06.018] [PMID: 15615650]

[85]   Ittner LM, Götz J. Amyloid-β and taua toxic pas de deux in Alzheimer's disease. Nat Rev Neurosci 2011; 12(2): 65-72.
[http://dx.doi.org/10.1038/nrn2967] [PMID: 21193853]

[86]   Gómez-Isla T, Hollister R, West H, *et al.* Neuronal loss correlates with but exceeds neurofibrillary tangles in Alzheimer's disease. Ann Neurol 1997; 41(1): 17-24.
[http://dx.doi.org/10.1002/ana.410410106] [PMID: 9005861]

[87]   Li X, Alafuzoff I, Soininen H, Winblad B, Pei JJ. Levels of mTOR and its downstream targets 4E-BP1, eEF2, and eEF2 kinase in relationships with tau in Alzheimer's disease brain. FEBS J 2005; 272(16): 4211-20.
[http://dx.doi.org/10.1111/j.1742-4658.2005.04833.x] [PMID: 16098202]

[88]   Pei JJ, Hugon J. mTOR-dependent signalling in Alzheimer's disease. J Cell Mol Med 2008; 12(6B): 2525-32.
[http://dx.doi.org/10.1111/j.1582-4934.2008.00509.x] [PMID: 19210753]

[89]   Pei J-J, Björkdahl C, Zhang H, Zhou X, Winblad B. p70 S6 kinase and tau in Alzheimer's disease. J Alzheimers Dis 2008; 14(4): 385-92.
[http://dx.doi.org/10.3233/JAD-2008-14405] [PMID: 18688088]

[90]   Oddo S. The role of mTOR signaling in Alzheimer disease. Front Biosci (Schol Ed) 2012; 4: 941-52.
[http://dx.doi.org/10.2741/s310] [PMID: 22202101]

[91]   Congdon EE, Wu JW, Myeku N, *et al.* Methylthioninium chloride (methylene blue) induces autophagy and attenuates tauopathy *in vitro* and *in vivo*. Autophagy 2012; 8(4): 609-22.
[http://dx.doi.org/10.4161/auto.19048] [PMID: 22361619]

[92]   Ozcelik S, Fraser G, Castets P, *et al.* Rapamycin attenuates the progression of tau pathology in P301S tau transgenic mice. PLoS One 2013; 8(5): e62459.
[http://dx.doi.org/10.1371/journal.pone.0062459] [PMID: 23667480]

[93]   Schaeffer V, Lavenir I, Ozcelik S, Tolnay M, Winkler DT, Goedert M. Stimulation of autophagy reduces neurodegeneration in a mouse model of human tauopathy. Brain 2012; 135(Pt 7): 2169-77.
[http://dx.doi.org/10.1093/brain/aws143] [PMID: 22689910]

[94]   Siman R, Cocca R, Dong Y. The mTOR inhibitor rapamycin mitigates perforant pathway neurodegeneration and synapse loss in a mouse model of early-stage Alzheimer-type tauopathy. PLoS One 2015; 10(11): e0142340.
[http://dx.doi.org/10.1371/journal.pone.0142340] [PMID: 26540269]

[95]   Bruch J, Xu H, Rösler TW, *et al.* PERK activation mitigates tau pathology *in vitro* and *in vivo*. EMBO Mol Med 2017; 9(3): 371-84.
[http://dx.doi.org/10.15252/emmm.201606664] [PMID: 28148553]

[96]   Hetz C, Mollereau B. Disturbance of endoplasmic reticulum proteostasis in neurodegenerative diseases. Nat Rev Neurosci 2014; 15(4): 233-49.
[http://dx.doi.org/10.1038/nrn3689] [PMID: 24619348]

[97]   Hetz C, Chevet E, Harding HP. Targeting the unfolded protein response in disease. Nat Rev Drug Discov 2013; 12(9): 703-19.
[http://dx.doi.org/10.1038/nrd3976] [PMID: 23989796]

[98]   Ohno M. PERK as a hub of multiple pathogenic pathways leading to memory deficits and neurodegeneration in Alzheimer's disease. Brain Res Bull 2017.
[PMID: 28804008]

[99]   Hamanaka RB, Bennett BS, Cullinan SB, Diehl JA. PERK and GCN2 contribute to eIF2α phosphorylation and cell cycle arrest after activation of the unfolded protein response pathway. Mol Biol Cell 2005; 16(12): 5493-501.
[http://dx.doi.org/10.1091/mbc.e05-03-0268] [PMID: 16176978]

[100]   Raven JF, Koromilas AE. PERK and PKR: old kinases learn new tricks. Cell Cycle 2008; 7(9): 1146-50.
[http://dx.doi.org/10.4161/cc.7.9.5811] [PMID: 18418049]

[101]   Donnelly N, Gorman AM, Gupta S, Samali A. The eIF2α kinases: their structures and functions. Cell Mol Life Sci 2013; 70(19): 3493-511.
[http://dx.doi.org/10.1007/s00018-012-1252-6] [PMID: 23354059]

[102]   Richter J, Klann E. Translational control of synaptic plasticity and learning and memory. Cold spring harbor: Cold spring harbor laboratory press 2007.

[103]   Alberini CM. The role of protein synthesis during the labile phases of memory: revisiting the skepticism. Neurobiol Learn Mem 2008; 89(3): 234-46.
[http://dx.doi.org/10.1016/j.nlm.2007.08.007] [PMID: 17928243]

[104]   Abisambra JF, Jinwal UK, Blair LJ, *et al.* Tau accumulation activates the unfolded protein response by impairing endoplasmic reticulum-associated degradation. J Neurosci 2013; 33(22): 9498-507.
[http://dx.doi.org/10.1523/JNEUROSCI.5397-12.2013] [PMID: 23719816]

[105]   Devi L, Ohno M. PERK mediates eIF2α phosphorylation responsible for BACE1 elevation, CREB dysfunction and neurodegeneration in a mouse model of Alzheimer's disease. Neurobiol Aging 2014; 35(10): 2272-81.
[http://dx.doi.org/10.1016/j.neurobiolaging.2014.04.031] [PMID: 24889041]

[106]   Moreno JA, Radford H, Peretti D, *et al.* Sustained translational repression by eIF2α-P mediates prion neurodegeneration. Nature 2012; 485(7399): 507-11.
[http://dx.doi.org/10.1038/nature11058] [PMID: 22622579]

[107]   Köhler C, Dinekov M, Götz J. Granulovacuolar degeneration and unfolded protein response in mouse models of tauopathy and Aβ amyloidosis. Neurobiol Dis 2014; 71: 169-79.
[http://dx.doi.org/10.1016/j.nbd.2014.07.006] [PMID: 25073087]

[108]   Atkin JD, Farg MA, Walker AK, McLean C, Tomas D, Horne MK. Endoplasmic reticulum stress and induction of the unfolded protein response in human sporadic amyotrophic lateral sclerosis. Neurobiol Dis 2008; 30(3): 400-7.
[http://dx.doi.org/10.1016/j.nbd.2008.02.009] [PMID: 18440237]

[109]   Saxena S, Cabuy E, Caroni P. A role for motoneuron subtype-selective ER stress in disease manifestations of FALS mice. Nat Neurosci 2009; 12(5): 627-36.
[http://dx.doi.org/10.1038/nn.2297] [PMID: 19330001]

[110]   Axten JM, Medina JR, Feng Y, *et al.* Discovery of 7-methyl-5-(1-{[3-(trifluoromethyl)phenyl]ac-tyl}-2,3-dihydro-1H-indol-5-yl)-7H-pyrrolo[2,3-d]pyrimidin-4-amine (GSK2606414), a potent and selective first-in-class inhibitor of protein kinase R (PKR)-like endoplasmic reticulum kinase (PERK). J Med Chem 2012; 55(16): 7193-207.
[http://dx.doi.org/10.1021/jm300713s] [PMID: 22827572]

[111]   Halliday M, Radford H, Sekine Y, *et al.* Partial restoration of protein synthesis rates by the small molecule ISRIB prevents neurodegeneration without pancreatic toxicity. Cell Death Dis 2015; 6(3): e1672.
[http://dx.doi.org/10.1038/cddis.2015.49] [PMID: 25741597]

[112]   Radford H, Moreno JA, Verity N, Halliday M, Mallucci GR. PERK inhibition prevents tau-mediated

neurodegeneration in a mouse model of frontotemporal dementia. Acta Neuropathol 2015; 130(5): 633-42.
[http://dx.doi.org/10.1007/s00401-015-1487-z] [PMID: 26450683]

[113] Ho Y-S, Yang X, Lau JC, *et al.* Endoplasmic reticulum stress induces tau pathology and forms a vicious cycle: implication in Alzheimer's disease pathogenesis. J Alzheimers Dis 2012; 28(4): 839-54.
[http://dx.doi.org/10.3233/JAD-2011-111037] [PMID: 22101233]

[114] Lin L, Yang SS, Chu J, *et al.* Region-specific expression of tau, amyloid-β protein precursor, and synaptic proteins at physiological condition or under endoplasmic reticulum stress in rats. J Alzheimers Dis 2014; 41(4): 1149-63.
[http://dx.doi.org/10.3233/JAD-140207] [PMID: 24787918]

[115] van der Harg JM, Nölle A, Zwart R, *et al.* The unfolded protein response mediates reversible tau phosphorylation induced by metabolic stress. Cell Death Dis 2014; 5(8): e1393.
[http://dx.doi.org/10.1038/cddis.2014.354] [PMID: 25165879]

[116] Nijholt DA, Nölle A, van Haastert ES, *et al.* Unfolded protein response activates glycogen synthase kinase-3 *via* selective lysosomal degradation. Neurobiol Aging 2013; 34(7): 1759-71.
[http://dx.doi.org/10.1016/j.neurobiolaging.2013.01.008] [PMID: 23415837]

[117] Santacruz K, Lewis J, Spires T, *et al.* Tau suppression in a neurodegenerative mouse model improves memory function. Science 2005; 309(5733): 476-81.
[http://dx.doi.org/10.1126/science.1113694] [PMID: 16020737]

[118] Ron D, Walter P. Signal integration in the endoplasmic reticulum unfolded protein response. Nat Rev Mol Cell Biol 2007; 8(7): 519-29.
[http://dx.doi.org/10.1038/nrm2199] [PMID: 17565364]

[119] Moreno JA, Halliday M, Molloy C, *et al.* Oral treatment targeting the unfolded protein response prevents neurodegeneration and clinical disease in prion-infected mice. Science translational medicine 2013; 5(206): 206ra138.
[http://dx.doi.org/10.1126/scitranslmed.3006767]

[120] Ma T, Klann E. PERK: a novel therapeutic target for neurodegenerative diseases? Alzheimers Res Ther 2014; 6(3): 30.
[http://dx.doi.org/10.1186/alzrt260] [PMID: 25031640]

[121] Ma T, Trinh MA, Wexler AJ, *et al.* Suppression of eIF2α kinases alleviates Alzheimer's disease-related plasticity and memory deficits. Nat Neurosci 2013; 16(9): 1299-305.
[http://dx.doi.org/10.1038/nn.3486] [PMID: 23933749]

[122] Trinh MA, Kaphzan H, Wek RC, Pierre P, Cavener DR, Klann E. Brain-specific disruption of the eIF2α kinase PERK decreases ATF4 expression and impairs behavioral flexibility. Cell Reports 2012; 1(6): 676-88.
[http://dx.doi.org/10.1016/j.celrep.2012.04.010] [PMID: 22813743]

[123] Zhang P, McGrath B, Li S, *et al.* The PERK eukaryotic initiation factor 2 α kinase is required for the development of the skeletal system, postnatal growth, and the function and viability of the pancreas. Mol Cell Biol 2002; 22(11): 3864-74.
[http://dx.doi.org/10.1128/MCB.22.11.3864-3874.2002] [PMID: 11997520]

[124] Mitchell P. Coupling of phosphorylation to electron and hydrogen transfer by a chemi-osmotic type of mechanism. Nature 1961; 191(4784): 144-8.
[http://dx.doi.org/10.1038/191144a0] [PMID: 13771349]

[125] Mitchell P. Chemiosmotic coupling in oxidative and photosynthetic phosphorylation. Biol Rev Camb Philos Soc 1966; 41(3): 445-502.
[http://dx.doi.org/10.1111/j.1469-185X.1966.tb01501.x] [PMID: 5329743]

[126] Mitchell P. Vectorial chemistry and the molecular mechanics of chemiosmotic coupling: power transmission by proticity. Portland Press Limited 1976.

[127] Rich P. Chemiosmotic coupling: The cost of living. Nature 2003; 421(6923): 583.
       [http://dx.doi.org/10.1038/421583a] [PMID: 12571574]

[128] Saraste M. Oxidative phosphorylation at the fin de siècle. Science 1999; 283(5407): 1488-93.
       [http://dx.doi.org/10.1126/science.283.5407.1488] [PMID: 10066163]

[129] Sun F, Huo X, Zhai Y, *et al.* Crystal structure of mitochondrial respiratory membrane protein complex
       II. Cell 2005; 121(7): 1043-57.
       [http://dx.doi.org/10.1016/j.cell.2005.05.025] [PMID: 15989954]

[130] Fiedorczuk K, Letts JA, Degliesposti G, Kaszuba K, Skehel M, Sazanov LA. Atomic structure of the
       entire mammalian mitochondrial complex I. Nature 2016; 538(7625): 406-10.
       [http://dx.doi.org/10.1038/nature19794] [PMID: 27595392]

[131] Lazarou M, *et al.* Assembly of mitochondrial complex I and defects in disease. Biochimica et
       Biophysica Acta (BBA)-. Molecular Cell Research 2009; 1793(1): 78-88.

[132] Sánchez-Caballero L, Guerrero-Castillo S, Nijtmans L. Unraveling the complexity of mitochondrial
       complex I assembly: A dynamic process. Biochim Biophys Acta 2016; 1857(7): 980-90.
       [http://dx.doi.org/10.1016/j.bbabio.2016.03.031] [PMID: 27040506]

[133] Wirth C, Brandt U, Hunte C, Zickermann W, *et al.* Structure and function of mitochondrial complex I.
       Biochimica et Biophysica Acta (BBA). Bioenergetics 2016; 1857(7): 902-14.
       [http://dx.doi.org/10.1016/j.bbabio.2016.02.013]

[134] Bhat AH, Dar KB, Anees S, *et al.* Oxidative stress, mitochondrial dysfunction and neurodegenerative
       diseases; a mechanistic insight. Biomed Pharmacother 2015; 74: 101-10.
       [http://dx.doi.org/10.1016/j.biopha.2015.07.025] [PMID: 26349970]

[135] Raha S, Robinson BH. Mitochondria, oxygen free radicals, disease and ageing. Trends Biochem Sci
       2000; 25(10): 502-8.
       [http://dx.doi.org/10.1016/S0968-0004(00)01674-1] [PMID: 11050436]

[136] Robinson BH. Human complex I deficiency: clinical spectrum and involvement of oxygen free
       radicals in the pathogenicity of the defect. Biochim Biophys Acta 1998; 1364(2): 271-86.
       [http://dx.doi.org/10.1016/S0005-2728(98)00033-4] [PMID: 9593934]

[137] Adrain C, Martin SJ. The mitochondrial apoptosome: a killer unleashed by the cytochrome seas.
       Trends Biochem Sci 2001; 26(6): 390-7.
       [http://dx.doi.org/10.1016/S0968-0004(01)01844-8] [PMID: 11406413]

[138] Wang C, Youle RJ. The role of mitochondria in apoptosis. Annu Rev Genet 2009; 43: 95-118.
       [http://dx.doi.org/10.1146/annurev-genet-102108-134850] [PMID: 19659442]

[139] Perier C, Tieu K, Guégan C, *et al.* Complex I deficiency primes Bax-dependent neuronal apoptosis
       through mitochondrial oxidative damage. Proc Natl Acad Sci USA 2005; 102(52): 19126-31.
       [http://dx.doi.org/10.1073/pnas.0508215102] [PMID: 16365298]

[140] Ahmadi FA, Linseman DA, Grammatopoulos TN, *et al.* The pesticide rotenone induces caspase--
       -mediated apoptosis in ventral mesencephalic dopaminergic neurons. J Neurochem 2003; 87(4): 914-
       21.
       [http://dx.doi.org/10.1046/j.1471-4159.2003.02068.x] [PMID: 14622122]

[141] Li J, Spletter ML, Johnson DA, Wright LS, Svendsen CN, Johnson JA. Rotenone-induced caspase
       9/3-independent and -dependent cell death in undifferentiated and differentiated human neural stem
       cells. J Neurochem 2005; 92(3): 462-76.
       [http://dx.doi.org/10.1111/j.1471-4159.2004.02872.x] [PMID: 15659217]

[142] Abolaji AO, Ojo M, Afolabi TT, Arowoogun MD, Nwawolor D, Farombi EO. Protective properties of
       6-gingerol-rich fraction from Zingiber officinale (Ginger) on chlorpyrifos-induced oxidative damage
       and inflammation in the brain, ovary and uterus of rats. Chem Biol Interact 2017; 270: 15-23.
       [http://dx.doi.org/10.1016/j.cbi.2017.03.017] [PMID: 28373059]

[143] Park JH, Ko J, Hwang J, Koh HC. Dynamin-related protein 1 mediates mitochondria-dependent apoptosis in chlorpyrifos-treated SH-SY5Y cells. Neurotoxicology 2015; 51: 145-57.
[http://dx.doi.org/10.1016/j.neuro.2015.10.008] [PMID: 26598294]

[144] Singh N, Lawana V, Luo J, *et al.* Organophosphate pesticide chlorpyrifos impairs STAT1 signaling to induce dopaminergic neurotoxicity: Implications for mitochondria mediated oxidative stress signaling events. Neurobiol Dis 2018; 117: 82-113.
[http://dx.doi.org/10.1016/j.nbd.2018.05.019] [PMID: 29859868]

[145] Venkatesan R, Park YU, Ji E, Yeo EJ, Kim SY. Malathion increases apoptotic cell death by inducing lysosomal membrane permeabilization in N2a neuroblastoma cells: a model for neurodegeneration in Alzheimer's disease. Cell Death Discov 2017; 3: 17007.
[http://dx.doi.org/10.1038/cddiscovery.2017.7] [PMID: 28487766]

[146] Kaur P, Radotra B, Minz RW, Gill KD. Impaired mitochondrial energy metabolism and neuronal apoptotic cell death after chronic dichlorvos (OP) exposure in rat brain. Neurotoxicology 2007; 28(6): 1208-19.
[http://dx.doi.org/10.1016/j.neuro.2007.08.001] [PMID: 17850875]

[147] Wani WY, Gudup S, Sunkaria A, *et al.* Protective efficacy of mitochondrial targeted antioxidant MitoQ against dichlorvos induced oxidative stress and cell death in rat brain. Neuropharmacology 2011; 61(8): 1193-201.
[http://dx.doi.org/10.1016/j.neuropharm.2011.07.008] [PMID: 21784090]

[148] Alberts B, Johnson A, Lewis J, *et al.* Molecular biology of the cell. New York: Garland Science; 2002.

[149] Bugger H, Abel ED. Molecular mechanisms for myocardial mitochondrial dysfunction in the metabolic syndrome. London, England : 1979Clinical science 2008; 114: pp. (3)195-210.
[http://dx.doi.org/10.1042/CS20070166]

[150] Esposti DD, Hamelin J, Bosselut N, *et al.* Mitochondrial roles and cytoprotection in chronic liver injury. Biochemistry research international 2012.
[http://dx.doi.org/10.1155/2012/387626]

[151] Diebold I, Hennigs JK, Miyagawa K, *et al.* BMPR2 preserves mitochondrial function and DNA during reoxygenation to promote endothelial cell survival and reverse pulmonary hypertension. Cell Metab 2015; 21(4): 596-608.
[http://dx.doi.org/10.1016/j.cmet.2015.03.010] [PMID: 25863249]

[152] Duchen MR. Contributions of mitochondria to animal physiology: from homeostatic sensor to calcium signalling and cell death. J Physiol 1999; 516(Pt 1): 1-17.
[http://dx.doi.org/10.1111/j.1469-7793.1999.001aa.x] [PMID: 10066918]

[153] Ernster L, Ikkos D, Luft R. Enzymic activities of human skeletal muscle mitochondria: a tool in clinical metabolic research. Nature 1959; 184(4702): 1851-4.
[http://dx.doi.org/10.1038/1841851a0] [PMID: 13820680]

[154] van Wijngaarden GK, Bethlem J, Meijer AE, Hülsmann WC, Feltkamp CA. Skeletal muscle disease with abnormal mitochondria. Brain 1967; 90(3): 577-92.
[http://dx.doi.org/10.1093/brain/90.3.577] [PMID: 6058143]

[155] Morgan-Hughes JA, Darveniza P, Kahn SN, *et al.* A mitochondrial myopathy characterized by a deficiency in reducible cytochrome b. Brain 1977; 100(4): 617-40.
[http://dx.doi.org/10.1093/brain/100.4.617] [PMID: 608115]

[156] Morgan-Hughes JA, Darveniza P, Landon DN, Land JM, Clark JB. A mitochondrial myopathy with a deficiency of respiratory chain NADH-CoQ reductase activity. J Neurol Sci 1979; 43(1): 27-46.
[http://dx.doi.org/10.1016/0022-510X(79)90071-6] [PMID: 521828]

[157] Schotland DL, DiMauro S, Bonilla E, Scarpa A, Lee CP. Neuromuscular disorder associated with a defect in mitochondrial energy supply. Arch Neurol 1976; 33(7): 475-9.
[http://dx.doi.org/10.1001/archneur.1976.00500070017003] [PMID: 180936]

[158] Holt IJ, Harding AE, Morgan-Hughes JA. Deletions of muscle mitochondrial DNA in patients with mitochondrial myopathies. Nature 1988; 331(6158): 717-9.
[http://dx.doi.org/10.1038/331717a0] [PMID: 2830540]

[159] Zeviani M, Moraes CT, DiMauro S, *et al.* Deletions of mitochondrial DNA in Kearns-Sayre syndrome. Neurology 1988; 38(9): 1339-46.
[http://dx.doi.org/10.1212/WNL.38.9.1339] [PMID: 3412580]

[160] Burns L. World Drug Report 2013 By United Nations Office on Drugs and Crime New York: United Nations, 2013 ISBN: 9789210561686, 151 pp. Grey literature. Drug Alcohol Rev 2014; 33(2): 216-6.
[http://dx.doi.org/10.1111/dar.12110]

[161] Schapira AH, Cooper JM, Dexter D, Jenner P, Clark JB, Marsden CD. Mitochondrial complex I deficiency in Parkinson's disease. Lancet 1989; 1(8649): 1269-9.
[http://dx.doi.org/10.1016/S0140-6736(89)92366-0] [PMID: 2566813]

[162] Schapira AHV, Mann VM, Cooper JM, *et al.* Anatomic and disease specificity of NADH CoQ1 reductase (complex I) deficiency in Parkinson's disease. J Neurochem 1990; 55(6): 2142-5.
[http://dx.doi.org/10.1111/j.1471-4159.1990.tb05809.x] [PMID: 2121905]

[163] Krige D, Carroll MT, Cooper JM, Marsden CD, Schapira AH. Platelet mitochondrial function in Parkinson's disease. Ann Neurol 1992; 32(6): 782-8.
[http://dx.doi.org/10.1002/ana.410320612] [PMID: 1471869]

[164] Parker WD Jr, Boyson SJ, Parks JK. Abnormalities of the electron transport chain in idiopathic Parkinson's disease. Ann Neurol 1989; 26(6): 719-23.
[http://dx.doi.org/10.1002/ana.410260606] [PMID: 2557792]

[165] Parker WD Jr, Filley CM, Parks JK. Cytochrome oxidase deficiency in Alzheimer's disease. Neurology 1990; 40(8): 1302-3.
[http://dx.doi.org/10.1212/WNL.40.8.1302] [PMID: 2166249]

[166] Kish SJ, Bergeron C, Rajput A, *et al.* Brain cytochrome oxidase in Alzheimer's disease. J Neurochem 1992; 59(2): 776-9.
[http://dx.doi.org/10.1111/j.1471-4159.1992.tb09439.x] [PMID: 1321237]

[167] Parker WD Jr, Parks J, Filley CM, Kleinschmidt-DeMasters BK. Electron transport chain defects in Alzheimer's disease brain. Neurology 1994; 44(6): 1090-6.
[http://dx.doi.org/10.1212/WNL.44.6.1090] [PMID: 8208407]

[168] Parker WD Jr, Boyson SJ, Luder AS, Parks JK. Evidence for a defect in NADH: ubiquinone oxidoreductase (complex I) in Huntington's disease. Neurology 1990; 40(8): 1231-4.
[http://dx.doi.org/10.1212/WNL.40.8.1231] [PMID: 2143271]

[169] Gu M, Gash MT, Mann VM, Javoy-Agid F, Cooper JM, Schapira AH. Mitochondrial defect in Huntington's disease caudate nucleus. Ann Neurol 1996; 39(3): 385-9.
[http://dx.doi.org/10.1002/ana.410390317] [PMID: 8602759]

[170] Holt IJ, Harding AE, Petty RK, Morgan-Hughes JA. A new mitochondrial disease associated with mitochondrial DNA heteroplasmy. Am J Hum Genet 1990; 46(3): 428-33.
[PMID: 2137962]

[171] Puddu P, Barboni P, Mantovani V, *et al.* Retinitis pigmentosa, ataxia, and mental retardation associated with mitochondrial DNA mutation in an Italian family. Br J Ophthalmol 1993; 77(2): 84-8.
[http://dx.doi.org/10.1136/bjo.77.2.84] [PMID: 8435424]

[172] Begley JG, Duan W, Chan S, Duff K, Mattson MP. Altered calcium homeostasis and mitochondrial dysfunction in cortical synaptic compartments of presenilin-1 mutant mice. J Neurochem 1999; 72(3): 1030-9.
[http://dx.doi.org/10.1046/j.1471-4159.1999.0721030.x] [PMID: 10037474]

[173] Farrer MJ. Genetics of Parkinson disease: paradigm shifts and future prospects. Nat Rev Genet 2006;

7(4): 306-18.
[http://dx.doi.org/10.1038/nrg1831] [PMID: 16543934]

[174] Hauptmann S, Keil U, Scherping I, Bonert A, Eckert A, Müller WE. Mitochondrial dysfunction in sporadic and genetic Alzheimer's disease. Exp Gerontol 2006; 41(7): 668-73.
[http://dx.doi.org/10.1016/j.exger.2006.03.012] [PMID: 16677790]

[175] Keller JN, Guo Q, Holtsberg FW, Bruce-Keller AJ, Mattson MP. Increased sensitivity to mitochondrial toxin-induced apoptosis in neural cells expressing mutant presenilin-1 is linked to perturbed calcium homeostasis and enhanced oxyradical production. J Neurosci 1998; 18(12): 4439-50.
[http://dx.doi.org/10.1523/JNEUROSCI.18-12-04439.1998] [PMID: 9614221]

[176] Martinat C, Shendelman S, Jonason A, *et al.* Sensitivity to oxidative stress in DJ-1-deficient dopamine neurons: an ES- derived cell model of primary Parkinsonism. PLoS Biol 2004; 2(11): e327-7.
[http://dx.doi.org/10.1371/journal.pbio.0020327] [PMID: 15502868]

[177] Rosen KM, Veereshwarayya V, Moussa CE, *et al.* Parkin protects against mitochondrial toxins and beta-amyloid accumulation in skeletal muscle cells. J Biol Chem 2006; 281(18): 12809-16.
[http://dx.doi.org/10.1074/jbc.M512649200] [PMID: 16517603]

[178] Lasagna-Reeves CA, Castillo-Carranza DL, Sengupta U, Clos AL, Jackson GR, Kayed R. Tau oligomers impair memory and induce synaptic and mitochondrial dysfunction in wild-type mice. Mol Neurodegener 2011; 6(1): 39-9.
[http://dx.doi.org/10.1186/1750-1326-6-39] [PMID: 21645391]

[179] Camilleri A, Zarb C, Caruana M, *et al.* Mitochondrial membrane permeabilisation by amyloid aggregates and protection by polyphenols. Biochim Biophys Acta 2013; 1828(11): 2532-43.
[http://dx.doi.org/10.1016/j.bbamem.2013.06.026] [PMID: 23817009]

[180] Kopeikina KJ, Carlson GA, Pitstick R, *et al.* Tau accumulation causes mitochondrial distribution deficits in neurons in a mouse model of tauopathy and in human Alzheimer's disease brain. Am J Pathol 2011; 179(4): 2071-82.
[http://dx.doi.org/10.1016/j.ajpath.2011.07.004] [PMID: 21854751]

[181] Gilley J, Seereeram A, Ando K, *et al.* Age-dependent axonal transport and locomotor changes and tau hypophosphorylation in a "P301L" tau knockin mouse. Neurobiol Aging 2012; 33(3): 621.e1-621.e15.
[http://dx.doi.org/10.1016/j.neurobiolaging.2011.02.014] [PMID: 21492964]

[182] Rodríguez-Martín T, Pooler AM, Lau DHW, *et al.* Reduced number of axonal mitochondria and tau hypophosphorylation in mouse P301L tau knockin neurons. Neurobiol Dis 2016; 85: 1-10.
[http://dx.doi.org/10.1016/j.nbd.2015.10.007] [PMID: 26459111]

[183] Amadoro G, Corsetti V, Florenzano F, *et al.* AD-linked, toxic NH2 human tau affects the quality control of mitochondria in neurons. Neurobiol Dis 2014; 62: 489-507.
[http://dx.doi.org/10.1016/j.nbd.2013.10.018] [PMID: 24411077]

[184] Manczak M, Reddy PH. Abnormal interaction between the mitochondrial fission protein Drp1 and hyperphosphorylated tau in Alzheimer's disease neurons: implications for mitochondrial dysfunction and neuronal damage. Hum Mol Genet 2012; 21(11): 2538-47.
[http://dx.doi.org/10.1093/hmg/dds072] [PMID: 22367970]

[185] Salama M, Mohamed WMY. NDUFA12L mitochondrial complex-I assembly factor: Implications for taupathies. Appl Transl Genomics 2015; 5: 37-9.
[http://dx.doi.org/10.1016/j.atg.2015.05.003] [PMID: 26937358]

[186] Betarbet R, Sherer TB, MacKenzie G, Garcia-Osuna M, Panov AV, Greenamyre JT. Chronic systemic pesticide exposure reproduces features of Parkinson's disease. Nat Neurosci 2000; 3(12): 1301-6.
[http://dx.doi.org/10.1038/81834] [PMID: 11100151]

[187] Cannon JR, Tapias V, Na HM, Honick AS, Drolet RE, Greenamyre JT. A highly reproducible rotenone model of Parkinson's disease. Neurobiol Dis 2009; 34(2): 279-90.

[http://dx.doi.org/10.1016/j.nbd.2009.01.016] [PMID: 19385059]

[188] Pouchieu C, Piel C, Carles C, *et al.* Pesticide use in agriculture and Parkinson's disease in the AGRICAN cohort study. Int J Epidemiol 2018; 47(1): 299-310.
[http://dx.doi.org/10.1093/ije/dyx225] [PMID: 29136149]

[189] Salama M, Ellaithy A, Helmy B, *et al.* Colchicine protects dopaminergic neurons in a rat model of Parkinson's disease. CNS & Neurological Disorders-Drug Targets (Formerly Current Drug Targets-CNS & Neurological Disorders) 2012; 11(7): 836-43.
[http://dx.doi.org/10.2174/1871527311201070836]

[190] Tanner CM, Kamel F, Ross GW, *et al.* Rotenone, paraquat, and Parkinson's disease. Environ Health Perspect 2011; 119(6): 866-72.
[http://dx.doi.org/10.1289/ehp.1002839] [PMID: 21269927]

[191] Höglinger GU, Lannuzel A, Khondiker ME, *et al.* The mitochondrial complex I inhibitor rotenone triggers a cerebral tauopathy. J Neurochem 2005; 95(4): 930-9.
[http://dx.doi.org/10.1111/j.1471-4159.2005.03493.x] [PMID: 16219024]

[192] Hongo H, Kihara T, Kume T, *et al.* Glycogen synthase kinase-3β activation mediates rotenone-induced cytotoxicity with the involvement of microtubule destabilization. Biochem Biophys Res Commun 2012; 426(1): 94-9.
[http://dx.doi.org/10.1016/j.bbrc.2012.08.042] [PMID: 22922102]

[193] Caparros-Lefebvre D, Elbaz A. Possible relation of atypical parkinsonism in the French West Indies with consumption of tropical plants: a case-control study. Lancet 1999; 354(9175): 281-6.
[http://dx.doi.org/10.1016/S0140-6736(98)10166-6] [PMID: 10440304]

[194] Champy P, Höglinger GU, Féger J, *et al.* Annonacin, a lipophilic inhibitor of mitochondrial complex I, induces nigral and striatal neurodegeneration in rats: possible relevance for atypical parkinsonism in Guadeloupe. J Neurochem 2004; 88(1): 63-9.
[http://dx.doi.org/10.1046/j.1471-4159.2003.02138.x] [PMID: 14675150]

[195] Escobar-Khondiker M, Höllerhage M, Muriel MP, *et al.* Annonacin, a natural mitochondrial complex I inhibitor, causes tau pathology in cultured neurons. J Neurosci 2007; 27(29): 7827-37.
[http://dx.doi.org/10.1523/JNEUROSCI.1644-07.2007] [PMID: 17634376]

[196] Yamada ES, Respondek G, Müssner S, *et al.* Annonacin, a natural lipophilic mitochondrial complex I inhibitor, increases phosphorylation of tau in the brain of FTDP-17 transgenic mice. Exp Neurol 2014; 253: 113-25.
[http://dx.doi.org/10.1016/j.expneurol.2013.12.017] [PMID: 24389273]

[197] Salama M, Elhussiny M, Magdy A, *et al.* Dual mTORC1/mTORC2 blocker as a possible therapy for tauopathy in cellular model. Metab Brain Dis 2018; 33(2): 583-7.
[http://dx.doi.org/10.1007/s11011-017-0137-7] [PMID: 29080085]

[198] Höllerhage M, Matusch A, Champy P, *et al.* Natural lipophilic inhibitors of mitochondrial complex I are candidate toxins for sporadic neurodegenerative tau pathologies. Exp Neurol 2009; 220(1): 133-42.
[http://dx.doi.org/10.1016/j.expneurol.2009.08.004] [PMID: 19682988]

[199] Yuan XL, Guo JF, Shi ZH, *et al.* R492X mutation in PTEN-induced putative kinase 1 induced cellular mitochondrial dysfunction and oxidative stress. Brain Res 2010; 1351: 229-37.
[http://dx.doi.org/10.1016/j.brainres.2010.06.005] [PMID: 20547144]

[200] Ye M, Zhou D, Zhou Y, Sun C. Parkinson's disease-associated PINK1 G309D mutation increases abnormal phosphorylation of Tau. IUBMB Life 2015; 67(4): 286-90.
[http://dx.doi.org/10.1002/iub.1367] [PMID: 25899925]

[201] Mortiboys H, Johansen KK, Aasly JO, Bandmann O. Mitochondrial impairment in patients with Parkinson disease with the G2019S mutation in LRRK2. Neurology 2010; 75(22): 2017-20.
[http://dx.doi.org/10.1212/WNL.0b013e3181ff9685] [PMID: 21115957]

[202] Ujiie S, Hatano T, Kubo S, *et al.* LRRK2 I2020T mutation is associated with tau pathology. Parkinsonism Relat Disord 2012; 18(7): 819-23.
[http://dx.doi.org/10.1016/j.parkreldis.2012.03.024] [PMID: 22525366]

CHAPTER 2

The Therapy of Alzheimer's Disease: Towards a New Generation of Drugs

Luca Piemontese[1,*], Fulvio Loiodice[1], Sílvia Chaves[2] and Maria Amélia Santos[2]

[1] *Dipartimento di Farmacia–Scienze del Farmaco, Università degli Studi di Bari "Aldo Moro", Via E. Orabona 4, 70125 Bari, Italy*

[2] *Centro de Química Estrutural, Instituto Superior Técnico, Universidade de Lisboa, Av. Rovisco Pais 1, 1049-001 Lisboa, Portugal*

Abstract: The treatment of neurodegenerative diseases is one of the most urgent challenges for pharmaceutical industry and public institutions. Alzheimer's disease (AD), in particular, is a severe age-dependent dementia, currently affecting 44.4 million people worldwide, and this number is estimated to rise to 131.5 million in 2050.

Only a few drugs are currently available for AD therapy, but these molecules are just able to temporarily improve the symptoms of the patients. Recently, important advancements have been achieved about the knowledge of this complex disease, even if, unfortunately, every attempt to obtain new efficient drugs for its therapy has failed.

Following the theory of the multifactorial origin of the disease, in the last decade, researchers mainly focused on the development of multi-target agents, acting on the classical features recognized as important against the onset of AD, such as NMDA receptor antagonism, inhibition of cholinesterases (ChEs) and beta-Secretase (BACE), as well as inhibition of beta amyloid peptide ($A\beta$) aggregation and antioxidant activity.

More recently, the modulation of dyshomeostasis of metal ions (*i.e.* copper, zinc and iron) in the AD patient brains, has been proposed as a disease-modifying therapeutic (DMT) strategy, due to their involvement in $A\beta$ aggregation and in the formation of Reactive Oxygen Species (ROS). Noteworthy is the role of Peroxisome Proliferator-Activated Receptors (PPARs) in the onset of neurodegenerative diseases as well. $PPAR\alpha$ expression levels have been reported to significantly decline in central nervous system (CNS) during the aging process. $PPAR\gamma$, instead, is reported to have a neuroprotective effect, with a different mechanism that influences the $A\beta$ precursor protein (APP) cleavage and the inflammatory response. The overexpression of certain types of ApoE also seems to increase the risk of developing AD. Therefore, the modulation of both PPAR subtypes seems to be a new interesting target to be explored.

[*] **Corresponding author Luca Piemontese**: Dipartimento di Farmacia–Scienze del Farmaco, Università degli Studi di Bari "Aldo Moro", Via E. Orabona 4, 70125 Bari, Italy; Tel: +39-080-5442732; E-mail: luca.piemontese@uniba.it

Atta-ur-Rahman (Ed.)

Other innovative targets as well are currently studied to find the final breakthrough for the therapy of AD: a number of years have passed since the approval of the last active drug in the treatment of this pathology and people now need a new hope.

Keywords: Alzheimer's Disease, Aβ Amyloid Plaques, Acetylcholinesterase Inhibitors, Disease-Modifying Agents, Donepezil, Innovative Targets, Insulin, Multi-Functional Drugs, Multitarget Therapy, Neurodegenerative Disease, Pioglitazone, PPARs, Type 2 Diabetes Mellitus, Type 3 Diabetes, Tacrine.

INTRODUCTION

Alzheimer's disease (AD) is a widely recognized socioeconomic problem. Indeed, the average annual cost per patient is estimated in $15,000–20,000, firstly due to the assistance expenses for not auto-sufficient people [1], taking into account that the annual incidence is around 34/1000 persons over 60 years old, and the prevalence is up to 42.1% at age > 95 years [1 - 4]. The impact of the pathology will be higher considering the wide increase of life expectancy in most of the world countries: it is estimated that, in the absence of effective therapies, the number of people with this dementia will reach 131.5 million by 2050 [5].

The pathology has been described for the first time by Aloïs Alzheimer in 1907 as an "unusual illness of the cerebral cortex". His short communication, published in German language, was recently translated [6] to English and it reported the experience of the physician with a 51 years old woman, who suffered from uncommon cognitive impairment, characterized by disorientation in space and time, hallucinations, and progressive memory loss [6]. The post-mortem exam of the patient's neuronal tissues showed a uniformly atrophic brain without macroscopic focal degeneration. However, further studies revealed many fibrils located next to each other, corresponding to the first observation of beta amyloid (Aβ) plaques [6]. This occurred around one century ago, but now this pathology is the most widespread neurodegenerative disease. It has a name (Alzheimer's disease) but its etiology continues to be unknown [7, 8].

AD is characterized by a progressive deterioration of cognitive functions that can be linked to a significant reduction of the volume of the brain, the degeneration of synapses and the death of neurons, especially in hippocampus (this being the reason of progressive memory loss) [8 - 10]. The Aβ plaques are located in the extracellular space while another kind of aggregates, very frequently found inside neuronal cells, are constituted by the so-called neurofibrillary tangles. These aggregates, which are also observed in Parkinson's disease (PD), are composed of hyper-phosphorylated Tau proteins [8]. The presence of both structures is very often correlated in AD patients [8].

In the last years, many routes have been suggested for understanding AD pathogenesis and addressing relevant drug strategies to fight this pathology. The most commonly pursued approaches are the cholinergic and the amyloid hypotheses [11].

According to numerous research works, the damage of cholinergic neurons or simply the cholinergic activity deficit, frequently linked with AD onset, leads to an overall loss of acetylcholine activity [12]. Moreover, acetylcholinesterase (AChE) seems to play also a certain role in several secondary non-cholinergic functions and particularly in the deposition of Aβ in the extracellular environment of AD diagnosed brains [12]. A recent research has shown that, in addition to AChE, also butyrylcholinesterase (BChE) plays an important role in cholinergic neurotransmission, especially in normal central nervous system (CNS) [12, 13]. Based on this consideration, many studies suggest that a non-selective ChE inhibitor should lead to better clinical results [12, 14].

The Aβ peptides are produced by cleavage operated by β- and γ-secretase enzymes in the inter-synaptic environment on the membrane-anchored APP (β-Amyloid Precursor Protein). The activity of these proteases leads to oligomeric fragments that, following a natural aggregation process, become themselves active neurotoxins, therefore causing neuronal dysfunction, loss of synaptic connections and cell death [15]. Aβ peptide aggregates lead to the formation of amyloid plaques that include in their structures metal ions such as copper-(II) and zinc-(II) [16]. Their cytotoxicity has been associated by several authors with the production of reactive oxygen species (ROS), with increased intracellular calcium level and consequent mitochondrial dysfunction, lipid peroxidation and neuronal inflammation [15, 17]. Therefore, considering that Aβ toxicity begins outside the cells and makes its way inside to disrupt basic cellular processes [15], the ideal therapeutic interventions should be directed both at the early stages of the process and at the end of it, through the inhibition of the aggregation of amyloid plaques and their disaggregation, respectively.

Although numerous clinical trials have been projected and accomplished over the last ten years for the evaluation of potential therapies [5, 18, 19], unfortunately only four acetyl-cholinesterase inhibitors (AChEi, namely tacrine, donepezil, rivastigmine, galantamine) and memantine (that acts by blocking NMDA receptors) have been approved for the use as anti-AD agents at international level, even if tacrine has been discontinued from the market in several countries due to its side effects at therapeutic doses [20]. These drugs are simply for symptomatic treatments, leading to temporary amelioration of memory and cognitive problems. They neither fight the cause of the pathology nor slow down neurodegeneration. Therefore, their clinical effects can be defined as modest [5, 21].

Researchers from Public Health Systems, as well as from private companies, are now involved in the identification of new disease-modifying agents (DMAs). Considering the wide number of patients, it would be a big breakthrough. However, to approve a DMA, a long phase of preclinical development and initial characterization of the molecule (up to several years), as well as three phases of clinical studies (around three years), are necessary [5, 22]. So, to have a realistic chance to be approved within 2025, the potential new drugs need to be currently at least in Phase 2 of clinical studies [5]. Including preclinical development, the total development time reaches up to ten years and the cost, together with that related to the failures, is estimated at around $5.7 billion [5, 22]. The main targets pursued are amyloid, followed by inflammatory mediators/factors and tau aggregation inhibitors, while the dominant innovative method is passive immunization targeting amyloid [5].

Another approach that recently was followed by numerous pharmaceutical companies is the so-called "drug repositioning" (also known as "drug re-profiling" or "drug repurposing"). It is the procedure of recycling, through a new development process, a drug or a past candidate drug for its use in the treatment of an alternative pathology [23]. In this way, the first steps of pre-clinical phases are faster, considering that these molecules have been tested in humans, and the most important pharmacokinetic and toxicological-kinetic information is already available and can be used as a fundamental starting point for further development [23]. Therefore, drug repositioning has many advantages over traditional *de novo* drug discovery approaches in that it can significantly reduce the cost and development time of new chemical entities and it is a milestone in the recent Alzheimer's disease drug development pipeline [23, 24]. Following this convenient approach, Pioglitazone and Insulin are among the compounds included in phase III clinical trials of 2017 [19]. Telmisartan, instead, is encompassed in repositioned drugs in Phase II of clinical trials, as reported in the Alzheimer's disease drug development pipeline of 2018 [24 - 26] (Table **1**).

The investigation about the possibility of using, in the therapy of Alzheimer's disease, drugs developed for the treatment of metabolic diseases is a new intriguing approach in the tentative comprehension of this complex pathology.

In fact, recent evidence in clinical and basic research has shown that there are common pathological and physiological changes which determine a relationship between AD and Type 2 Diabetes Mellitus (T2DM). This condition is currently termed as Type 3 Diabetes (T3D) [27] and this is probably due to the presence of several common pathways in both pathologies. This fact can justify their co-occurrence in > 65 years old people. If these suppositions will be confirmed, a common pharmacological treatment would be desirable both for prevention and therapy of the oldest sections of the population [28], according to an innovative therapeutic approach based on the patient's phenotype.

Table 1. A selection of potential drugs included in the Alzheimer's disease drug development pipeline: 2018 [25].

s Name of Molecule	Target(s)	Phase	Company
Anavex 2-73	Tau-protein, Metabolism	II	Anavex Life Sciences
Benfotiamine	Metabolic	II	Burke Medical Research Institute
BIIB092	Tau-protein	II	Biogen
Bryostatin 1	β-amyloid, Metabolism	II	Neurotrope Bioscience
E2609	BACE Inhibitor	II	Eisai, Biogen
GV1001	β-amyloid, Metabolism	II	GemVax & Kael
ID1201	β-amyloid, Metabolism	II	IlDong Pharmaceutical Co
Levetiracetam	Metabolism	II	University of California
LY3202626	BACE Inhibitor	II	Eli Lilly
Nicotinamide	Tau-protein	II	University of California
Telmisartan	Metabolism	II	Sunnybrook Health Sciences Centre
Aducanumab	β-amyloid	III	Biogen
AZD3293	BACE Inhibitor	III	AstraZeneca
CNP520	BACE Inhibitor	III	Novartis
Crenezumab	β-amyloid	III	Roche
Elenbecestat	BACE Inhibitor	III	Eisai, Biogen
Gantenerumab	β-amyloid	III	Roche
Insulin intranasal	Metabolism	III	University of Southern California
Solanezumab	β-amyloid	III	Eli Lilly
TRx0237	Tau-protein	III	TauRx Therapeutics
Verubecestat	BACE Inhibitor	III	Merck
Zolpidem	GABA-A	III	Brasilia University Hospital

Although most of the recent literature is focused on the relationships between AD and T2DM, it was recently demonstrated, in *in vivo* experiments, that type 1 diabetes negatively affects cognitive performance [29]. Therefore, an insulin unbalance could also be important in the onset of the neurological deficit typical of AD [30].

These new targets, together with those classically pursued in AD therapy research, can be the basis for the design and synthesis of new multifunctional molecules

[31]. In fact, due to the multifactorial origin of this pathology, the multi-target-directed ligand (MTDL) approach is a hopeful method in the search for new drugs [32 - 35]. Even if the common structures developed in the last two decades are AChE inhibitors, very often the pharmacological profile of the compounds was enriched with other biological activities, oriented to the identification of DMAs and able to treat the symptoms as well [32]. Aβ anti-aggregation activity, BACE and MAO inhibition, as well as antioxidant and metal chelating activities are the most reported targets [16, 32 - 35]. However, the breakthrough has not yet been reached and many innovative therapeutic goals have been identified over the years while new ones are continuously being discovered.

The model based on the "one molecule/multiple targets" paradigm has led to the design of novel molecules possessing heterodimeric structures, able to interact with different targets by combining different pharmacophores in natural products, synthetic molecules and/or existing drugs already used in therapy (*i.e.* tacrine and donepezil) [32, 36 - 38]. Particularly attractive at the industrial level is the use of natural compounds contained in food supplements [35, 39], which can be also a source of inspiration for the synthesis of more complex molecules or can be developed and commercialized following a nutraceutical approach [39].

The design of multifunctional drugs is not for sure an easy task, but numerous examples are reported in the literature [32, 36 - 38, 40 - 45] (Table **2**). The potential of this type of approach will be detailed in the text below, as well as the introduction of potential targets of innovative biological activities, such as PPAR agonism (Table **2**).

Table 2. Summary of the AD targets.

Compound(s)	Ref	Target(s)
Tacrine	Figure 2	AChE
Donepezil	Figure 2	AChE
Galantamine	Figure 2	AChE
Rivastigmine	Figure 2	AChE
Memantine	Figure 2	NMDA receptor
Tacrine-Hydroxybenzymidazol hybrids	Figure 3	AChE, Aβ, ROS, Metal Ions, BACE
WMY90	Figure 3	AChE, MAO B, ROS, Metal Ions
Rivastigmine-Lipoic acid hybrids	Figure 3	AChE, Aβ, ROS
Galantamine-Memantine hybrids	Figure 3	AChE, NMDA receptor
PBT2	Figure 4	Aβ, ROS, Metal Ions
N,N-donors	Figure 4	Aβ, ROS, Metal Ions

(Table 2) cont.....

Compound(s)	Ref	Target(s)
O,O-donors	Figure 4	Aβ, ROS, Metal Ions
3,4-Hydroxypyridinones	Figure 4	Aβ, ROS, Metal Ions
Flavonoids	Figure 4	Aβ, ROS, Metal Ions
Verubecestat	Figure 5	BACE
Elenbecestat	Figure 5	BACE
AZD3293	Figure 5	BACE
Rosiglitazone	Figure 7	PPARγ
Pioglitazone	Figure 7	PPARγ
INT131	Figure 8	PPARγ
K-877	Figure 9	PPARα

MULTITARGETING DRUG DEVELOPMENT

Despite the enormous and continuing research efforts to discover new anti-AD drugs [9, 10], the multifaceted nature of AD led to the proposal by Melchiorre *et al.*'s pioneering group, about one decade ago, of using multitargeting ligands (MTDLs) [46]. Accordingly, by enabling the simultaneous interference with two or more pathological dysfunctions of AD, the MTDLs could achieve better therapeutic efficacy than a single-target drug ("one-molecule/one-target" paradigm) [44, 47]. Since then, the MTDL's strategy has been extensively adopted in drug development approaches and so, along the last 10 years, a huge number of compounds have been reported for potential AD treatment [32, 48 - 55]. Notwithstanding all the efforts made in this area, up to now, none of the MTDL drug candidates has yet been approved. However, the multi-targeting approach still goes on as a growing challenge towards the discovery of a correct and efficient therapy.

Multi-Targeting Approach in Anti-AD Drugs: Past and Present

The main targets that have challenged the medicinal chemists in drug development (Fig. **1**) are related with the different factors that interfere in the AD pathophysiology and its most relevant hallmarks, such as the massive loss of cholinergic activity, the deposition of extracellular Aβ plaques and intracellular neurofibrillary tangles, the oxidative stress, and metal dyshomeostasis, although, more recently, several other receptors and enzymes have also been considered.

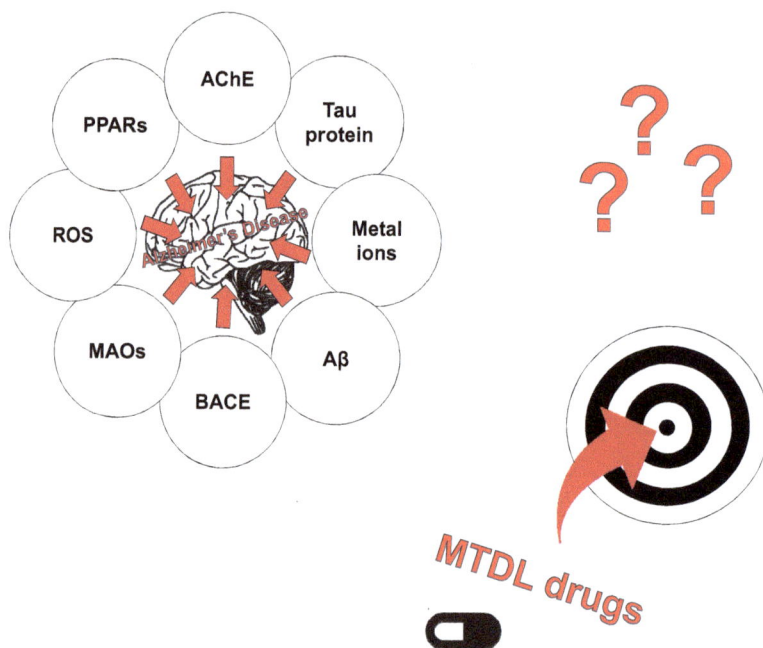

Fig. (1). Schematic illustration of the multiple targets of AD pathogenesis.

Therefore, at least theoretically, an efficient multi-target drug should include several pharmacophore groups able to interfere with some of the classical targets, namely: inhibition of enzymes (cholinesterases, secretases, monoamineoxidases), inhibition of amyloid-β aggregation, inhibition of tau-hyperphosphorylation, modulation of oxidative stress, chelation of misplaced metal ions [32, 48 - 55] and modulation of several disrupted neurotransmitter systems, such as the NMDA receptor [56], as well as the regulation of dopaminergic and serotoninergic systems [57, 58].

Importantly, such a multifunctional molecule should also possess drug-like properties, such as appropriate ADMET molecular descriptors. Therefore, the discovery of small molecules able to provide effective pharmacological responses for diverse receptors or enzymatic systems is recognized as a huge challenge but not easy to be accomplished [35, 59].

Furthermore, both the still limited knowledge of the mechanisms underlying the AD process and the high number of proposed disease targets have led the medicinal chemists to face the problem of selecting the most relevant AD targets. Also, ideally, the right combination should lead to additive or synergistic effects, turning out the task even more difficult [60].

In this context and in a simplistic way, the strategies for the development of multi-targeting anti-AD drugs have pursued two main approaches [61]: one gives higher emphasis to the treatment of the ongoing AD process, namely to control symptomatic effects (mostly associated to loss of neurotransmitter systems), together with some potential process reversion [62, 63]; another one gives higher relevance to disease modifying effects, namely to avoid or delay the AD onset and the upstream disease events (mostly connected with Aβ and Tau aggregates production) [64 - 66]. Herein, a review of the most relevant current developments (last three years) of multitargeting anti-AD drug candidates will be made, aiming at both the symptoms and the causes of the disease.

MTDLs with Symptomatic Effects: Multifunctional Cholinesterase Inhibitors

Given the association of impairment of cognitive function with the failure of cholinergic neurotransmission in AD [67], along the last two decades, several strategies have been developed to compensate that loss. In particular, since the discovery of tacrine (1993), five drugs were FDA approved for AD treatment and, among those, four cholinesterase (ChE) inhibitors [tacrine (Cognex®), donepezil (Aricept®), rivastigmine (Exelon®), galantamine (Razadyne®)] and one NMDA receptor antagonist [memantine (Namenda®)] (see Fig. **2**). Excluding tacrine, that was withdrawn from the market due to its hepatotoxicity, the other drugs are still under current use, while recently the drug combination of AChE inhibitors with memantine has also been recommended [68, 69].

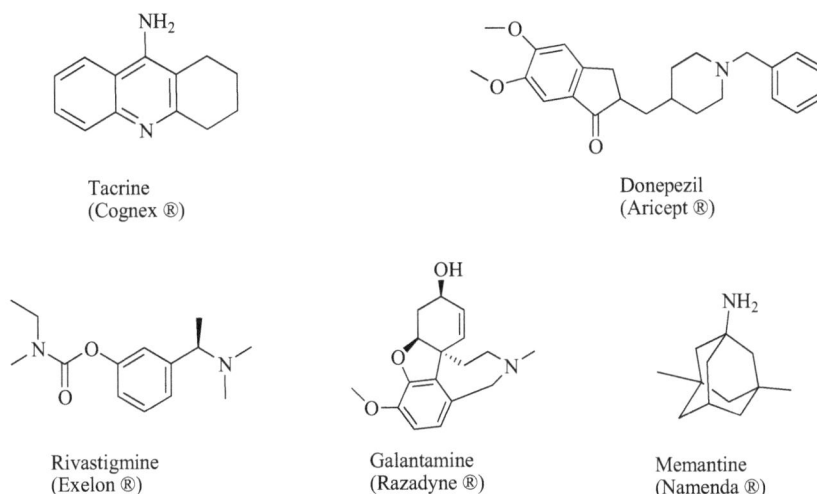

Tacrine
(Cognex ®)

Donepezil
(Aricept ®)

Rivastigmine
(Exelon ®)

Galantamine
(Razadyne ®)

Memantine
(Namenda ®)

Fig. (2). Structural formulae of FDA-approved therapeutic agents in AD (and corresponding market names®).

However, the fact that only symptomatic ameliorations can result from these treatments led to an important search for compounds that, besides the cholinesterase inhibitory capacity, could also provide other pharmacological responses, namely through: inhibition of Aβ aggregation and/or inhibition of other enzymes such as β-secretase (BACE-1, responsible for the synthesis of Aβ) or monoamine oxidase A and B (MAO-A and MAO-B, that catalyze the oxidation of monoamines, including several neurotransmitters), anti-oxidant and/or metal modulating capacity [33], or even interference with other receptors such as the serotonergic ones.

Typically, the MTDLs result from combining pharmacophores (of selected molecule backbones already known as drugs or drug candidates) in the same molecular entity by linking, fusing or merging strategies, taking in mind that a higher integration should correspond to smaller molecules and should lead to better drug-like properties [40, 70]. Therefore, before starting the expensive experimental process of synthesis and activity assessments, the rational design of multi-target compounds can provide an essential aid to find the right target combination, but it is not an easy task [50, 71, 72]. For that purpose, computational simulations, such as the cheminformatics, virtual screening, molecular docking and many other computational approaches, are very important tools for predicting the best compound/association options to interact with the drug targets and also fulfill druggability requirements.

Multitargeting Approach Based in Repositioning "Old" Anti-AD Drugs

Under the MTDLs strategy for anti-AD drugs, a huge number of hybrid molecules has been based on the derivatization of already known "old" anti-AD drugs, namely AChE inhibitors (AChEi), to provide them with capacity to interact with other important disease targets. This strategy of repositioning "old" drugs is based on the recognized paradigm "if it works, let´s make it working even better". Also, the use of drugs involving approved or already validated combinations may facilitate therapeutic outcomes and translational strategies.

Usually, one AChEi drug molecule (*e.g.* tacrine) or one important molecular segment (*e.g.* benzylpiperidine from donepezil or phenyl-carbamate from rivastigmine) is selected as a template for the generation of multifunctional hybrids. Although tacrine is not under current usage for AD treatment, it is still one of the most used AChEi drugs in MTDL approach, due to its easy derivatization; also, by targeting complementary pathways, its hybrid is ultimately expected to have therapeutic effects at lower doses than tacrine and consequently a better safety profile.

The recognized interest for this drug development strategy is reflected by the huge amount of reviews published along the last three years, which have been mainly focused on the discovery of multipotent hybrids bearing mostly tacrine- or donepezil-like structural segments [48, 50, 59, 73, 74]. However, some authors have given higher emphasis to derivatives with specific targets, besides the AChE, such as the MAO enzymes [75] or oxidative stress/metal dyshomeostasis [54, 55]. The current continuing interest for this approach is also illustrated by the number of very recent publications (last two years) focused on combining molecular segments for ChE inhibition with other moieties to hit multiple targets [37, 38, 76, 77] or otherwise mostly addressing one extra-target (dual targeting), such as anti-oxidant/metal dyshomeostasis [49, 78, 79] or receptor antagonism in neurodegenerative diseases, such as the serotonergic [80, 81] or histamine systems [82] as well as melatonin receptors [83]. It is noteworthy that some mimetics of other AChE inhibitory drugs, such as rivastigmine [55] or galantamine [84], have also been recently used for multitargeting strategies by hybridization (Fig. **3**).

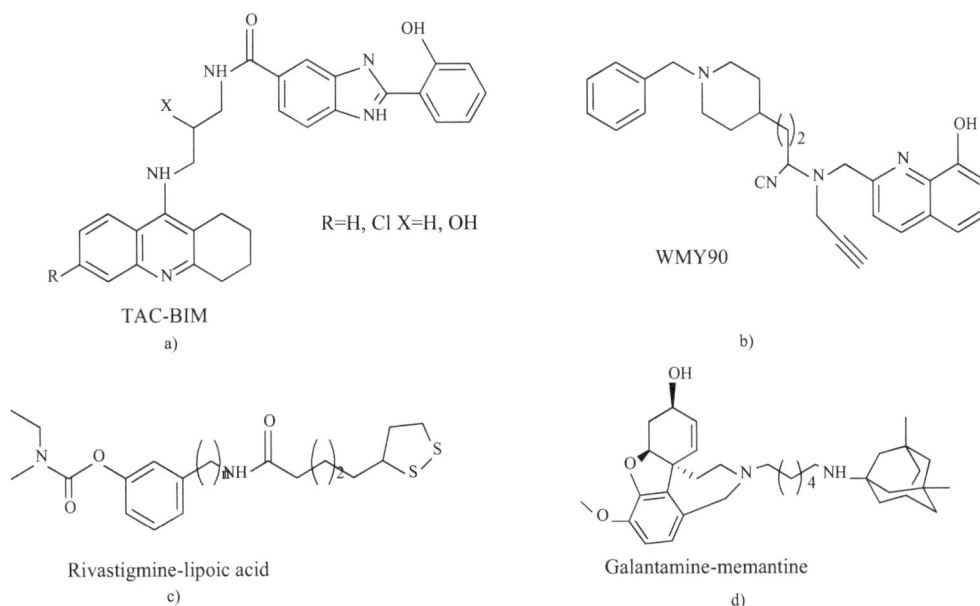

Fig. (3). Examples of multitarget-drugs based on repurposing "old" AChE inhibitory drugs: a) tacrine-hydroxybenzymidazole hybrids, targeting: AChE, Aβ, ROS, biometals and potential BACE-1 (X = OH) [38]; b) donepezil segment hybridized, targeting: AChE, MAO B, ROS, biometals [48]; c) rivastigmine-lipoic acid hybrids, targeting: AChE, Aβ, ROS [55]; d) galantamine-memantine hybrids, targeting AChE and NMDA receptor [84].

Multitargeting Approach with Major Emphasis on Disease Modifying Agents

Notwithstanding the recognized great significance given to the development of multitargeting anti-AD drugs able to combat the symptoms of the disease, together with the tackling of other targets to delay its progress (previous section), there has been a lot of interest on engineering new multifunctional compounds with high emphasis on disease modifying effects to avoid or delay the AD onset and the upstream disease events. They mostly aim to reduce the amyloid plaque formation by inhibition of Aβ aggregation (and related features as metal dysregulation and oxidative stress) and, more recently, to reduce the production of the neurotoxic $A\beta_{1-42}$ peptide through the inhibition of secretases, especially β-secretases type 1 (BACE-1) that cleave the membrane-associated APP.

In fact, although the amyloid cascade hypothesis has long been considered the main cause of AD pathology [85], only a few small molecules have been developed as inhibitors of Aβ aggregation for potential therapeutic treatment of AD, but without significant results [86]. In spite of the lack of positive clinical trials associated to that strategy [87], in the last years there was a renaissance of potential multitarget drugs centered in Aβ anti-aggregation and/or BACE-1 inhibition. Regarding the multi-target drug developing strategy with main emphasis on the inhibition of Aβ aggregation, quite a number of polyfunctional potential drugs have also included molecular moieties for cholinesterase inhibition [38, 79, 83]. On the other hand, a huge amount of research has also been focused on the development of small molecules for the inhibition of Aβ aggregation and the modulation of associated metal dyshomeostasis and oxidative stress.

Alzheimer′s Disease and "Neuro-Metals"

Since a long time, there has been increasing evidences indicating that the disruption of "neuro-metal" homeostasis contributes to various neurodegenerative diseases, including AD. In fact, it is well known that some bio-metals, including copper (Cu), zinc (Zn) and iron (Fe), play significant roles in the conformations, functions and toxicity of amyloidogenic proteins, particularly at the synapses, and so they have been associated with the pathogenesis of some neurodegenerative diseases such as AD [88].

Among the mentioned metal ions, Fe is the most abundant in the brain, as well as in the whole body. Although it is essential for several important biological functions, it has critical roles in specialized brain functions, including synthesis of the neurotransmitter dopamine [89]. However, when it is in excess, this redox-active metal ion can generate reactive oxygen species (ROS) through Fenton and Haber-Weiss reactions, that can damage DNA, proteins and lipids, and thus be toxic to neurons. Zinc is the second most abundant element in the brain, existing mostly bound to metalloproteins but also in part as free ion, particularly in synaptic vesicles [88, 90]. The secreted Zn^{2+} modulates overall brain excitability by binding *N*-methyl-*D*-aspartate (NMDA) type glutamate receptor and other important receptors, which are critical to information processing and synaptic plasticity [91]. Copper, the third most abundant element in the brain, is a cofactor of several enzymes, such as cytochrome C and superoxide dismutase, and it is essential for brain functions. However, when in excess, similarly to Fe, free Cu is toxic because it produces ROS which can bind to functional proteins and cause mutations [88]. Different levels of metal cations can coexist in various brain regions and interfere with each other. For example, Cu and Zn can interact with one another and bind the same metal-binding proteins and the Aβ peptide, changing the Aβ aggregation profile [92]. Aluminum is an environmental risk factor for AD because, by sharing similarities with Fe, it can interfere with the Fe functions and be highly toxic to brain [93].

Besides the above referred recognition that extracellular Aβ plaque deposits and intracellular neurofibrillary tangles within the brains of patients are primary markers of AD, dyshomeostasis of bio-metals is another important biomarker. This conclusion is based on the comparison between the concentrations of bio-metals within the brains of AD patients and those of healthy individuals [94]. In particular, in Aβ plaque deposits of AD brains, the metal concentrations were found about 3-fold higher for Zn^{2+} and Fe^{2+}/Fe^{3+}, and 5-6-fold higher for Cu^{2+}, being especially greater close to the N-terminus of the Aβ peptide [95]. Furthermore, the dyshomeostasis of redox active metal ions, including those unbounded and bounded to Aβ peptides, can also strengthen the generation of reactive oxygen species (ROS) that endanger DNA, proteins, lipids and can cause neuronal damage in the CNS of AD patients. Therefore, the "neuro-metals" can be taken as biomarkers of AD and their modulation has been admitted as an important disease target in anti-AD multi-target drug approach, namely in an integrated strategy of controlling the amyloid plaque formation [54].

Multi-Target Drugs with Anti-Aβ Aggregation and Metal Modulation Capacity

Based on the inherent complexities of AD, specifically associated with the deleterious effects of brain bio-metal dyshomeostasis in the misfolded protein aggregation in the brains of AD patients, and also with the serious injury of related ROS in biological tissues of CNS, a number of novel multi-functional drugs has been recently developed in an integrative way. In particular, there is a huge amount of scientific publications based on the rational design of molecules able to target and react with specific distinct pathological targets associated with

AD, aimed to understand and identify the disease mechanisms implicated in upstream causative events leading to the onset and progression of AD. In this context, several new small molecules have been recently developed and explored for their capacity to interact with metal ions, to modulate metal-free and metal-bound Aβ aggregation *in vitro*, and also to have regulatory activities towards ROS and free organic radicals. Their structure is mainly based on the conjugation of mono- or poly-aromatic or -heteroaromatic systems (for the interaction/ intercalation with the Aβ oligomers to avoid/destroy the formation of amyloid fibrils) with metal chelating systems, mostly bidentate systems, containing X,Y-donor atoms ($X = N, O$; $Y = O, N$) [54]. The selection of the chelating systems and corresponding donor atoms depends on the relevance given to the modulation of a specific metal ion or on some degree of selectivity in the set of cations. The metal of major concern in AD has been copper followed by zinc, which are more frequently bound to APP and Aβ than iron, although the role of iron in AD pathogenesis is well substantiated and mostly associated with iron-dependent ROS [95, 96]. Therefore, the most recent anti-Aβ aggregation drugs integrating metal chelating functions have been addressed for binding *hard-soft* metal ions (Cu^{2+}, Zn^{2+}). So, they involve mostly N,O-donor atoms as in the series of 8-hydroxyquinoline hybrids [54] represented by PBT2 (Fig. **4a**). Despite its very promising properties, it was discontinued in phase II clinical trials by the Australian company "Prana Biotechnology" in April 2014 [97], apparently due to complications with large-scale manufacturing of the compound. Analogous compounds, namely hydroxyquinoline conjugates with indol, have been reported to accomplish ameliorations of the cognitive and spatial memory deficits in AD mice as well as reduction of overall cerebral β-amyloid deposits [98]; also conjugates with triazole seem to rescue the neurotoxicity associated with Cu-Aβ, so warranting further investigation in animal models [99]. Quite a number of reported N,O-donor ligands is based on betaphenol-imines or -amines involving different type of aromatic moieties, namely benzothiazole (Fig. **4b**) [100] or resveratrol (Fig. **4c**); these compounds have been shown to improve, respectively, the anti-Aβ aggregation or the anti-oxidant properties. Even an amino-propargyl substituent has been introduced to target MAO (Fig. **4c**) [101]. Regarding hybrids integrating metal-chelating moieties with the *soft* N,N-donor ability (Fig. **4d**) [102], quite a number of compounds has been also rationally designed to tune the targeting and regulation of distinct pathological factors linked to AD pathology (*e.g.* metal ions, metal-free Aβ, metal–Aβ, ROS, free organic radicals), involving studies *in vitro* and also some in mouse models [53]. Regarding the hybrids integrating chelating moieties with O,O-donors, there are several derivatives of the marketed iron-chelating drug Desferal®, used to treat iron overload disorders, that include in their structure the 3-hydroxy-4-pyridinone (3,4-HP) moiety [103] and flavonoid derivatives as mimetics of naturally occurring flavonoids existent in fruits and vegetables (*e.g.* myricetin from berries, quercetin from onions and apples, epigallocatechin (EGCG) from green tea extract) [104]. Although all this type of ligands has high capacity for chelation of the redox-active iron and copper, they can have a broad range of pharmacological properties, such as anti-inflammatory and neuroprotection capacity and even anti-amyloidogenic properties, especially depending on the type of substituents [16, 103, 105].

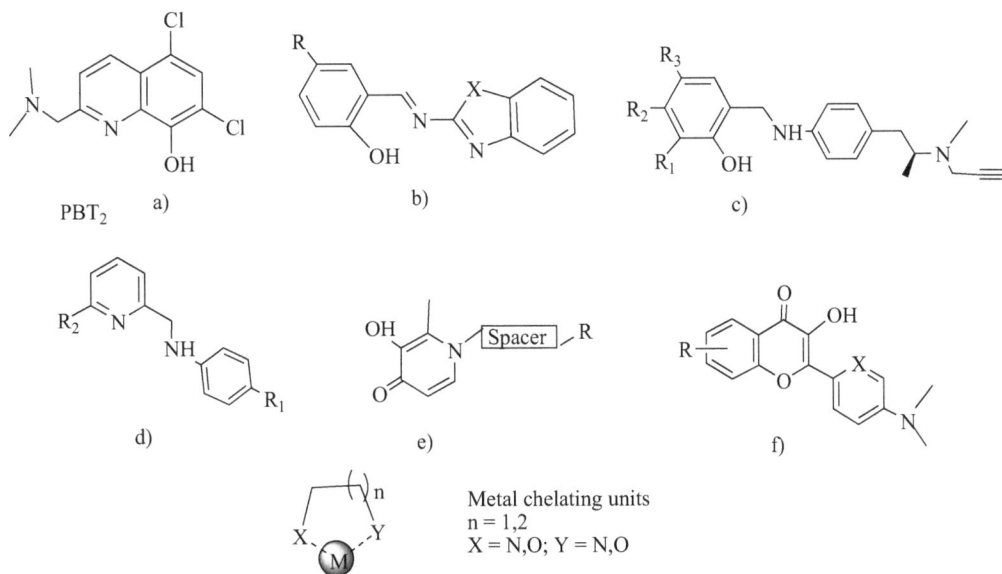

Fig. (4). Structural formulae for representative examples of small size hybrids integrating anti-Aβ aggregation, metal (M) modulation and anti-oxidant properties; *N,O*-donors: a) hydroxyquinolines (PBT2), b) and c) beta-phenol-imines or –amines; *N,N*-donors: d) beta-aminopyridines; *O,O*-donors: e) 3,4-hydroxypyridinones, f) flavonoids.

For example, to increase the capacity of these compounds to interact with Aβ oligomers and act as anti-Aβ aggregators, 3,4-HP derivatives (Fig. **4e**) have been functionalized with benzothiazole (as analogues of the thioflavin AD imaging agent) [106] or benzofuran moieties [107], and some of these compounds showed capacity to dissolve Aβ from Cu(II)- and Zn(II)-promoted aggregation *in vitro*. For the flavonoids, the improvement of Aβ interaction was induced by the 4'-$N(CH_3)_2$ group in lateral aromatic B ring (Fig. **4f**) [108]. Overall these are representative examples of small sized hybrids with capacity to protect neurons from oxidative species (by modulating the redox active metal ions, *e.g.* Cu^{2+} and Fe^{3+}, and by capturing radicals), as well as reduce the formation of senile plaques and clear them from the brain.

Strategies Based on BACE-1 Inhibition

The accumulation of amyloid plaques, composed of Aβ peptides, in hippocampal and cortical regions of the brain of AD patients has long been considered one of the major histopathological hallmarks of AD, and so, the inhibition of Aβ peptide formation has been considered a key strategy for disease-modifying therapy of AD [109 - 111]. Since the discovery in 1999 of the role of the transmembrane aspartic protease β-secretase (β-site APP-cleaving enzyme 1, BACE-1) in the proteolytic cleavage of APP, as one of the first steps in the cascade of events leading to toxic Aβ production [112], extensive research has been conducted by pharmaceutical companies and academia. These studies have led to a number of BACE-1 inhibitors with remarkable potency, several of which are subjected to ongoing clinical studies [113 - 117]. Along almost two decades, there has been a great evolution about the variety of structural chemotypes of β-secretase inhibitors. As with other proteases, the first strategy was based on peptide-mimetics, by substrate homology, but soon, to overcome several issues, particularly related with their molecular weight and other drawbacks like BBB penetration, they were substituted by small size non-peptide inhibitors using a set of different main scaffolds such as acylguanidine, 2-aminopyridine, aminoimidazole, amino- or imino-hydantoin, and aminothia- or aminooxa-zoline [113, 114].

Therefore, these scaffolds and many other analogues have been recently used as main tools for chemical derivatizations in many medicinal chemistry labs, following different drug development approaches and design strategies to guarantee not only a selective inhibitory activity by strong interaction with enzyme´s catalytic machinery (aspartic dyad) within the large BACE-1 active site and subsites (through hydrogen, hydrophilic and lipophilic bindings), but also drug-like properties [114 - 117]. Furthermore, several of the lately reported BACE-1 inhibitors appear as drug candidates (Fig. **5**) and some of them have been enrolled in clinical trials: *e.g.* Verubecestat (MK-8931, removed in February 2018) [118, 119] and Elenbecestat, (E2609, advanced Elsai´s and Biogen´s Phase 3 clinical trials) [120]. Other molecules (e.g. AZD3293 [115], JNJ-54861911 and CNP520) are still in clinical trials [114]. Another possibility for BACE inhibitors is their use in pre-symptomatic phase of Alzheimer's disease. In order to evaluate it, two large trials are running and will finish in 2023, aimed to investigate the effects of the candidate drugs in individuals with normal cognition but who carry genetic risk factor for AD [120].

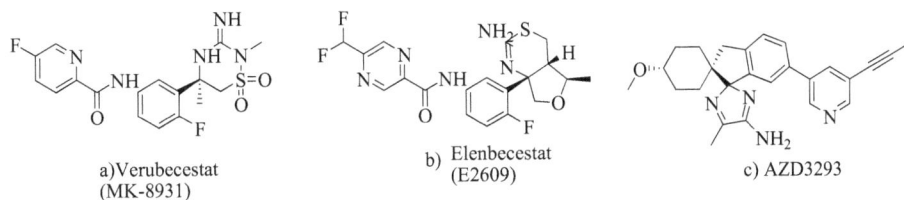

Fig. (5). Representative known structures of BACE-1 clinical drug candidates.

Multitargeting Approach Centered on BACE-1 Inhibition

Notwithstanding the extensive worldwide recent research on the inhibitors of BACE-1 and its account against amyloidogenesis, unfortunately, the discovery of an effective drug to cure AD keeps being a mirage. Therefore, once more, the recognized multi-factorial nature of AD and the relevant role of BACE-1 in the onset of amyloid cascade led medicinal chemists to give a step forward in the search for new anti-AD therapies, aimed to discover new chemical entities to be potentially brought to AD clinical trials.

As above reported, some of the recent approaches for new multi-targeting drug candidates integrate BACE-1 inhibition with other targets (symptomatic and/or disease-modifying agents) *e.g.* through the dual inhibition of AChE and BACE-1 activity, as well as the dual Aβ and tau anti-aggregation [121, 122]. This wide combination of functionalities is expected to confer to the drugs a broad and well-balanced spectrum of biological activities, and so maybe a good starting point for further developments.

The difficulty of targeting BACE-1 along with several other disease targets led to recent drug developing strategies based on a fine-tuned modulation of a lower number of selected pathways, instead of a more complete blockade [123]. Therefore, there have been recent reports on promising MTDLs which can combine the alleviation of the symptoms in the early symptomatic phase of the disease, by inhibiting AChE, and the fight of amyloidogenesis, by inhibiting BACE-1. The strategy of dual AChE/BACE-1 inhibitors started to be followed in 2008 [124], and since then it keeps being the object of intensive research. Some compounds revealed also other important properties as inhibitors of Aβ aggregation or as antagonists of histamine H3 receptor [125] (section 4). It is noteworthy that most of the strategies used for AChE inhibition were based on the repurposing of already known anti-AD drugs, namely tacrine [126] and donepezil [127] (Fig. **6a,b**).

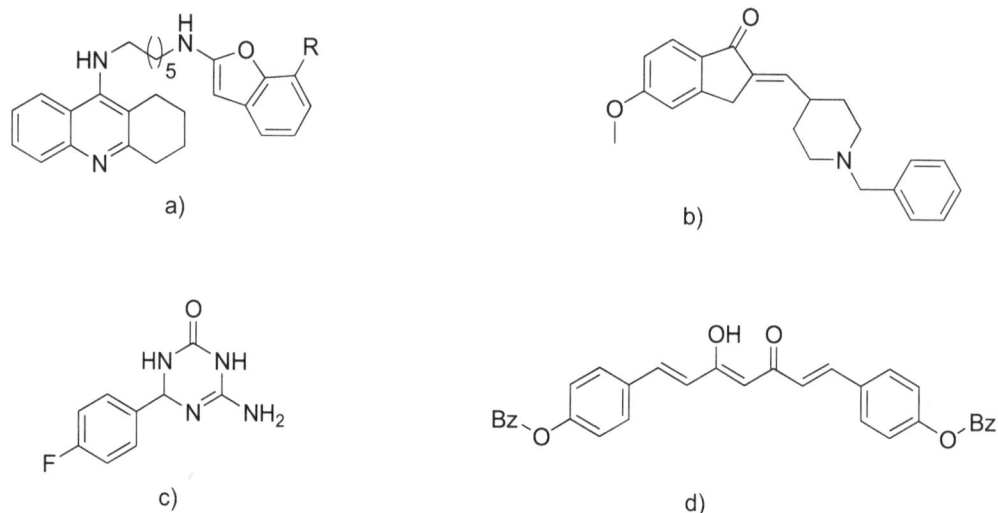

Fig. (6). Multitarget compounds centered on BACE-1: dual BACE-1/AChE inhibitors (**a,b**) and dual BACE-1/GSK-3β inhibitors (**c,d**).

Another important BACE-1 related dual targeting strategy for anti-AD therapeutics has been recently pursued aimed to further reinforce the targeting of upstream AD events, namely *via* the dual inhibition of two key enzymes, BACE-1/GSK-3β, which are involved in two main pathological cascades of AD [123] (section 4 and Fig. **6c,d**).

Compound c) (Fig. **6**) is a small size molecule, bearing a triazinone skeleton resulting from combining two pharmacophore functions, and demonstrated a balanced micromolar range of activity for both enzyme targets [128]. Regarding compound d), resulting from derivatization of the natural compound curcumin, as compared with the previous compound, it has a higher molecular weight, higher and lower activity against BACE-1 and GSK-3β, respectively, but it is further endowed with antioxidant activity [129]. However, none of these compounds has been enrolled in clinical trials.

PEROXISOME PROLIFERATOR-ACTIVATED RECEPTORS

Peroxisome Proliferator-Activated Receptors (PPARs) are ligand-activated transcriptional factors that belong to the nuclear hormone receptor superfamily. They form heterodimers with the retinoid X receptor (RXR) and bind to specific PPAR response elements which are specific regions on the DNA of target genes [130 - 132]. Under the inactivated conditions, these heterodimers are associated with corepressors which inhibit the gene expression. However, binding of the ligand induces conformational changes which facilitate the release of corepressors and promotes the recruitment of coactivators thus initiating the gene transcription [133]. There are three distinct PPAR subtypes, PPARα, PPARγ, and PPARδ, with different tissue distributions and physiological functions. The modulation of these nuclear receptors plays a key role in the regulation of a large number of genes whose products are directly or indirectly involved in energy homeostasis and lipid and carbohydrate metabolism. For this reason, they have been considered suitable targets for the treatment of metabolic disorders. The lipid-lowering fibrates (*e.g.*, fenofibrate and gemfibrozil) are PPARα ligands [134, 135]. The marketed thiazolidinedione (TZD) class of antidiabetic agents (rosiglitazone and pioglitazone, Fig. (**7a**) and (**b**) activates PPARγ [136].

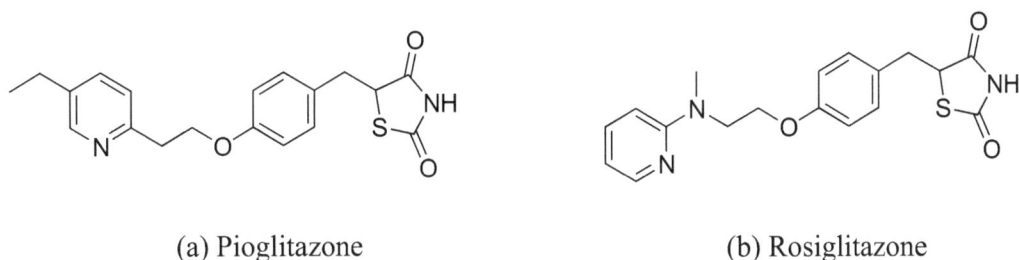

(a) Pioglitazone   (b) Rosiglitazone

Fig. (7). Marketed thiazolidinedione (TZD) class of antidiabetic agents.

They enhance insulin sensitivity in target tissues and lower glucose and fatty acid levels in type 2 diabetic patients. In the last 10-15 years, experimental evidence has revealed the presence of all PPAR subtypes in the CNS and suggested an important role of PPAR in brain function and neuronal development. *In vitro* and *in vivo* experiments have shown that PPARδ is the dominant form present in the rat brain and that it is expressed in all neural cell types, whereas the distributions of PPARα and γ are more restricted [137, 138]. The presence of human PPARs is difficult to verify *in vivo*. However, several clinical trials have demonstrated similar PPAR-mediated effects in both animal models and human cells suggesting that these receptors are present and fully active within the human CNS [139]. PPARs regulate many neurophysiological processes, such as energy metabolism, redox homeostasis, cell cycle and differentiation [139]. PPARα seems to be involved in the control of brain inflammation [140, 141], whereas PPARγ has an important role in the CNS by regulating insulin-sensitization, anti-inflammatory activity, brain metabolite balance, neuronal differentiation and neurite outgrowth [142 - 146]. The role of PPARδ has been less addressed and several studies have indicated that antioxidant and anti-inflammatory modulations are the most important PPARδ-mediated effects [147]; in addition, recent studies have established a link between metabolic syndrome and neurodegeneration [148] suggesting that PPARδ-mediated hypolipidemia has a critical role in CNS physiology [149].

PPARs and Alzheimer's Disease

PPARs have recently emerged as viable targets in palliative treatments for AD. Indeed, diseases such as T2DM and obesity significantly increase the risk of AD [150, 151]. Consistent with these epidemiological studies, the hypothesis of brain insulin resistance as a possible etiological factor of sporadic AD, has gained more attention [152]. In fact, inhibition of insulin signaling in the brain is associated with memory impairment, tau-hyperphosphorylation, and Aβ accumulation [153, 154]. T2DM and obesity are usually treated with PPARγ and PPARα agonists, such as TZDs and fibrates, respectively; therefore, these drugs could represent a promising therapeutic approach for AD. A large body of experimental evidence shows the beneficial effects of PPAR agonists that probably do not rely on one particular mechanism, but rather on an array of different pathways which is in line with the fact that AD is a multifactorial disease. The mechanisms that have been described, so far, as underlying the beneficial effects of these nuclear receptors can be summarized as resulting basically from three categories: Aβ metabolism, neuroinflammation, and mitochondrial dysfunction.

Aβ Metabolism

PPAR activation has been shown to exert neuroprotective effects against Aβ-induced neuronal damage by modulating the generation or clearance of this peptide. Santos and co-workers were the first to demonstrate that neurotoxic damage caused by β-amyloid was reduced by pretreatment with the PPARα agonist Wy-14,463 [155]. In particular, hippocampal neuronal death and neuritic network loss were decreased, concomitant with an increase of PPARα-mediated peroxisomes and catalase activity. A later *in vivo* investigation from the same group in a double transgenic mouse model of AD treated with a set of synthetic PPARα activators showed lower levels of Aβ aggregates, decreased tau-phosphorylation, a reduction of learning and memory impairment, synaptic failure and neurodegeneration [156]. The anti-amyloidogenic effects were also reported for some other PPARα agonists: fenofibrate and GW7647 were shown to modulate the APP processing *in vitro* and *in vivo*, suggesting the involvement of PI3-K pathway [157, 158]; Corbett *et al.* demonstrated that gemfibrozil stimulates the non-amyloidogenic pathway in a PPARα-dependent manner, by upregulating the α-secretase ADAM metallopeptidase domain 10 (ADAM10) expression in hippocampal cultures and decreasing Aβ levels [159]. The role played by PPARα in mediating this effect was confirmed in PPARα-/- mice, which exhibited increased levels of endogenous Aβ.

It has also been demonstrated that the activation of PPARγ reduces amyloid deposition by repressing BACE-1 promoter activity and expression [160]. The role of this receptor in the modulation of Aβ production was further supported by the ability of PPARγ agonists to reduce BACE-1 expression and Aβ levels through a mechanism mediated by the PPARγ co-activator-1α (PGC-1α) [161] whose decreased expression in the brains of AD patients has been recently demonstrated [162].

The regulatory role of PPARs is not limited to production of Aβ but may be extended also to its clearance. The stimulation of Aβ clearance, in fact, is one of the most popular strategies to ameliorate AD pathology. PPARα action on this pathway seems to involve lysosomal biogenesis as demonstrated by the increased expression of the lysosomal biogenesis master regulatory gene TFEB in brain cells induced by the PPARα agonist gemfibrozil [163]. In addition, PPARα activation has been reported to stimulate Aβ clearance through regulation of autophagy by increased peroxisome proliferation as well as by downregulating mTOR pathway [164, 165].

The clearance of soluble forms of Aβ from the brain is regulated by apolipoprotein E (ApoE). This lipoprotein is principally expressed in liver and brain and it has been shown to enhance both the degradation and phagocytosis of Aβ in the microglia and astrocytes [166, 167]. Interestingly, PPARγ is able to increase expression of ApoE as well as its transporter protein, ATP-binding cassette submemberA1 (ABCA1) in the brains of AD animal models [168]. Moreover, proteolytic degradation of Aβ is stimulated by ApoE through extracellular insulin-degrading enzyme (IDE) whose expression is induced by PPARγ in neurons [169, 170]. In the AD mice model 5XFAD, also, PPARβ/δ activation has been shown to drive the expression of IDE which can also contribute to an increased clearance of Aβ [171]. Additionally, PPARγ activation has been shown to stimulate microglial Aβ phagocytosis by increasing the expression of the scavenger receptor CD36 and, interestingly, the combined treatment with PPARγ and retinoid X receptor (RXR) agonists was shown to have an additive effect on Aβ uptake by myeloid cells [172].

Neuroinflammation

Inflammation is basically a defensive mechanism leading to repair of the damaged area. However, chronic inflammatory reaction leads to neuronal degeneration and to the promotion of the initial changes that lead to the development of AD [173, 174]. According to this model, the identification of the cellular pathways involved in the inflammatory response may represent a new approach for the development of therapeutic agents for the treatment or prevention of neurodegenerative disorders. The anti-inflammatory action of PPARs in neurodegenerative disorders is well established [175]. Concerning PPARα, several natural and synthetic agonists proved effective in lowering the expression levels of neuroinflammatory mediators in degenerative diseases, including AD [176]. Palmitoylethanolamide (PEA) is an endocannabinoid-like compound which is essentially inactive at cannabinoid CB1/CB2 receptors but recognized as a PPARα ligand [177]. In particular, PEA is abundant in the CNS, where is conspicuously released by microglial cells and overproduced in response to cellular stress [178, 179]. This ligand was shown to attenuate Aβ-induced upregulation of a wide range of pro-inflammatory molecules, in a PPARα-dependent manner [180]. More recently, it has been demonstrated that PEA treatment of Aβ-loaded rat astroglioma cells inhibits pro-angiogenic cytokine release by a PPARα-mediated mechanism [181]. This effect is especially relevant if one considers the role of angiogenesis in AD, which is thought to alter blood-brain barrier permeability, thus promoting passage of inflammatory/immune-competent cells into the brain. As regards *in vivo* effects of PEA, two rodent AD experimental models have been studied, leading to similar conclusions. In the first investigation, chronic subcutaneous injections of PEA prevented the amyloid-$\beta_{25\text{-}35}$-induced learning and memory impairment in mice with a concomitant decrease of inflammatory mediators, oxidative damage markers and apoptotic activation [182]. More recently, Scuderi and colleagues reported that PEA systemic administration to adult male rats intrahippocampally injected with $A\beta_{1\text{-}42}$ reduced behavioral deficits and biochemical markers by controlling reactive gliosis and exerting neuroprotective functions [183]. Concerning PPARγ, the potent anti-inflammatory effect resulting from its activation prompted several groups to investigate *in vitro* and *in vivo* the effects of PPARγ agonists treatment against the inflammatory response in AD. After PPARγ activation, expression of several nuclear factors, such as NF-κB, AP-1, STAT-1, NFAT, Egr-1, Jun and GATA-3, was inhibited in both astrocytes and microglia. Under these conditions, expression of pro-inflammatory cytokines such as IL-8, IL-12 and TNFα, and chemokines such as MCP-1 and MCP-3, that have been linked to cognitive impairment, was also reduced [184 - 186]. More recently, studies conducted *in vivo* have demonstrated that these PPARγ-mediated effects are responsible for improved cognitive performance in animal models. For example, the treatment with pioglitazone reduced astrocytes and microglial activation in the brain of A/T mice that overproduce Aβ and TGF-β1 [187]. At the same time, injection of rosiglitazone into the brain of Aβ oligomers-treated Wistar rats, prevented microglia activation and improved cognitive decline by decreasing the levels of inflammatory cytokines [188]. Interestingly, the same effects occurred in the AD transgenic mouse models J20 and APP/PS1 with previous oral administration with rosiglitazone or pioglitazone [189, 190]. All these studies indicate that inhibition of inflammatory response is involved in the beneficial roles of TZD treatment in AD, and this response is mediated by microglia and astrocytes.

Although, among the three subtypes, PPARδ is the most widely expressed in various brain areas, until recently, few reports have been published regarding its anti-inflammatory functions. Noteworthy, a recent study has shown that oral treatment of 5XFAD mice with the selective PPARδ agonist GW0742 decreased overall microglial activation and reduced pro-inflammatory mediators suggesting that PPARδ agonists may have therapeutic utility in treating AD [191].

Mitochondrial Dysfunction

Several reports have showed that the exposure to Aβ affects different aspects of mitochondrial function contributing to the AD pathology in the brain. An impaired mitochondrial function induces a negative energy balance leading to altered cell physiology and ultimately to cell death. Neurons are highly specialized cells requiring a constant energy supply derived from mitochondrial aerobic respiration of glucose; therefore, any disruption or alteration to these organelles, will lead to a reduction in the energy available for neuronal physiology [192, 193]. Given that a compromised transcription of proteins has been suggested within the damaged mitochondria, a drug able to regulate the mitochondrial genes, with a key role as essential initiators for mitochondrial biogenesis and respiration may help or assist in treating AD disease. Interestingly, PPAR stimulation can rescue mitochondrial failure through an increased expression of several nuclear and mitochondrial genes that are necessary for mitochondrial biogenesis. Studies on N2A cells showed that the treatment with rosiglitazone increased the mitochondrial mass and function through the activation of PPAR coactivator 1α (PGC1α) whose expression is diminished in AD [146, 162, 194]. Additionally, it was proposed that rosiglitazone prevented $Aβ_{1-42}$-induced deficits in synapse formation and plasticity by increasing the number of mitochondria in neuronal dendrites and spines [195]. In another study, the chronic treatment with pioglitazone of Aβ-injected Wistar rats attenuated oxidative damage, restored mitochondrial respiratory activity and promoted mitochondrial biogenesis [190]. Interestingly, there is growing evidence that not all the effects observed in mitochondria could be mediated by the direct activation of PPARγ; in fact, according to Colca and colleagues, the pleiotropic effects elicited by TZD might be induced, at least in part, through specific mitochondrial targets and a direct regulation of mitochondrial metabolism of the neurons [193]. In addition, more recently, the anti-diabetes drug pioglitazone has been shown to inhibit the activity of Cyclin-dependent kinase 5 (Cdk5), a serine/threonine kinase that is activated by the neuron specific activators p35/p39 and plays many important roles in neuronal development. Aberrant activation of Cdk5 is believed to be associated with the pathogenesis of several neurodegenerative diseases, including AD [196]. Interestingly, pioglitazone suppressed Cdk5 activity both *in vitro* and *in vivo* by decreasing p35 protein level in a proteasome dependent manner. The mechanism by which pioglitazone can effectively promote p35 degradation remains unknown; however, these findings suggest that this drug can be a promising one to prevent AD progression [197].

New PPAR Activators and Potential Use in Alzheimer's Disease

Pathologies such as diabetes and dyslipidemia are usually treated with PPARγ and PPARα agonists, such as TZDs and fibrates, respectively. However, their clinical use has led to complications. Several studies and meta-analyses of patients with T2DM treated with TZDs have demonstrated that the use of these compounds is associated with unwanted effects, such as weight gain, fluid retention, increased incidence of cardiovascular events, and bone fractures [198 - 200]. On the other hand, some fibrates have been reported to increase the risk of cardiovascular, renal, and hepatic dysfunction and rhabdomyolysis [201, 202]. Different strategies have been conceived to obtain compounds with good therapeutic potential, trying to avoid the known side effects. In particular, a series of PPAR dual activators (α/γ, α/δ, γ/δ) and pan-agonists (α/γ/δ) have been designed to investigate their combined hypolipidemic and insulin-sensitizing effects [203 - 207]. Some of these molecules are partial agonists, which means that they have lower affinity for each PPAR subtype and induce a partial expression of genes compared to full agonists [208, 209]. Another set of selective compounds can activate a specific PPAR subtype but only affects a subset of the genes whose expression is usually induced with the full agonists. These activators are known as selective PPAR modulators (SPPARMs) and show an improved therapeutic profile as a consequence of peculiar ligand-receptor interactions resulting in a different structural conformation of the complex that gives rise to different PPAR transcriptional signatures [210 - 213]. Some of these new-generation PPAR activators are currently under investigation for treatment of neurodegenerative diseases including AD.

Bezafibrate is a PPAR pan-agonist which lowers levels of triglycerides and increases HDL cholesterol. It can also lower blood glucose and glycosylated hemoglobin A1C (HbA1C) levels and has therefore been proposed for use in patients with insulin resistance and atherogenic dyslipidemia [214]. In a recent study, bezafibrate has been reported to induce the expression of PCG1α in the cortex and hippocampus of a mouse model of mitochondrial encephalopathy together with increases in mitochondrial mass and enhanced cytochrome c oxidase (COX) activity [215]. These results suggest that bezafibrate may be useful in the treatment of neurodegenerative diseases due to its effects on mitochondrial function.

Also, a recent experimental work showed that the novel potent PPAR pan-agonist

GFT1803 exerted beneficial effects on an AD mouse model. GFT1803 produced both quantitatively superior and qualitatively different therapeutic effects with respect to amyloid plaque burden, insoluble Aβ content, and neuroinflammation compared to the sole use of the selective PPARγ agonist pioglitazone [216]. The effects of this new experimental compound on carbohydrate and lipid metabolism were not evaluated; however, these results seem to confirm that targeting all three PPAR subtypes may represent a successful strategy to suppress cerebral amyloidosis and chronic inflammation.

INT131 is a potent selective modulator of PPARγ and in humans and in animal models of T2DM this drug has shown to be well tolerated and significantly improve insulin sensitivity and glucose tolerance without any undesirable effects [217, 218]. A recent study has demonstrated that INT131 (Fig. **8**) increased dendritic branching, promoted neuronal survival against Aβ amyloid, increased expression of PGC1α and modulated neuronal mitochondrial dynamics. These results suggest that INT131, a drug that has been shown to be safe and effective in metabolic disorders, may constitute a new therapeutic alternative for AD [219].

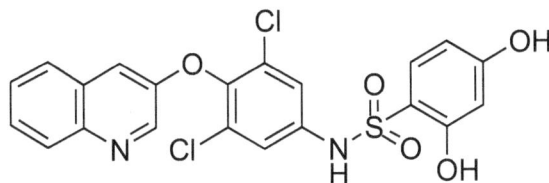

Fig. (8). Chemical structure of INT131.

Telmisartan is an angiotensin II receptor blocker used in the treatment of hypertension. It has been reported to be a partial PPARγ agonist with a specific coactivator recruitment profile that translated to a different gene expression pattern compared to full PPARγ agonists of the TZD class [220]. In *in vivo* studies, telmisartan showed a TZD-like therapeutic profile by lowering levels of insulin, glucose and plasma triglycerides, thereby improving insulin sensitivity, but simultaneously avoiding the full agonist's side effects [221, 222]. In patients with AD and hypertension, telmisartan was shown to improve the cognitive function by decreasing levels of $A\beta_{1-42}$, IL-1β and TNF in cerebrospinal fluid [223]. This additional benefit in comparison with common anti-hypertensive drugs may offer a novel therapeutic strategy of AD.

Iso-α-acids are the bitter component of beer and act as dual agonists of PPARα and PPARγ [224, 225]. These compounds have antioxidant and anti-metabolic syndrome activities and are also reported to prevent diet-induced obesity in rodents and to improve hyperglycemia, which has also been confirmed in humans

[226]. In a recent study, iso-α-acids have been shown to increase microglial phagocytosis of β-amyloid by increasing CD36 expression and suppressing inflammation in neuronal tissue. In Alzheimer's model 5xFAD mice, oral administration of iso-α-acids resulted in a 21% reduction in β-amyloid in the cerebral cortex, a significant reduction in inflammatory cytokines and chemokines and a significant improvement of cognitive function. These results suggest that these acids may represent a new useful therapeutic approach for the prevention of dementia [227].

More recently, a highly potent, specific PPARα-agonist (K-877, Fig. **9**) has emerged with SPPARM characteristics. Compared to fenofibrate, K-877 has more potent PPARα-activating efficacy *in vitro*, greater effects on triglycerides- and HDL-C levels in humans, and a reduced risk of adverse effects [213].

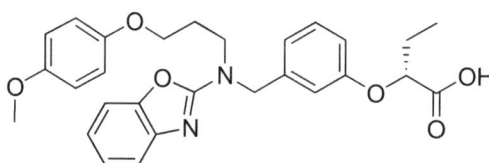

Fig. (9). Chemical structure of K-877.

In pre-clinical studies in animal models for obesity, it induced a dramatic decrease of plasma triglycerides, an effect that was accompanied by higher levels of plasma fibroblast growth factor 21 (FGF-21) [228]. These observations are important because FGF-21 expression in white adipose tissue increases in response to feeding and is generally associated with weight loss, antidiabetic and hypolipidemic effects in animal models of T2DM and obesity [229, 230]. Interestingly, a recent study showed that FGF-21 is expressed in primary rat brain neurons, and that its expression levels are boosted by combined treatment with the mood stabilizers lithium and valproic acid. The increased expression of the FGF-21 gene was associated with increased Akt-1 activity and inhibition of GSK-3β [231] and is responsible, at least in part, for the neuroprotective effects of co-treatment with lithium and valproic acid against glutamate-induced excitotoxicity. Growing evidence has linked glutamate excitotoxicity in discrete brain areas to diverse neurodegenerative conditions, including AD. Therefore, the results of this study suggest that FGF-21 could represent a potential new target for the treatment of brain disorders.

In this regard, also a new dual PPARα/γ agonist, LT175, with a partial agonist profile against PPARγ, has been reported to induce an increase in FGF-21 levels. In addition to this effect, LT175 differentially activated PPARγ target genes involved in fatty acid esterification and storage in 3T3-L1-derived adipocytes.

This resulted in a less severe lipid accumulation compared with that triggered by rosiglitazone, suggesting that LT175 may have a lower adipogenic activity. Consistent with this hypothesis, *in vivo* administration of LT175 to mice fed a high-fat diet decreased body weight, adipocyte size, and white adipose tissue mass. Furthermore, LT175 significantly reduced plasma glucose, insulin, non-esterified fatty acids, triglycerides, and cholesterol [232]. This new PPAR ligand, which is able to modulate lipid and glucose metabolism with reduced adipogenic activity, may represent an interesting model for a series of novel molecules with an improved pharmacological profile for the treatment of dyslipidemia and T2DM. Besides, the increased expression of FGF-21 induced by this ligand, as well as K-877, could suggest investigating their possible neuroprotective effects and therapeutic potential for the treatment of neurodegenerative diseases including AD.

Clinical Trials

Although the animal studies are very promising, the clinical efficacy of PPAR activation in AD has yet to be demonstrated. In addition, even though PPARα and PPARγ agonists seem to be promising drugs for treatment of AD, so far, only PPARγ agonists TZDs have been employed in clinical trials. A placebo-controlled, double-blind clinical trial of rosiglitazone, which included subjects with mild AD was encouraging [233]. The study found that subjects receiving rosiglitazone exhibited better delayed recall, selective attention, and unchanged plasma Aβ levels from baseline compared to subjects receiving placebo. Another trial with pioglitazone in patients with mild AD accompanied with T2DM was also positive [234]. Pioglitazone treatment resulted in improved cognition and regional cerebral blood flow in the parietal lobe together with a decrease in fasting plasma insulin levels, indicating enhanced insulin sensitivity. However, a phase III trial to test the PPARγ agonist rosiglitazone as monotherapy in mild to moderate AD failed to show clinical efficacy [235], which may be related to the poor brain penetrance of this drug. Recently, a phase III trial was initiated to test pioglitazone efficacy in AD, and this trial is currently underway. Additionally, Heneka and colleagues found that chronic treatment with pioglitazone is associated with a lower risk of dementia in diabetic patients when these are compared to patients treated with metformin or insulin [236]. Recently, meta-analysis studies showed that pioglitazone treatment might be therapeutically beneficial in early stages of AD and for patients with mild-to-moderate AD [237]. The ongoing clinical trials will hopefully determine whether TZDs can be used for treating AD.

OTHER INNOVATIVE TARGETS

Several other innovative targets have been identified over the years with the aim to obtain new chemical entities to be used for the therapy and the diagnosis of AD.

A hot topic is the study of the role of ATP-binding cassette (ABC) transporter in the onset and progression of the pathology. This protein superfamily, designated as ABCA-ABCG, comprises 49 human polypeptides, classified in 7 sub-families. Their role in the brain is to mediate the transport of a wide typology of substances, including $A\beta$ at the BBB. Non-invasive molecular imaging methods, such as PET, possess a great potential to directly study the functional activity of ABC transporters in the brains of patients with neurodegenerative disorders. Indeed, an insufficient $A\beta$ export, which these proteins have been demonstrated to be involved in, can be for sure connected with AD and seems to play a fundamental role in disease initiation and progression of the pathology [238]. Therefore, activation of these transporters could be also used in new preventive and/or therapeutic treatments, as well as be useful to develop innovative early diagnostic techniques [238].

Asparagine Endopeptidase, also known as delta-secretase, seems as well to be a promising protein to be targeted, due to its involvement in the cleavage of APP and in the processes leading to tau-hyperphosphorylation. Moreover, this enzyme seems to have a pivotal role in neuronal cell death in several neurodegenerative diseases. Therefore, the inhibition of its catalytic activity can be defined as a potential DMT for AD [239 - 242]. As mentioned in the previous sections, another enzyme, the Glycogen Synthase Kinase-3β (GSK-3β), has been found to be a potential target for the effective treatment of AD due to its potential to unusually phosphorylate tau proteins and lead to neurodegeneration, through the disconnection from microtubules and formation of the neurofibrillary tangles [243 - 245]. In particular, AZD1080, an ATP competitive inhibitor of GSK-3β tested in Phase I clinical studies, was recently identified as a potential drug for the treatment of the tauopathies [244], even if its development was stopped because of a poor therapeutic window [243].

Other recent studies have demonstrated that the Phosphatase 2A (PP2A) inhibition joined to the neuronal Ca^{2+} overload are two important early physio-pathological events in AD that have been recently studied using 3-aminomethylindole derivatives as multi-target agents [246]. The high recovery of the PP2A activity and blockade of voltage-gated Ca^{2+} channels have been linked to good neuroprotective profiles of these molecules, and further preclinical assays could lead to the development of new drugs for the treatment of AD with attractive innovative pharmacological profiles [246].

Interesting findings have been obtained studying the block of H3-type histamine receptors [247] and the activation of σ_1 receptor [248]. Recently, Ciproxifan, an H3 antagonist, was demonstrated to alleviate the hyperactivity and cognitive deficits observed in a transgenic mouse model able to simulate AD, suggesting that the modulation of these receptors could represent an innovative therapeutic strategy in the treatment of the pathology [249]. Moreover, new lipoic acid-based σ_1 agonists, instead, showed multi-target properties acting as neurogenic, antioxidant, anticholinergic and Aβ aggregation inhibitors, and are now candidates to become new disease-modifying drugs [248].

Finally, recent studies have permitted to identify druggable molecular targets and signaling pathways involved in neurogenic processes and consequently different drug types have been developed and tested in neuronal plasticity [250]. The most promising chemical entities act by modulating serotonergic and/or melatonergic systems (as mentioned above), as well as Wnt/β-catenin pathway (that can be modulated by using PPAR agonists as well [204]), nuclear erythroid 2-related factor and nicotinamide phosphoribosyl-transferase [250].

CONCLUDING REMARKS

The therapeutic options available for the treatment of Alzheimer's disease are not able to permit the early diagnosis and/or the regression of the pathology. Considering the expected increase in the number of patients, due to the higher life expectancy all over the world and the long and painful progression of the pathology, AD represents an urgent emergency. Researchers are now focused on different strategies: it is important the discovery of new therapeutic targets and ligands acting as disease-modifying agents, as well as including different chemical framework in the same molecular entity to obtain new multi-functional compounds. The MDT approach is a reality for other pathologies and seems to be the correct strategy for AD as well [251]. However, the major part of the already reported multi-functional compounds appear just as chemical tools to provide venue for discovering effective therapeutics and not more, considering the poor results in the clinical studies.

On the other hand, a different approach of combinatorial active and/or passive immunotherapies against different amyloidogenic proteins, at distinct levels of the disease progression, might offer an effective alternative therapy in AD [19, 25, 252].

Maybe a combination of multi-target and immunotherapy approaches can lead to synergistic effects and could even be more effective in the prevention and the treatment of AD. Moreover, a new fascinating theory links this neurological disorder to metabolic dysfunction: further studies in this field could address the researcher toward a new generation of drugs, able not only to control the symptoms, but finally with the ability to cure this terrible pathology.

CONSENT FOR PUBLICATION

Not applicable.

CONFLICT OF INTEREST

The authors declare no conflict of interest.

ACKNOWLEDGEMENTS

L.P. would like to acknowledge Fondo di Sviluppo e Coesione 2007–2013, APQ Ricerca Regione Puglia "Programma regionale a sostegno della specializzazione intelligente e della sostenibilità sociale ed ambientale–FutureInResearch", Project ID: I2PCTF6 and Arch. Cinzia Tarantino for the Figure editing. M.A.S. and S.C. acknowledge the Portuguese *Fundação para a Ciência e Tecnologia* (FCT) for the project UID/QUI/00100/2013.

REFERENCES

[1]   Cacabelos R. Have there been improvements in Alzheimer's disease drug discovery over the past 5 years? Expert Opin Drug Discov 2018; 13(6): 523-38.
[http://dx.doi.org/10.1080/17460441.2018.1457645] [PMID: 29607687]

[2]   Chan KY, Wang W, Wu JJ, *et al.* Global Health Epidemiology Reference Group (GHERG). Epidemiology of Alzheimer's disease and other forms of dementia in China, 1990-2010: a systematic review and analysis. Lancet 2013; 381(9882): 2016-23.
[http://dx.doi.org/10.1016/S0140-6736(13)60221-4] [PMID: 23746902]

[3]   Fiest KM, Roberts JI, Maxwell CJ, *et al.* The prevalence and incidence of dementia due to Alzheimer's disease: a systematic review and meta-analysis. Can J Neurol Sci 2016; 43(S1) (Suppl. 1): S51-82.
[http://dx.doi.org/10.1017/cjn.2016.36] [PMID: 27307128]

[4]   GBD 2015 Neurological Disorders Collaborator Group. Global, regional, and national burden of neurological disorders during 1990-2015: a systematic analysis for the Global Burden of Disease Study 2015. Lancet Neurol 2017; 16(11): 877-97.
[http://dx.doi.org/10.1016/S1474-4422(17)30299-5] [PMID: 28931491]

[5]   Cummings J, Aisen PS, DuBois B, *et al.* Drug development in Alzheimer's disease: the path to 2025. Alzheimers Res Ther 2016; 8: 39.
[http://dx.doi.org/10.1186/s13195-016-0207-9] [PMID: 27646601]

[6]   Alzheimer A, Stelzmann RA, Schnitzlein HN, Murtagh FR. An English translation of Alzheimer's 1907 paper, "Uber eine eigenartige Erkankung der Hirnrinde". Clin Anat 1995; 8(6): 429-31.
[http://dx.doi.org/10.1002/ca.980080612] [PMID: 8713166]

[7]   Goedert M, Spillantini MG. A century of Alzheimer's disease. Science 2006; 314(5800): 777-81.
        [http://dx.doi.org/10.1126/science.1132814] [PMID: 17082447]

[8]   Cheignon C, Tomas M, Bonnefont-Rousselot D, Faller P, Hureau C, Collin F. Oxidative stress and the
        amyloid beta peptide in Alzheimer's disease. Redox Biol 2018; 14: 450-64.
        [http://dx.doi.org/10.1016/j.redox.2017.10.014] [PMID: 29080524]

[9]   Mattson MP. Pathways towards and away from Alzheimer's disease. Nature 2004; 430(7000): 631-9.
        [http://dx.doi.org/10.1038/nature02621] [PMID: 15295589]

[10]   Soto-Rojas LO, de la Cruz-López F, Torres MAO, *et al.* Neuro-inflammation and alteration of the
        blood-brain barrier in Alzheimers disease. Alzheimer's Disease – Challenges for the Future, InTech
        2015.

[11]   Orhan IE, Senol FS. Designing multi-targeted therapeutics for the treatment of alzheimer's disease.
        Curr Top Med Chem 2016; 16(17): 1889-96.
        [http://dx.doi.org/10.2174/1568026616666160204121832] [PMID: 26845553]

[12]   Daoud I, Melkemi N, Salah T, Ghalem S. Combined QSAR, molecular docking and molecular
        dynamics study on new Acetylcholinesterase and Butyrylcholinesterase inhibitors. Comput Biol Chem
        2018; 74: 304-26.
        [http://dx.doi.org/10.1016/j.compbiolchem.2018.03.021] [PMID: 29747032]

[13]   Darvesh S, Hopkins DA, Geula C. Neurobiology of butyrylcholinesterase. Nat Rev Neurosci 2003;
        4(2): 131-8.
        [http://dx.doi.org/10.1038/nrn1035] [PMID: 12563284]

[14]   Greig NH, Lahiri DK, Sambamurti K, Kumar S. Butyrylcholinesterase: an important new target in
        Alzheimer's disease therapy. Int Psychogeriatr 2002; 14(1) (Suppl. 1): 77-91.
        [http://dx.doi.org/10.1017/S1041610203008676] [PMID: 12636181]

[15]   Rivera I, Capone R, Cauvi DM, Arispe N, De Maio A. Modulation of Alzheimer's amyloid β peptide
        oligomerization and toxicity by extracellular Hsp70. Cell Stress Chaperones 2018; 23(2): 269-79.
        [http://dx.doi.org/10.1007/s12192-017-0839-0] [PMID: 28956268]

[16]   Chaves S, Piemontese L, Hiremathad A, Santos MA. Hydroxypyridinone Derivatives: A fascinating
        class of chelators with therapeutic applications - an update. Curr Med Chem 2018; 25(1): 97-112.
        [http://dx.doi.org/10.2174/0929867324666170330092304] [PMID: 28359230]

[17]   Crews L, Masliah E. Molecular mechanisms of neurodegeneration in Alzheimer's disease. Hum Mol
        Genet 2010; 19(R1): R12-20.
        [http://dx.doi.org/10.1093/hmg/ddq160] [PMID: 20413653]

[18]   Cummings JL, Morstorf T, Zhong K. Alzheimer's disease drug-development pipeline: few candidates,
        frequent failures. Alzheimers Res Ther 2014; 6(4): 37.
        [http://dx.doi.org/10.1186/alzrt269] [PMID: 25024750]

[19]   Cummings J, Lee G, Mortsdorf T, Ritter A, Zhong K. Alzheimer's disease drug development pipeline:
        2017. Alzheimers Dement (N Y) 2017; 3(3): 367-84.
        [http://dx.doi.org/10.1016/j.trci.2017.05.002] [PMID: 29067343]

[20]   Dudley J, Berliocchi L. Drug Repositioning: Approaches and applications for neurotherapeutic. 2017.
        CRC Press, Taylor&Francis Group. ISBN: 9781482220834

[21]   Schneider LS, Sano M. Current Alzheimer's disease clinical trials: methods and placebo outcomes.
        Alzheimers Dement 2009; 5(5): 388-97.
        [http://dx.doi.org/10.1016/j.jalz.2009.07.038] [PMID: 19751918]

[22]   Scott TJ, O'Connor AC, Link AN, Beaulieu TJ. Economic analysis of opportunities to accelerate
        Alzheimer's disease research and development. Ann N Y Acad Sci 2014; 1313: 17-34.
        [http://dx.doi.org/10.1111/nyas.12417] [PMID: 24673372]

[23]   Astin JW, Keerthisinghe P, Du L, *et al.* Innate immune cells and bacterial infection in zebrafish.

Methods Cell Biol 2017; 138: 31-60.
[http://dx.doi.org/10.1016/bs.mcb.2016.08.002] [PMID: 28129850]

[24]   Mei H, Feng G, Zhu J, *et al.* A practical guide for exploring opportunities of repurposing drugs for CNS diseases in systems biology. Methods Mol Biol 2016; 1303: 531-47.
[http://dx.doi.org/10.1007/978-1-4939-2627-5_33] [PMID: 26235090]

[25]   Cummings J, Lee G, Ritter A, Zhong K. Alzheimer's disease drug development pipeline: 2018. Alzheimers Dement (NY) 2018; 4: 195-214.
[http://dx.doi.org/10.1016/j.trci.2018.03.009] [PMID: 29955663]

[26]   Torika N, Asraf K, Danon A, Apte RN, Fleisher-Berkovich S. Telmisartan modulates glial activation: *In vitro* and *In vivo* studies. PLoS one 2016; 11(5): e0155823.
[http://dx.doi.org/10.1371/journal.pone.0155823] [PMID: 27187688]

[27]   Kandimalla R, Thirumala V, Reddy PH. Is Alzheimer's disease a Type 3 Diabetes? A critical appraisal. Biochim Biophys Acta Mol Basis Dis 2017; 1863(5): 1078-89.
[http://dx.doi.org/10.1016/j.bbadis.2016.08.018] [PMID: 27567931]

[28]   Akter K, Lanza EA, Martin SA, Myronyuk N, Rua M, Raffa RB. Diabetes mellitus and Alzheimer's disease: shared pathology and treatment? Br J Clin Pharmacol 2011; 71(3): 365-76.
[http://dx.doi.org/10.1111/j.1365-2125.2010.03830.x] [PMID: 21284695]

[29]   Jolivalt CG, Hurford R, Lee CA, Dumaop W, Rockenstein E, Masliah E. Type 1 diabetes exaggerates features of Alzheimer's disease in APP transgenic mice. Exp Neurol 2010; 223(2): 422-31.
[http://dx.doi.org/10.1016/j.expneurol.2009.11.005] [PMID: 19931251]

[30]   Morris JK, Burns JM. Insulin: an emerging treatment for Alzheimer's disease dementia? Curr Neurol Neurosci Rep 2012; 12(5): 520-7.
[http://dx.doi.org/10.1007/s11910-012-0297-0] [PMID: 22791280]

[31]   Piemontese L. An innovative approach for the treatment of Alzheimer's disease: the role of peroxisome proliferator-activated receptors and their ligands in development of alternative therapeutic interventions. Neural Regen Res 2019; 14(1): 43-5.
[http://dx.doi.org/10.4103/1673-5374.241043] [PMID: 30531068]

[32]   Guzior N, Wieckowska A, Panek D, Malawska B. Recent development of multifunctional agents as potential drug candidates for the treatment of Alzheimer's disease. Curr Med Chem 2015; 22(3): 373-404.
[http://dx.doi.org/10.2174/0929867321666141106122628] [PMID: 25386820]

[33]   Piemontese L. New approaches for prevention and treatment of Alzheimer's disease: a fascinating challenge. Neural Regen Res 2017; 12(3): 405-6.
[http://dx.doi.org/10.4103/1673-5374.202942] [PMID: 28469653]

[34]   Hiremathad A, Piemontese L. Heterocyclic compounds as key structures for the interaction with old and new targets in Alzheimer's disease therapy Neur Regen Res 2017; 12(8): 1256-61.

[35]   Piemontese L, Vitucci G, Catto M, *et al.* Natural scaffolds with multi-target activity for the potential treatment of alzheimer's disease. Molecules 2018; 23(9): 2182.
[http://dx.doi.org/10.3390/molecules23092182] [PMID: 30158491]

[36]   Das S, Basu S. Multi-targeting strategics for alzheimer's disease therapeutics: pros and cons. Curr Top Med Chem 2017; 17(27): 3017-61.
[http://dx.doi.org/10.2174/1568026617666170707130652] [PMID: 28685694]

[37]   Piemontese L, Tomás D, Hiremathad A, *et al.* Donepezil structure-based hybrids as potential multifunctional anti-Alzheimer's drug candidates. J Enzyme Inhib Med Chem 2018; 33(1): 1212-24.
[http://dx.doi.org/10.1080/14756366.2018.1491564] [PMID: 30160188]

[38]   Hiremathad A, Keri RS, Esteves AR, Cardoso SM, Chaves S, Santos MA. Novel Tacrine-Hydroxyphenylbenzimidazole hybrids as potential multitarget drug candidates for Alzheimer's disease. Eur J Med Chem 2018; 148: 255-67.

[http://dx.doi.org/10.1016/j.ejmech.2018.02.023] [PMID: 29466775]

[39]   Piemontese L. Plant food supplements with antioxidant properties for the treatment of chronic and neurodegenerative diseases: Benefits or risks? J Diet Suppl 2017; 14(4): 478-84.
[http://dx.doi.org/10.1080/19390211.2016.1247936] [PMID: 27893282]

[40]   Prati F, Cavalli A, Bolognesi ML. Navigating the chemical space of multitarget-directed ligands: From hybrids to fragments in alzheimer's disease. Molecules 2016; 21(4): 466.
[http://dx.doi.org/10.3390/molecules21040466] [PMID: 27070562]

[41]   Hopkins AL, Mason JS, Overington JP. Can we rationally design promiscuous drugs? Curr Opin Struct Biol 2006; 16(1): 127-36.
[http://dx.doi.org/10.1016/j.sbi.2006.01.013] [PMID: 16442279]

[42]   Morphy R, Rankovic Z. Designing multiple ligands - medicinal chemistry strategies and challenges. Curr Pharm Des 2009; 15(6): 587-600.
[http://dx.doi.org/10.2174/138161209787315594] [PMID: 19199984]

[43]   Ali MY, Seong SH, Reddy MR, Seo SY, Choi JS, Jung HA. Kinetics and molecular docking studies of 6-formyl-umbelliferone isolated from *angelica decursiva* as an inhibitor of cholinesterase and BACE1. Molecules 2017; 22: 1604.
[http://dx.doi.org/10.3390/molecules22101604]

[44]   Morphy R, Kay C, Rankovic Z. From magic bullets to designed multiple ligands. Drug Discov Today 2004; 9(15): 641-51.
[http://dx.doi.org/10.1016/S1359-6446(04)03163-0] [PMID: 15279847]

[45]   Morphy R, Rankovic Z. The physicochemical challenges of designing multiple ligands. J Med Chem 2006; 49(16): 4961-70.
[http://dx.doi.org/10.1021/jm0603015] [PMID: 16884308]

[46]   Cavalli A, Bolognesi ML, Minarini A, *et al.* Multi-target-directed ligands to combat neurodegenerative diseases. J Med Chem 2008; 51(3): 347-72.
[http://dx.doi.org/10.1021/jm7009364] [PMID: 18181565]

[47]   Morphy R, Rankovic Z. Designed multiple ligands. An emerging drug discovery paradigm. J Med Chem 2005; 48(21): 6523-43.
[http://dx.doi.org/10.1021/jm058225d] [PMID: 16220969]

[48]   Ismaili L, Refouvelet B, Benchekroun M, *et al.* Multitarget compounds bearing tacrine- and donepezil-like structural and functional motifs for the potential treatment of Alzheimer's disease. Prog Neurobiol 2017; 151: 4-34.
[http://dx.doi.org/10.1016/j.pneurobio.2015.12.003] [PMID: 26797191]

[49]   Mezeiova E, Spilovska K, Nepovimova E, *et al.* Profiling donepezil template into multipotent hybrids with antioxidant properties. J Enzyme Inhib Med Chem 2018; 33(1): 583-606.
[http://dx.doi.org/10.1080/14756366.2018.1443326] [PMID: 29529892]

[50]   Ramsay RR, Popovic-Nikolic MR, Nikolic K, Uliassi E, Bolognesi ML. A perspective on multi-target drug discovery and design for complex diseases. Clin Transl Med 2018; 7(1): 3.
[http://dx.doi.org/10.1186/s40169-017-0181-2] [PMID: 29340951]

[51]   Savelieff MG, Nam G, Kang J, Lee HJ, Lee M, Lim MH. Development of multifunctional moleculesas potential therapeutic candidates for Alzheimer's disease, Parkinson's disease and amyothrophic lateral sclerosis in the last decade. Chem Rev 2018.
[http://dx.doi.org/10.1021/acs.chemrev.8b00138] [PMID: 30095897]

[52]   Rampa A, Tarozzi A, Mancini F, *et al.* Naturally inspired molecules as multifunctional agents for Alzheimer's disease treatment. Molecules 2016; 21(5): 643.
[http://dx.doi.org/10.3390/molecules21050643] [PMID: 27196880]

[53]   Beck MW, Derrick JS, Kerr RA, *et al.* Structure-mechanism-based engineering of chemical regulators targeting distinct pathological factors in Alzheimer's disease. Nat Commun 2016; 7: 13115.

[http://dx.doi.org/10.1038/ncomms13115] [PMID: 27734843]

[54]   Santos MA, Chand K, Chaves S. Recent progress in multifunctional metal chelators as potential drugs for Alzheimer's disease. Coord Chem Rev 2016; 327-328: 287-303.
[http://dx.doi.org/10.1016/j.ccr.2016.04.013]

[55]   Nesi G, Sestito S, Digiacomo M, Rapposelli S. Oxidative stress, mitochondria abnormalities, and proteins deposition: multitarget approaches in Alzheimer's disease. Curr Top Med Chem 2017; 17(27): 3062-79.
[PMID: 28595557]

[56]   Francis PT. The interplay of neurotransmitters in Alzheimer's disease. CNS Spectr 2005; 10(11) (Suppl. 18): 6-9.
[http://dx.doi.org/10.1017/S1092852900014164] [PMID: 16273023]

[57]   Mann DM, Yates PO, Marcyniuk B. Dopaminergic neurotransmitter systems in Alzheimer's disease and in Down's syndrome at middle age. J Neurol Neurosurg Psychiatry 1987; 50(3): 341-4.
[http://dx.doi.org/10.1136/jnnp.50.3.341] [PMID: 2951499]

[58]   Rodríguez JJ, Noristani HN, Verkhratsky A. The serotonergic system in ageing and Alzheimer's disease. Prog Neurobiol 2012; 99(1): 15-41.
[http://dx.doi.org/10.1016/j.pneurobio.2012.06.010] [PMID: 22766041]

[59]   Oset-Gasque MJ, Marco-Contelles J. Alzheimer's disease, the "one-molecule, one-target" paradigm, and the multitarget directed ligand approach. ACS Chem Neurosci 2018; 9(3): 401-3.
[http://dx.doi.org/10.1021/acschemneuro.8b00069] [PMID: 29465220]

[60]   Bottegoni G, Favia AD, Recanatini M, Cavalli A. The role of fragment-based and computational methods in polypharmacology. Drug Discov Today 2012; 17(1-2): 23-34.
[http://dx.doi.org/10.1016/j.drudis.2011.08.002] [PMID: 21864710]

[61]   Grill JD, Cummings JL. Current therapeutic targets for the treatment of Alzheimer's disease. Expert Rev Neurother 2010; 10(5): 711-28.
[http://dx.doi.org/10.1586/ern.10.29] [PMID: 20420492]

[62]   Hardy J, Allsop D. Amyloid deposition as the central event in the aetiology of Alzheimer's disease. Trends Pharmacol Sci 1991; 12(10): 383-8.
[http://dx.doi.org/10.1016/0165-6147(91)90609-V] [PMID: 1763432]

[63]   Frisardi V, Solfrizzi V, Imbimbo PB, *et al.* Towards disease-modifying treatment of Alzheimer's disease: drugs targeting beta-amyloid. Curr Alzheimer Res 2010; 7(1): 40-55.
[http://dx.doi.org/10.2174/156720510790274400] [PMID: 19939231]

[64]   Rampa A, Gobbi S, Belluti F, Bisi A. Emerging targets in neurodegeneration: new opportunities for Alzheimer's disease treatment? Curr Top Med Chem 2013; 13(15): 1879-904.
[http://dx.doi.org/10.2174/15680266113139990143] [PMID: 23931436]

[65]   Caraci F, Bosco P, Leggio GM, *et al.* Clinical pharmacology of novel anti-Alzheimer disease modifying medications. Curr Top Med Chem 2013; 13(15): 1853-63.
[http://dx.doi.org/10.2174/15680266113139990141] [PMID: 23931438]

[66]   Bennett DA, Yu L, De Jager PL. Building a pipeline to discover and validate novel therapeutic targets and lead compounds for Alzheimer's disease. Biochem Pharmacol 2014; 88(4): 617-30.
[http://dx.doi.org/10.1016/j.bcp.2014.01.037] [PMID: 24508835]

[67]   Terry AV Jr, Buccafusco JJ. The cholinergic hypothesis of age and Alzheimer's disease-related cognitive deficits: recent challenges and their implications for novel drug development. J Pharmacol Exp Ther 2003; 306(3): 821-7.
[http://dx.doi.org/10.1124/jpet.102.041616] [PMID: 12805474]

[68]   Raina P, Santaguida P, Ismaila A, *et al.* Effectiveness of cholinesterase inhibitors and memantine for treating dementia: evidence review for a clinical practice guideline. Ann Intern Med 2008; 148(5): 379-97.

[http://dx.doi.org/10.7326/0003-4819-148-5-200803040-00009] [PMID: 18316756]

[69] Rodda J, Carter J. Cholinesterase inhibitors and memantine for symptomatic treatment of dementia. BMJ 2012; 344: e2986.
[http://dx.doi.org/10.1136/bmj.e2986] [PMID: 22550350]

[70] Morphy R, Rankovic Z. Design of multitarget ligands.Lead generation approaches in drug discovery. Hoboken: Wiley 2010; pp. 141-64.
[http://dx.doi.org/10.1002/9780470584170.ch5]

[71] Hughes RE, Nikolic K, Ramsay RR. One for all? Hitting multiple Alzheimer's disease targets with one drug. Front Neurosci 2016; 10: 177.
[http://dx.doi.org/10.3389/fnins.2016.00177] [PMID: 27199640]

[72] Nikolic K, Mavridis L, Bautista-Aguilera OM, *et al.* Predicting targets of compounds against neurological diseases using cheminformatic methodology. J Comput Aided Mol Des 2015; 29(2): 183-98.
[http://dx.doi.org/10.1007/s10822-014-9816-1] [PMID: 25425329]

[73] Decker M, Muñoz-Torrero D. Special Issue: "Molecules against Alzheimer". Molecules 2016; 21(12): 1736.
[http://dx.doi.org/10.3390/molecules21121736] [PMID: 27999295]

[74] Santos MA, Chand K, Chaves S. Recent progress in repositioning Alzheimer's disease drugs based on a multitarget strategy. Future Med Chem 2016; 8(17): 2113-42.
[http://dx.doi.org/10.4155/fmc-2016-0103] [PMID: 27774814]

[75] Unzeta M, Esteban G, Bolea I, *et al.* Multi-target directed donepezil-like ligands for Alzheimer's disease. Front Neurosci 2016; 10: 205.
[http://dx.doi.org/10.3389/fnins.2016.00205] [PMID: 27252617]

[76] Ma F, Du H. Novel deoxyvasicinone derivatives as potent multitarget-directed ligands for the treatment of Alzheimer's disease: Design, synthesis, and biological evaluation. Eur J Med Chem 2017; 140: 118-27.
[http://dx.doi.org/10.1016/j.ejmech.2017.09.008] [PMID: 28923380]

[77] Dias KST, de Paula CT, Dos Santos T, *et al.* Design, synthesis and evaluation of novel feruloyl-donepezil hybrids as potential multitarget drugs for the treatment of Alzheimer's disease. Eur J Med Chem 2017; 130: 440-57.
[http://dx.doi.org/10.1016/j.ejmech.2017.02.043] [PMID: 28282613]

[78] Jeřábek J, Uliassi E, Guidotti L, *et al.* Tacrine-resveratrol fused hybrids as multi-target-directed ligands against Alzheimer's disease. Eur J Med Chem 2017; 127: 250-62.
[http://dx.doi.org/10.1016/j.ejmech.2016.12.048] [PMID: 28064079]

[79] Chand K, Rajeshwari, Candeias E, Cardoso SM, Chaves S, Santos MA. Tacrine-deferiprone hybrids as multi-target-directed metal chelators against Alzheimer's disease: a two-in-one drug. Metallomics 2018; 10(10): 1460-75.
[http://dx.doi.org/10.1039/C8MT00143J] [PMID: 30183790]

[80] Rochais C, Lecoutey C, Gaven F, *et al.* Novel multitarget-directed ligands (MTDLs) with acetylcholinesterase (AChE) inhibitory and serotonergic subtype 4 receptor (5-HT$_4$R) agonist activities as potential agents against Alzheimer's disease: the design of donecopride. J Med Chem 2015; 58(7): 3172-87.
[http://dx.doi.org/10.1021/acs.jmedchem.5b00115] [PMID: 25793650]

[81] Więckowska A, Wichur T, Godyń J, *et al.* Novel multi-target-directed ligands aiming at symptoms and causes of Alzheimer's disease. ACS Chem Neurosci 2018; 9(5): 1195-214.
[http://dx.doi.org/10.1021/acschemneuro.8b00024] [PMID: 29384656]

[82] Bautista-Aguilera OM, Hagenow S, Palomino-Antolin A, *et al.* Multitarget-directed ligands combining cholinesterase and monoamine oxidase inhibition with histamine H3R antagonism for

neurodegenerative diseases. Angew Chem Int Ed Engl 2017; 56(41): 12765-9.
[http://dx.doi.org/10.1002/anie.201706072] [PMID: 28861918]

[83] Benchekroun M, Romero A, Egea J, *et al.* The antioxidant additive approach for alzheimer's disease therapy: New ferulic (lipoic) acid plus melatonin modified tacrines as cholinesterases inhibitors, direct antioxidants, and nuclear factor (erythroid-derived 2)-like 2 activators. J Med Chem 2016; 59(21): 9967-73.
[http://dx.doi.org/10.1021/acs.jmedchem.6b01178] [PMID: 27736061]

[84] Vezenkov L, Sevalle J, Danalev D, *et al.* Galantamine-based hybrid molecules with acetylcholinesterase, butyrylcholinesterase and γ-secretase inhibition activities. Curr Alzheimer Res 2012; 9(5): 600-5.
[http://dx.doi.org/10.2174/156720512800618044] [PMID: 22211487]

[85] Hardy JA, Higgins GA. Alzheimer's disease: the amyloid cascade hypothesis. Science 1992; 256(5054): 184-5.
[http://dx.doi.org/10.1126/science.1566067] [PMID: 1566067]

[86] Nie Q, Du XG, Geng MY. Small molecule inhibitors of amyloid β peptide aggregation as a potential therapeutic strategy for Alzheimer's disease. Acta Pharmacol Sin 2011; 32(5): 545-51.
[http://dx.doi.org/10.1038/aps.2011.14] [PMID: 21499284]

[87] Hardy J, Selkoe DJ. The amyloid hypothesis of Alzheimer's disease: progress and problems on the road to therapeutics. Science 2002; 297(5580): 353-6.
[http://dx.doi.org/10.1126/science.1072994] [PMID: 12130773]

[88] Kawahara M, Kato-Negishi M, Tanaka K. Cross talk between neurometals and amyloidogenic proteins at the synapse and the pathogenesis of neurodegenerative diseases. Metallomics 2017; 9(6): 619-33.
[http://dx.doi.org/10.1039/C7MT00046D] [PMID: 28516990]

[89] Mills E, Dong XP, Wang F, Xu H. Mechanisms of brain iron transport: insight into neurodegeneration and CNS disorders. Future Med Chem 2010; 2(1): 51-64.
[http://dx.doi.org/10.4155/fmc.09.140] [PMID: 20161623]

[90] Uttara B, Singh AV, Zamboni P, Mahajan RT. Oxidative stress and neurodegenerative diseases: a review of upstream and downstream antioxidant therapeutic options. Curr Neuropharmacol 2009; 7(1): 65-74.
[http://dx.doi.org/10.2174/157015909787602823] [PMID: 19721819]

[91] Tamano H, Koike Y, Nakada H, Shakushi Y, Takeda A. Significance of synaptic Zn^{2+} signaling in zincergic and non-zincergic synapses in the hippocampus in cognition. J Trace Elem Med Biol 2016; 38: 93-8.
[http://dx.doi.org/10.1016/j.jtemb.2016.03.003] [PMID: 26995290]

[92] (a). Faller P, Hureau C. Bioinorganic chemistry of copper and zinc ions coordinated to amyloid-beta peptide. Dalton Trans 2009; 21(7): 1080-94.
[http://dx.doi.org/10.1039/B813398K] [PMID: 19322475]
(b). Atrián-Blasco E, Gonzalez P, Santoro A, Alies B, Faller P, Hureau C. Cu and Zn coordination to amyloid peptides: From fascinating chemistry to debated pathological relevance. Coord Chem Rev 2018; 375: 38-55.
[http://dx.doi.org/10.1016/j.ccr.2018.04.007] [PMID: 30262932]

[93] (a). Vasudevaraju P, Govindaraju M, Palanisamy AP, Sambamurti K, Rao KS. Molecular toxicity of aluminium in relation to neurodegeneration. Indian J Med Res 2008; 128(4): 545-56.
[PMID: 19106446]
(b). Maya S, Prakash T, Madhu KD, Goli D. Multifaceted effects of aluminium in neurodegenerative diseases: A review. Biomed Pharmacother 2016; 83: 746-54.
[http://dx.doi.org/10.1016/j.biopha.2016.07.035] [PMID: 27479193]

[94] Li D-D, Zhang W, Wang Z-Y, Zhao P. Serum copper, zinc, and iron levels in patients with alzheimer's disease: A meta-analysis of case-control studies. Front Aging Neurosci 2017; 9: 300.

[http://dx.doi.org/10.3389/fnagi.2017.00300] [PMID: 28966592]

[95] Barnham KJ, Bush AI. Biological metals and metal-targeting compounds in major neurodegenerative diseases. Chem Soc Rev 2014; 43(19): 6727-49.
[http://dx.doi.org/10.1039/C4CS00138A] [PMID: 25099276]

[96] Telpoukhovskaia MA, Orvig C. Werner coordination chemistry and neurodegeneration. Chem Soc Rev 2013; 42(4): 1836-46.
[http://dx.doi.org/10.1039/C2CS35236B] [PMID: 22952002]

[97] http://wwwalzforumorg/news/research-news/pbt2-takes-dive-phase-2-alzheimers-trial Last accessed on 10/10/2018 2018.

[98] Wang Z, Hu J, Yang X, *et al.* Design, synthesis, and evaluation of orally bioavailable quinoline-indole derivatives as innovative multitarget-directed ligands: promotion of cell proliferation in the adult murine hippocampus for the treatment of Alzheimer's disease. J Med Chem 2018; 61(5): 1871-94.
[http://dx.doi.org/10.1021/acs.jmedchem.7b01417] [PMID: 29420891]

[99] Jones MR, Mathieu E, Dyrager C, *et al.* Multi-target-directed phenol-triazole ligands as therapeutic agents for Alzheimer's disease. Chem Sci (Camb) 2017; 8(8): 5636-43.
[http://dx.doi.org/10.1039/C7SC01269A] [PMID: 28989601]

[100] Geng J, Li M, Wu L, Ren J, Qu X. Liberation of copper from amyloid plaques: making a risk factor useful for Alzheimer's disease treatment. J Med Chem 2012; 55(21): 9146-55.
[http://dx.doi.org/10.1021/jm3003813] [PMID: 22663067]

[101] Li SY, Wang XB, Kong LY. Design, synthesis and biological evaluation of imine resveratrol derivatives as multi-targeted agents against Alzheimer's disease. Eur J Med Chem 2014; 71: 36-45.
[http://dx.doi.org/10.1016/j.ejmech.2013.10.068] [PMID: 24269515]

[102] Hindo SS, Mancino AM, Braymer JJ, *et al.* Small molecule modulators of copper-induced Abeta aggregation. J Am Chem Soc 2009; 131(46): 16663-5.
[http://dx.doi.org/10.1021/ja907045h] [PMID: 19877631]

[103] Santos MA, Chaves S. 3-Hydroxypyridinone derivatives as metal-sequestering agents for therapeutic use. Future Med Chem 2015; 7(3): 383-410.
[http://dx.doi.org/10.4155/fmc.14.162] [PMID: 25826364]

[104] Solanki I, Parihar P, Mansuri ML, Parihar MS. Flavonoid-based therapies in the early management of neurodegenerative diseases. Adv Nutr 2015; 6(1): 64-72.
[http://dx.doi.org/10.3945/an.114.007500] [PMID: 25593144]

[105] Williams RJ, Spencer JP. Flavonoids, cognition, and dementia: actions, mechanisms, and potential therapeutic utility for Alzheimer disease. Free Radic Biol Med 2012; 52(1): 35-45.
[http://dx.doi.org/10.1016/j.freeradbiomed.2011.09.010] [PMID: 21982844]

[106] Telpoukhovskaia MA, Cawthray JF, Rodríguez-Rodríguez C, *et al.* 3-Hydroxy-4-pyridinone derivatives designed for fluorescence studies to determine interaction with amyloid protein as well as cell permeability. Bioorg Med Chem Lett 2015; 25(17): 3654-7.
[http://dx.doi.org/10.1016/j.bmcl.2015.06.059] [PMID: 26141772]

[107] Hiremathad A, Chand K, Tolayan L, *et al.* Hydroxypyridinone-benzofuran hybrids with potential protective roles for Alzheimer's disease therapy. J Inorg Biochem 2018; 179: 82-96.
[http://dx.doi.org/10.1016/j.jinorgbio.2017.11.015] [PMID: 29182921]

[108] Leuma Yona R, Mazères S, Faller P, Gras E. Thioflavin derivatives as markers for amyloid-β fibrils: insights into structural features important for high-affinity binding. ChemMedChem 2008; 3(1): 63-6.
[http://dx.doi.org/10.1002/cmdc.200700188] [PMID: 17926318]

[109] Alzheimer's Association 2017 Alzheimer's Disease Facts and Figures. Alzheimers Dement 2017; 13: 325-73.
[http://dx.doi.org/10.1016/j.jalz.2017.02.001]

[110]   World Health Organization. http://wwwwhoint/mediacentre/factsheets/fs362/en/ Last accessed on 10/10/2018 2018.

[111]   Querfurth HW, LaFerla FM. Alzheimer's disease. N Engl J Med 2010; 362(4): 329-44.
[http://dx.doi.org/10.1056/NEJMra0909142] [PMID: 20107219]

[112]   Vassar R, Bennett BD, Babu-Khan S, *et al.* β-secretase cleavage of Alzheimer's amyloid precursor protein by the transmembrane aspartic protease BACE. Science 1999; 286(5440): 735-41.
[http://dx.doi.org/10.1126/science.286.5440.735] [PMID: 10531052]

[113]   Ghosh AK, Osswald HL. BACE1 (β-secretase) inhibitors for the treatment of Alzheimer's disease. Chem Soc Rev 2014; 43(19): 6765-813.
[http://dx.doi.org/10.1039/C3CS60460H] [PMID: 24691405]

[114]   Ghosh AK, Cárdenas EL, Osswald HL. The design, development, and evaluation of BACE1 inhibitors for the treatment of alzheimer's disease. Top Med Chem 2017; 24: 27-86.
[http://dx.doi.org/10.1007/7355_2016_16]

[115]   Cebers G, Alexander RC, Haeberlein SB, *et al.* AZD3293: pharmacokinetic and pharmacodynamic effects in healthy subjects and patients with Alzheimer's disease. J Alzheimers Dis 2017; 55(3): 1039-53.
[http://dx.doi.org/10.3233/JAD-160701] [PMID: 27767991]

[116]   Johansson P, Kaspersson K, Gurrell IK, *et al.* Toward β-Secretase-1 Inhibitors with Improved Isoform Selectivity. J Med Chem 2018; 61(8): 3491-502.
[http://dx.doi.org/10.1021/acs.jmedchem.7b01716] [PMID: 29617572]

[117]   Nakahara K, Fuchino K, Komano K, *et al.* Discovery of potent and centrally active 6-substituted 5-fluoro-1,3-dihydro-oxazine β-secretase (BACE1) inhibitors *via* active conformation stabilization. J Med Chem 2018; 61(13): 5525-46.
[http://dx.doi.org/10.1021/acs.jmedchem.8b00011] [PMID: 29775538]

[118]   Kennedy ME, Stamford AW, Chen X, *et al.* The BACE1 inhibitor verubecestat (MK-8931) reduces CNS beta-amyloid in animal models and in Alzheimer's disease patients Sci Transl Med 2016; 8:363 ra150

[119]   Scott JD, Li SW, Brunskill AP, *et al.* Discovery of the 3-imino-1,2,4-thiadiazinane 1,1-dioxide derivative verubecestat (MK-8931)-Abeta-site smyloid precursor protein cleaving enzyme 1 inhibitor for the treatment of Alzheimer's disease. J Med Chem 2016; 59(23): 10435-50.
[http://dx.doi.org/10.1021/acs.jmedchem.6b00307] [PMID: 27933948]

[120]   Burki T. Alzheimer's disease research: the future of BACE inhibitors. Lancet 2018; 391(10139): 2486.
[http://dx.doi.org/10.1016/S0140-6736(18)31425-9] [PMID: 29976459]

[121]   Viayna E, Sola I, Bartolini M, *et al.* Synthesis and multitarget biological profiling of a novel family of rhein derivatives as disease-modifying anti-Alzheimer agents. J Med Chem 2014; 57(6): 2549-67.
[http://dx.doi.org/10.1021/jm401824w] [PMID: 24568372]

[122]   Panek D, Więckowska A, Jończyk J, *et al.* Design, synthesis and biological evaluation of 1-benzylamino-2-hydroxyalkyl derivatives as new potential disease-modifying multifunctional anti-Alzheimer's agents. ACS Chem Neurosci 2018; 9(5): 1074-94.
[http://dx.doi.org/10.1021/acschemneuro.7b00461] [PMID: 29345897]

[123]   Prati F, Bottegoni G, Bolognesi ML, Cavalli A. BACE-1 Inhibitors: From Recent Single-Target Molecules to Multitarget Compounds for Alzheimer's Disease. J Med Chem 2018; 61(3): 619-37.
[http://dx.doi.org/10.1021/acs.jmedchem.7b00393] [PMID: 28749667]

[124]   Piazzi L, Cavalli A, Colizzi F, *et al.* Multi-target-directed coumarin derivatives: hAChE and BACE1 inhibitors as potential anti-Alzheimer compounds. Bioorg Med Chem Lett 2008; 18(1): 423-6.
[http://dx.doi.org/10.1016/j.bmcl.2007.09.100] [PMID: 17998161]

[125]   Camps P, Formosa X, Galdeano C, *et al.* Pyrano[3,2-c]quinoline-6-chlorotacrine hybrids as a novel

family of acetylcholinesterase- and beta-amyloid-directed anti-Alzheimer compounds. J Med Chem 2009; 52(17): 5365-79.
[http://dx.doi.org/10.1021/jm900859q] [PMID: 19663388]

[126] Jain P, Wadhwai PK, Jadhavi HR. Design, synthesis and evaluation of acridin-9-yl hydrazide derivatives as BACE-1 inhibitors. Med Chem Res 2016; 25(7): 1507-13.
[http://dx.doi.org/10.1007/s00044-016-1581-3]

[127] Costanzo P, Cariati L, Desiderio D, *et al.* Design, synthesis, and evaluation of donepezil-like compounds as AChE and BACE-1 inhibitors. ACS Med Chem Lett 2016; 7(5): 470-5.
[http://dx.doi.org/10.1021/acsmedchemlett.5b00483] [PMID: 27190595]

[128] Prati F, De Simone A, Armirotti A, *et al.* 3,4-Dihydro-1,3,5-triazin-2(1H)-ones as the first dual BACE-1/GSK-3betafragment hits against Alzheimer's disease. ACS Chem Neurosci 2015; 6(10): 1665-82.
[http://dx.doi.org/10.1021/acschemneuro.5b00121] [PMID: 26171616]

[129] Di Martino RM, De Simone A, Andrisano V, *et al.* Versatility of the curcumin scaffold: discovery of potent and balanced dual BACE-1 and GSK-3beta inhibitors. J Med Chem 2016; 59(2): 531-44.
[http://dx.doi.org/10.1021/acs.jmedchem.5b00894] [PMID: 26696252]

[130] Kliewer SA, Umesono K, Noonan DJ, Heyman RA, Evans RM. Convergence of 9-cis retinoic acid and peroxisome proliferator signalling pathways through heterodimer formation of their receptors. Nature 1992; 358(6389): 771-4.
[http://dx.doi.org/10.1038/358771a0] [PMID: 1324435]

[131] Gearing KL, Göttlicher M, Teboul M, Widmark E, Gustafsson JA. Interaction of the peroxisome-proliferator-activated receptor and retinoid X receptor. Proc Natl Acad Sci USA 1993; 90(4): 1440-4.
[http://dx.doi.org/10.1073/pnas.90.4.1440] [PMID: 8381967]

[132] Keller H, Dreyer C, Medin J, Mahfoudi A, Ozato K, Wahli W. Fatty acids and retinoids control lipid metabolism through activation of peroxisome proliferator-activated receptor-retinoid X receptor heterodimers. Proc Natl Acad Sci USA 1993; 90(6): 2160-4.
[http://dx.doi.org/10.1073/pnas.90.6.2160] [PMID: 8384714]

[133] Feige JN, Gelman L, Michalik L, Desvergne B, Wahli W. From molecular action to physiological outputs: peroxisome proliferator-activated receptors are nuclear receptors at the crossroads of key cellular functions. Prog Lipid Res 2006; 45(2): 120-59.
[http://dx.doi.org/10.1016/j.plipres.2005.12.002] [PMID: 16476485]

[134] Elisaf M. Effects of fibrates on serum metabolic parameters. Curr Med Res Opin 2002; 18(5): 269-76.
[http://dx.doi.org/10.1185/030079902125000516] [PMID: 12240789]

[135] Staels B, Dallongeville J, Auwerx J, Schoonjans K, Leitersdorf E, Fruchart J-C. Mechanism of action of fibrates on lipid and lipoprotein metabolism. Circulation 1998; 98(19): 2088-93.
[http://dx.doi.org/10.1161/01.CIR.98.19.2088] [PMID: 9808609]

[136] Campbell IW. The clinical significance of PPAR gamma agonism. Curr Mol Med 2005; 5(3): 349-63.
[http://dx.doi.org/10.2174/1566524053766068] [PMID: 15892654]

[137] Moreno S, Farioli-Vecchioli S, Cerù MP. Immunolocalization of peroxisome proliferator-activated receptors and retinoid X receptors in the adult rat CNS. Neuroscience 2004; 123(1): 131-45.
[http://dx.doi.org/10.1016/j.neuroscience.2003.08.064] [PMID: 14667448]

[138] Warden A, Truitt J, Merriman M, *et al.* Localization of PPAR isotypes in the adult mouse and human brain. Sci Rep 2016; 6: 27618.
[http://dx.doi.org/10.1038/srep27618] [PMID: 27283430]

[139] Heneka MT, Landreth GE. PPARs in the brain. Biochim Biophys Acta 2007; 1771(8): 1031-45.
[http://dx.doi.org/10.1016/j.bbalip.2007.04.016] [PMID: 17569578]

[140] Inoue H, Jiang XF, Katayama T, Osada S, Umesono K, Namura S. Brain protection by resveratrol and fenofibrate against stroke requires peroxisome proliferator-activated receptor alpha in mice. Neurosci

Lett 2003; 352(3): 203-6.
[http://dx.doi.org/10.1016/j.neulet.2003.09.001] [PMID: 14625020]

[141] Besson VC, Chen XR, Plotkine M, Marchand-Verrecchia C. Fenofibrate, a peroxisome proliferator-activated receptor alpha agonist, exerts neuroprotective effects in traumatic brain injury. Neurosci Lett 2005; 388(1): 7-12.
[http://dx.doi.org/10.1016/j.neulet.2005.06.019] [PMID: 16087294]

[142] Chen Z, Zhong C. Decoding Alzheimer's disease from perturbed cerebral glucose metabolism: implications for diagnostic and therapeutic strategies. Prog Neurobiol 2013; 108: 21-43.
[http://dx.doi.org/10.1016/j.pneurobio.2013.06.004] [PMID: 23850509]

[143] Cramer PE, Cirrito JR, Wesson DW, *et al.* ApoE-directed therapeutics rapidly clear β-amyloid and reverse deficits in AD mouse models. Science 2012; 335(6075): 1503-6.
[http://dx.doi.org/10.1126/science.1217697] [PMID: 22323736]

[144] Gold PW, Licinio J, Pavlatou MG. Pathological parainflammation and endoplasmic reticulum stress in depression: potential translational targets through the CNS insulin, klotho and PPAR-γ systems. Mol Psychiatry 2013; 18(2): 154-65.
[http://dx.doi.org/10.1038/mp.2012.167] [PMID: 23183489]

[145] Kanakasabai S, Pestereva E, Chearwae W, Gupta SK, Ansari S, Bright JJ. PPARγ agonists promote oligodendrocyte differentiation of neural stem cells by modulating stemness and differentiation genes. PLoS One 2012; 7(11): e50500.
[http://dx.doi.org/10.1371/journal.pone.0050500] [PMID: 23185633]

[146] Chiang MC, Cheng YC, Chen HM, Liang YJ, Yen CH. Rosiglitazone promotes neurite outgrowth and mitochondrial function in N2A cells *via* PPARgamma pathway. Mitochondrion 2014; 14(1): 7-17.
[http://dx.doi.org/10.1016/j.mito.2013.12.003] [PMID: 24370585]

[147] Schnegg CI, Robbins ME. Neuroprotective mechanisms of PPARδ: modulation of oxidative stress and inflammatory processes. PPAR Res 2011; 2011: 373560.
[http://dx.doi.org/10.1155/2011/373560] [PMID: 22135673]

[148] Frisardi V. Impact of metabolic syndrome on cognitive decline in older age: protective or harmful, where is the pitfall? J Alzheimers Dis 2014; 41(1): 163-7.
[http://dx.doi.org/10.3233/JAD-140389] [PMID: 24595195]

[149] Kocalis HE, Turney MK, Printz RL, *et al.* Neuron-specific deletion of peroxisome proliferator-activated receptor delta (PPARδ) in mice leads to increased susceptibility to diet-induced obesity. PLoS One 2012; 7(8): e42981.
[http://dx.doi.org/10.1371/journal.pone.0042981] [PMID: 22916190]

[150] Williamson JD, Launer LJ, Bryan RN, *et al.* Action to Control Cardiovascular Risk in Diabetes Memory in Diabetes Investigators. Cognitive function and brain structure in persons with type 2 diabetes mellitus after intensive lowering of blood pressure and lipid levels: a randomized clinical trial. JAMA Intern Med 2014; 174(3): 324-33.
[http://dx.doi.org/10.1001/jamainternmed.2013.13656] [PMID: 24493100]

[151] Mayeux R, Stern Y. Epidemiology of Alzheimer disease. Cold Spring Harb Perspect Med 2012; 2(8): a006239.
[http://dx.doi.org/10.1101/cshperspect.a006239] [PMID: 22908189]

[152] Chen Y, Deng Y, Zhang B, Gong CX. Deregulation of brain insulin signaling in Alzheimer's disease. Neurosci Bull 2014; 30(2): 282-94.
[http://dx.doi.org/10.1007/s12264-013-1408-x] [PMID: 24652456]

[153] Chen Y, Liang Z, Blanchard J, *et al.* A non-transgenic mouse model (icv-STZ mouse) of Alzheimer's disease: similarities to and differences from the transgenic model (3xTg-AD mouse). Mol Neurobiol 2013; 47(2): 711-25.
[http://dx.doi.org/10.1007/s12035-012-8375-5] [PMID: 23150171]

[154]   Chen Y, Liang Z, Tian Z, *et al.* Intracerebroventricular streptozotocin exacerbates Alzheimer-like changes of 3xTg-AD mice. Mol Neurobiol 2014; 49(1): 547-62.
[http://dx.doi.org/10.1007/s12035-013-8539-y] [PMID: 23996345]

[155]   Santos MJ, Quintanilla RA, Toro A, *et al.* Peroxisomal proliferation protects from beta-amyloid neurodegeneration. J Biol Chem 2005; 280(49): 41057-68.
[http://dx.doi.org/10.1074/jbc.M505160200] [PMID: 16204253]

[156]   Inestrosa NC, Carvajal FJ, Zolezzi JM, *et al.* Peroxisome proliferators reduce spatial memory impairment, synaptic failure, and neurodegeneration in brains of a double transgenic mice model of Alzheimer's disease. J Alzheimers Dis 2013; 33(4): 941-59.
[http://dx.doi.org/10.3233/JAD-2012-120397] [PMID: 23109558]

[157]   Zhang H, Gao Y, Qiao PF, Zhao FL, Yan Y. Fenofibrate reduces amyloidogenic processing of APP in APP/PS1 transgenic mice *via* PPAR-α/PI3-K pathway. Int J Dev Neurosci 2014; 38: 223-31.
[http://dx.doi.org/10.1016/j.ijdevneu.2014.10.004] [PMID: 25447788]

[158]   Zhang H, Gao Y, Qiao PF, Zhao FL, Yan Y. PPAR-α agonist regulates amyloid-β generation *via* inhibiting BACE-1 activity in human neuroblastoma SH-SY5Y cells transfected with APPswe gene. Mol Cell Biochem 2015; 408(1-2): 37-46.
[http://dx.doi.org/10.1007/s11010-015-2480-5] [PMID: 26092426]

[159]   Corbett GT, Gonzalez FJ, Pahan K. Activation of peroxisome proliferator-activated receptor α stimulates ADAM10-mediated proteolysis of APP. Proc Natl Acad Sci USA 2015; 112(27): 8445-50.
[http://dx.doi.org/10.1073/pnas.1504890112] [PMID: 26080426]

[160]   Sastre M, Dewachter I, Rossner S, *et al.* Nonsteroidal anti-inflammatory drugs repress beta-secretase gene promoter activity by the activation of PPARgamma. Proc Natl Acad Sci USA 2006; 103(2): 443-8.
[http://dx.doi.org/10.1073/pnas.0503839103] [PMID: 16407166]

[161]   Katsouri L, Parr C, Bogdanovic N, Willem M, Sastre M. PPARγ co-activator-1α (PGC-1α) reduces amyloid-β generation through a PPARγ-dependent mechanism. J Alzheimers Dis 2011; 25(1): 151-62.
[http://dx.doi.org/10.3233/JAD-2011-101356] [PMID: 21358044]

[162]   Qin W, Haroutunian V, Katsel P, *et al.* PGC-1α expression decreases in the Alzheimer disease brain as a function of dementia. Arch Neurol 2009; 66(3): 352-61.
[http://dx.doi.org/10.1001/archneurol.2008.588] [PMID: 19273754]

[163]   Ghosh A, Jana M, Modi K, *et al.* Activation of peroxisome proliferator-activated receptor α induces lysosomal biogenesis in brain cells: implications for lysosomal storage disorders. J Biol Chem 2015; 290(16): 10309-24.
[http://dx.doi.org/10.1074/jbc.M114.610659] [PMID: 25750174]

[164]   Zhang J, Kim J, Alexander A, *et al.* A tuberous sclerosis complex signalling node at the peroxisome regulates mTORC1 and autophagy in response to ROS. Nat Cell Biol 2013; 15(10): 1186-96.
[http://dx.doi.org/10.1038/ncb2822] [PMID: 23955302]

[165]   Sarnelli G, D'Alessandro A, Iuvone T, *et al.* Palmitoylethanolamide Modulates Inflammation-Associated Vascular Endothelial Growth Factor (VEGF) Signaling *via* the Akt/mTOR Pathway in a Selective Peroxisome Proliferator-Activated Receptor Alpha (PPAR-α)-Dependent Manner. PLoS One 2016; 11(5): e0156198.
[http://dx.doi.org/10.1371/journal.pone.0156198] [PMID: 27219328]

[166]   Holtzman DM. Role of apoe/Abeta interactions in the pathogenesis of Alzheimer's disease and cerebral amyloid angiopathy. J Mol Neurosci 2001; 17(2): 147-55.
[http://dx.doi.org/10.1385/JMN:17:2:147] [PMID: 11816788]

[167]   Barage SH, Sonawane KD. Amyloid cascade hypothesis: Pathogenesis and therapeutic strategies in Alzheimer's disease. Neuropeptides 2015; 52: 1-18.
[http://dx.doi.org/10.1016/j.npep.2015.06.008] [PMID: 26149638]

[168] Mandrekar-Colucci S, Karlo JC, Landreth GE. Mechanisms underlying the rapid peroxisome proliferator-activated receptor-γ-mediated amyloid clearance and reversal of cognitive deficits in a murine model of Alzheimer's disease. J Neurosci 2012; 32(30): 10117-28.
[http://dx.doi.org/10.1523/JNEUROSCI.5268-11.2012] [PMID: 22836247]

[169] Jiang Q, Lee CY, Mandrekar S, *et al.* ApoE promotes the proteolytic degradation of Abeta. Neuron 2008; 58(5): 681-93.
[http://dx.doi.org/10.1016/j.neuron.2008.04.010] [PMID: 18549781]

[170] Du J, Zhang L, Liu S, *et al.* PPARgamma transcriptionally regulates the expression of insulin-degrading enzyme in primary neurons. Biochem Biophys Res Commun 2009; 383(4): 485-90.
[http://dx.doi.org/10.1016/j.bbrc.2009.04.047] [PMID: 19383491]

[171] Kalinin S, Richardson JC, Feinstein DL. A PPARdelta agonist reduces amyloid burden and brain inflammation in a transgenic mouse model of Alzheimer's disease. Curr Alzheimer Res 2009; 6(5): 431-7.
[http://dx.doi.org/10.2174/156720509789207949] [PMID: 19874267]

[172] Yamanaka M, Ishikawa T, Griep A, Axt D, Kummer MP, Heneka MT. PPARγ/RXRα-induced and CD36-mediated microglial amyloid-β phagocytosis results in cognitive improvement in amyloid precursor protein/presenilin 1 mice. J Neurosci 2012; 32(48): 17321-31.
[http://dx.doi.org/10.1523/JNEUROSCI.1569-12.2012] [PMID: 23197723]

[173] Morales I, Guzmán-Martínez L, Cerda-Troncoso C, Farías GA, Maccioni RB. Neuroinflammation in the pathogenesis of Alzheimer's disease. A rational framework for the search of novel therapeutic approaches. Front Cell Neurosci 2014; 8: 112.
[http://dx.doi.org/10.3389/fncel.2014.00112] [PMID: 24795567]

[174] Heppner FL, Ransohoff RM, Becher B. Immune attack: the role of inflammation in Alzheimer disease. Nat Rev Neurosci 2015; 16(6): 358-72.
[http://dx.doi.org/10.1038/nrn3880] [PMID: 25991443]

[175] Laganà AS, Vitale SG, Nigro A, *et al.* Pleiotropic Actions of Peroxisome Proliferator-Activated Receptors (PPARs) in Dysregulated Metabolic Homeostasis, Inflammation and Cancer: Current Evidence and Future Perspectives. Int J Mol Sci 2016; 17(7): E999.
[http://dx.doi.org/10.3390/ijms17070999] [PMID: 27347932]

[176] Fidaleo M, Fanelli F, Ceru MP, Moreno S. Neuroprotective properties of peroxisome proliferator-activated receptor alpha (PPARα) and its lipid ligands. Curr Med Chem 2014; 21(24): 2803-21.
[http://dx.doi.org/10.2174/0929867321666140303143455] [PMID: 24606520]

[177] Pertwee RG, Howlett AC, Abood ME, *et al.* International Union of Basic and Clinical Pharmacology. LXXIX. Cannabinoid receptors and their ligands: beyond CB$_1$ and CB$_2$. Pharmacol Rev 2010; 62(4): 588-631.
[http://dx.doi.org/10.1124/pr.110.003004] [PMID: 21079038]

[178] Mattace Raso G, Russo R, Calignano A, Meli R. Palmitoylethanolamide in CNS health and disease. Pharmacol Res 2014; 86: 32-41.
[http://dx.doi.org/10.1016/j.phrs.2014.05.006] [PMID: 24844438]

[179] Skaper SD, Facci L, Barbierato M, *et al.* N-Palmitoylethanolamine and Neuroinflammation: a Novel Therapeutic Strategy of Resolution. Mol Neurobiol 2015; 52(2): 1034-42.
[http://dx.doi.org/10.1007/s12035-015-9253-8] [PMID: 26055231]

[180] Scuderi C, Valenza M, Stecca C, Esposito G, Carratù MR, Steardo L. Palmitoylethanolamide exerts neuroprotective effects in mixed neuroglial cultures and organotypic hippocampal slices *via* peroxisome proliferator-activated receptor-α. J Neuroinflammation 2012; 9: 49.
[http://dx.doi.org/10.1186/1742-2094-9-49] [PMID: 22405189]

[181] Cipriano M, Esposito G, Negro L, *et al.* Palmitoylethanolamide Regulates Production of Pro-Angiogenic Mediators in a Model of β Amyloid-Induced Astrogliosis *In Vitro*. CNS Neurol Disord

Drug Targets 2015; 14(7): 828-37.
[http://dx.doi.org/10.2174/1871527314666150317224155] [PMID: 25801844]

[182] D'Agostino G, Russo R, Avagliano C, Cristiano C, Meli R, Calignano A. Palmitoylethanolamide protects against the amyloid-β25-35-induced learning and memory impairment in mice, an experimental model of Alzheimer disease. Neuropsychopharmacology 2012; 37(7): 1784-92.
[http://dx.doi.org/10.1038/npp.2012.25] [PMID: 22414817]

[183] Scuderi C, Stecca C, Valenza M, *et al.* Palmitoylethanolamide controls reactive gliosis and exerts neuroprotective functions in a rat model of Alzheimer's disease. Cell Death Dis 2014; 5: e1419.
[http://dx.doi.org/10.1038/cddis.2014.376] [PMID: 25210802]

[184] Cameron B, Landreth GE. Inflammation, microglia, and Alzheimer's disease. Neurobiol Dis 2010; 37(3): 503-9.
[http://dx.doi.org/10.1016/j.nbd.2009.10.006] [PMID: 19833208]

[185] Skerrett R, Malm T, Landreth G. Nuclear receptors in neurodegenerative diseases. Neurobiol Dis 2014; 72(Pt A): 104-16.
[http://dx.doi.org/10.1016/j.nbd.2014.05.019] [PMID: 24874548]

[186] Bastías-Candia S, Garrido A N, Zolezzi JM, Inestrosa NC. Recent advances in neuroinflammation therapeutics: PPARs/LXR as neuroinflammatory modulators. Curr Pharm Des 2016; 22(10): 1312-23.
[http://dx.doi.org/10.2174/1381612822666151223103038] [PMID: 26696410]

[187] Papadopoulos P, Rosa-Neto P, Rochford J, Hamel E. Pioglitazone improves reversal learning and exerts mixed cerebrovascular effects in a mouse model of Alzheimer's disease with combined amyloid-β and cerebrovascular pathology. PLoS One 2013; 8(7): e68612.
[http://dx.doi.org/10.1371/journal.pone.0068612] [PMID: 23874687]

[188] Xu S, Guan Q, Wang C, *et al.* Rosiglitazone prevents the memory deficits induced by amyloid-beta oligomers *via* inhibition of inflammatory responses. Neurosci Lett 2014; 578: 7-11.
[http://dx.doi.org/10.1016/j.neulet.2014.06.010] [PMID: 24933538]

[189] Escribano L, Simón A-M, Gimeno E, *et al.* Rosiglitazone rescues memory impairment in Alzheimer's transgenic mice: mechanisms involving a reduced amyloid and tau pathology. Neuropsychopharmacology 2010; 35(7): 1593-604.
[http://dx.doi.org/10.1038/npp.2010.32] [PMID: 20336061]

[190] Prakash A, Kumar A. Role of nuclear receptor on regulation of BDNF and neuroinflammation in hippocampus of β-amyloid animal model of Alzheimer's disease. Neurotox Res 2014; 25(4): 335-47.
[http://dx.doi.org/10.1007/s12640-013-9437-9] [PMID: 24277156]

[191] Malm T, Mariani M, Donovan LJ, Neilson L, Landreth GE. Activation of the nuclear receptor PPARδ is neuroprotective in a transgenic mouse model of Alzheimer's disease through inhibition of inflammation. J Neuroinflammation 2015; 12: 7.
[http://dx.doi.org/10.1186/s12974-014-0229-9] [PMID: 25592770]

[192] Cabezas-Opazo FA, Vergara-Pulgar K, Pérez MJ, Jara C, Osorio-Fuentealba C, Quintanilla RA. Mitochondrial dysfunction contributes to the pathogenesis of Alzheimer's disease. Oxid Med Cell Longev 2015; 2015: 509654.
[http://dx.doi.org/10.1155/2015/509654] [PMID: 26221414]

[193] Colca JR, McDonald WG, Kletzien RF. Mitochondrial target of thiazolidinediones. Diabetes Obes Metab 2014; 16(11): 1048-54.
[http://dx.doi.org/10.1111/dom.12308] [PMID: 24774061]

[194] Zolezzi JM, Silva-Alvarez C, Ordenes D, *et al.* Peroxisome proliferator-activated receptor (PPAR) γ and PPARα agonists modulate mitochondrial fusion-fission dynamics: relevance to reactive oxygen species (ROS)-related neurodegenerative disorders? PLoS One 2013; 8(5): e64019.
[http://dx.doi.org/10.1371/journal.pone.0064019] [PMID: 23675519]

[195] Xu S, Liu G, Bao X, *et al.* Rosiglitazone prevents amyloid-β oligomer-induced impairment of synapse

formation and plasticity *via* increasing dendrite and spine mitochondrial number. J Alzheimers Dis 2014; 39(2): 239-51.
[http://dx.doi.org/10.3233/JAD-130680] [PMID: 24150104]

[196] Cheung ZH, Ip NY. Cdk5: a multifaceted kinase in neurodegenerative diseases. Trends Cell Biol 2012; 22(3): 169-75.
[http://dx.doi.org/10.1016/j.tcb.2011.11.003] [PMID: 22189166]

[197] Chen J, Li S, Sun W, Li J. Anti-diabetes drug pioglitazone ameliorates synaptic defects in AD transgenic mice by inhibiting cyclin-dependent kinase 5 activity. PLoS One 2015; 10: 4.

[198] Guan Y, Hao C, Cha DR, *et al.* Thiazolidinediones expand body fluid volume through PPARgamma stimulation of ENaC-mediated renal salt absorption. Nat Med 2005; 11(8): 861-6.
[http://dx.doi.org/10.1038/nm1278] [PMID: 16007095]

[199] Home PD, Pocock SJ, Beck-Nielsen H, *et al.* RECORD Study Team. Rosiglitazone evaluated for cardiovascular outcomes in oral agent combination therapy for type 2 diabetes (RECORD): a multicentre, randomised, open-label trial. Lancet 2009; 373(9681): 2125-35.
[http://dx.doi.org/10.1016/S0140-6736(09)60953-3] [PMID: 19501900]

[200] Wei W, Wang X, Yang M, *et al.* PGC1beta mediates PPARgamma activation of osteoclastogenesis and rosiglitazone-induced bone loss. Cell Metab 2010; 11(6): 503-16.
[http://dx.doi.org/10.1016/j.cmet.2010.04.015] [PMID: 20519122]

[201] Davidson MH, Armani A, McKenney JM, Jacobson TA. Safety considerations with fibrate therapy. Am J Cardiol 2007; 99(6A): 3C-18C.
[http://dx.doi.org/10.1016/j.amjcard.2006.11.016] [PMID: 17368275]

[202] Rubenstrunk A, Hanf R, Hum DW, Fruchart JC, Staels B. Safety issues and prospects for future generations of PPAR modulators. Biochim Biophys Acta 2007; 1771(8): 1065-81.
[http://dx.doi.org/10.1016/j.bbalip.2007.02.003] [PMID: 17428730]

[203] Laghezza A, Piemontese L, Cerchia C, *et al.* Identification of the first PPARα/γ dual agonist able to bind to canonical and alternative sites of PPARγ and to inhibit its cdk5-mediated phosphorylation. J Med Chem 2018; 61(18): 8282-98.
[http://dx.doi.org/10.1021/acs.jmedchem.8b00835] [PMID: 30199253]

[204] Piemontese L, Cerchia C, Laghezza A, *et al.* New diphenylmethane derivatives as peroxisome proliferator-activated receptor alpha/gamma dual agonists endowed with anti-proliferative effects and mitochondrial activity. Eur J Med Chem 2017; 127: 379-97.
[http://dx.doi.org/10.1016/j.ejmech.2016.12.047] [PMID: 28076827]

[205] Laghezza A, Montanari R, Lavecchia A, *et al.* On the metabolically active form of metaglidasen: improved synthesis and investigation of its peculiar activity on peroxisome proliferator-activated receptors and skeletal muscles. ChemMedChem 2015; 10(3): 555-65.
[http://dx.doi.org/10.1002/cmdc.201402462] [PMID: 25641779]

[206] Piemontese L, Fracchiolla G, Carrieri A, *et al.* Design, synthesis and biological evaluation of a class of bioisosteric oximes of the novel dual peroxisome proliferator-activated receptor α/γ ligand LT175. Eur J Med Chem 2015; 90: 583-94.
[http://dx.doi.org/10.1016/j.ejmech.2014.11.044] [PMID: 25497132]

[207] Fracchiolla G, Laghezza A, Piemontese L, *et al.* Synthesis, biological evaluation and molecular investigation of fluorinated PPARalpha/gamma dual agonists. Bioorg Med Chem 2012; 20: 2141-51.
[http://dx.doi.org/10.1016/j.bmc.2012.01.025] [PMID: 22341573]

[208] Heald M, Cawthorne MA. Dual acting and pan-PPAR activators as potential anti-diabetic therapies. Handb Exp Pharmacol 2011; 203(203): 35-51.
[http://dx.doi.org/10.1007/978-3-642-17214-4_2] [PMID: 21484566]

[209] Agrawal R. The first approved agent in the Glitazar's Class: Saroglitazar. Curr Drug Targets 2014; 15(2): 151-5.

[http://dx.doi.org/10.2174/13894501113149990199] [PMID: 23906191]

[210] Yew T, Toh SA, Millar JS. Selective peroxisome proliferator-activated receptor-γ modulation to reduce cardiovascular risk in patients with insulin resistance. Recent Pat Cardiovasc Drug Discov 2012; 7(1): 33-41.
[http://dx.doi.org/10.2174/157489012799362359] [PMID: 22044303]

[211] Rosenson RS, Wright RS, Farkouh M, Plutzky J. Modulating peroxisome proliferator-activated receptors for therapeutic benefit? Biology, clinical experience, and future prospects. Am Heart J 2012; 164(5): 672-80.
[http://dx.doi.org/10.1016/j.ahj.2012.06.023] [PMID: 23137497]

[212] Higgins LS, Depaoli AM. Selective peroxisome proliferator-activated receptor γ (PPARgamma) modulation as a strategy for safer therapeutic PPARgamma activation. Am J Clin Nutr 2010; 91(1): 267S-72S.
[http://dx.doi.org/10.3945/ajcn.2009.28449E] [PMID: 19906796]

[213] Fruchart JC. Selective peroxisome proliferator-activated receptor α modulators (SPPARMα): the next generation of peroxisome proliferator-activated receptor α-agonists. Cardiovasc Diabetol 2013; 12: 82.
[http://dx.doi.org/10.1186/1475-2840-12-82] [PMID: 23721199]

[214] Tenenbaum A, Fisman EZ. Balanced pan-PPAR activator bezafibrate in combination with statin: comprehensive lipids control and diabetes prevention? Cardiovasc Diabetol 2012; 11: 140.
[http://dx.doi.org/10.1186/1475-2840-11-140] [PMID: 23150952]

[215] Noe N, Dillon L, Lellek V, *et al.* RETRACTED: Bezafibrate improves mitochondrial function in the CNS of a mouse model of mitochondrial encephalopathy. Mitochondrion 2013; 13(5): 417-26.
[http://dx.doi.org/10.1016/j.mito.2012.12.003] [PMID: 23261681]

[216] Kummer MP, Schwarzenberger R, Sayah-Jeanne S, *et al.* Pan-PPAR modulation effectively protects APP/PS1 mice from amyloid deposition and cognitive deficits. Mol Neurobiol 2015; 51(2): 661-71.
[http://dx.doi.org/10.1007/s12035-014-8743-4] [PMID: 24838579]

[217] Dunn FL, Higgins LS, Fredrickson J, DePaoli AM. INT131-004 study group. Selective modulation of PPARγ activity can lower plasma glucose without typical thiazolidinedione side-effects in patients with Type 2 diabetes. J Diabetes Complications 2011; 25(3): 151-8.
[http://dx.doi.org/10.1016/j.jdiacomp.2010.06.006] [PMID: 20739195]

[218] Taygerly JP, McGee LR, Rubenstein SM, *et al.* Discovery of INT131: a selective PPARγ modulator that enhances insulin sensitivity. Bioorg Med Chem 2013; 21(4): 979-92.
[http://dx.doi.org/10.1016/j.bmc.2012.11.058] [PMID: 23294830]

[219] Godoy JA, Zolezzi JM, Inestrosa NC. INT131 increases dendritic arborization and protects against Aβ toxicity by inducing mitochondrial changes in hippocampal neurons. Biochem Biophys Res Commun 2017; 490(3): 955-62.
[http://dx.doi.org/10.1016/j.bbrc.2017.06.146] [PMID: 28655613]

[220] Schupp M, Janke J, Clasen R, Unger T, Kintscher U. Angiotensin type 1 receptor blockers induce peroxisome proliferator-activated receptor-γ activity. Circulation 2004; 109(17): 2054-7.
[http://dx.doi.org/10.1161/01.CIR.0000127955.36250.65] [PMID: 15117841]

[221] Mori H, Okada Y, Arao T, Nishida K, Tanaka Y. Telmisartan at 80 mg/day increases high-molecula--weight adiponectin levels and improves insulin resistance in diabetic patients. Adv Ther 2012; 29(7): 635-44.
[http://dx.doi.org/10.1007/s12325-012-0032-x] [PMID: 22821644]

[222] Murakami K, Wada J, Ogawa D, *et al.* The effects of telmisartan treatment on the abdominal fat depot in patients with metabolic syndrome and essential hypertension: Abdominal fat Depot Intervention Program of Okayama (ADIPO). Diab Vasc Dis Res 2013; 10(1): 93-6.
[http://dx.doi.org/10.1177/1479164112444640] [PMID: 22561230]

[223] Li W, Zhang JW, Lu F, *et al.* Effects of telmisartan on the level of Aβ1-42, interleukin-1β, tumor

necrosis factor α and cognition in hypertensive patients with Alzheimer's disease. Zhonghua Yi Xue Za Zhi 2012; 92(39): 2743-6.
[PMID: 23290159]

[224] Yajima H, Ikeshima E, Shiraki M, *et al.* Isohumulones, bitter acids derived from hops, activate both peroxisome proliferator-activated receptor α and γ and reduce insulin resistance. J Biol Chem 2004; 279(32): 33456-62.
[http://dx.doi.org/10.1074/jbc.M403456200] [PMID: 15178687]

[225] Yajima H, Noguchi T, Ikeshima E, *et al.* Prevention of diet-induced obesity by dietary isomerized hop extract containing isohumulones, in rodents. Int J Obes 2005; 29(8): 991-7.
[http://dx.doi.org/10.1038/sj.ijo.0802965] [PMID: 15852044]

[226] Obara K, Mizutani M, Hitomi Y, Yajima H, Kondo K. Isohumulones, the bitter component of beer, improve hyperglycemia and decrease body fat in Japanese subjects with prediabetes. Clin Nutr 2009; 28(3): 278-84.
[http://dx.doi.org/10.1016/j.clnu.2009.03.012] [PMID: 19395131]

[227] Ano Y, Dohata A, Taniguchi Y, *et al.* Iso-α-acids, bitter components of beer, prevent inflammation and cognitive decline induced in a mouse model of alzheimer's disease. J Biol Chem 2017; 292(9): 3720-8.
[http://dx.doi.org/10.1074/jbc.M116.763813] [PMID: 28087694]

[228] Takizawa T, Murakami K, Yano W, *et al.* K-877, a highly potent and selective PPARα agonist, improves dyslipidemia and atherosclerosis in experimental animal models. The 80th EAS Congress. Milan; Italy. 2012.

[229] Cantó C, Auwerx J. Cell biology. FGF21 takes a fat bite. Science 2012; 336(6082): 675-6.
[http://dx.doi.org/10.1126/science.1222646] [PMID: 22582248]

[230] Ong KL, Rye KA, O'Connell R, *et al.* FIELD study investigators. Long-term fenofibrate therapy increases fibroblast growth factor 21 and retinol-binding protein 4 in subjects with type 2 diabetes. J Clin Endocrinol Metab 2012; 97(12): 4701-8.
[http://dx.doi.org/10.1210/jc.2012-2267] [PMID: 23144467]

[231] Leng Y, Wang Z, Tsai LK, *et al.* FGF-21, a novel metabolic regulator, has a robust neuroprotective role and is markedly elevated in neurons by mood stabilizers. Mol Psychiatry 2015; 20(2): 215-23.
[http://dx.doi.org/10.1038/mp.2013.192] [PMID: 24468826]

[232] Gilardi F, Giudici M, Mitro N, *et al.* LT175 is a novel PPARα/γ ligand with potent insulin-sensitizing effects and reduced adipogenic properties. J Biol Chem 2014; 289(10): 6908-20.
[http://dx.doi.org/10.1074/jbc.M113.506394] [PMID: 24451380]

[233] Watson GS, Cholerton BA, Reger MA, *et al.* Preserved cognition in patients with early Alzheimer disease and amnestic mild cognitive impairment during treatment with rosiglitazone: a preliminary study. Am J Geriatr Psychiatry 2005; 13(11): 950-8.
[PMID: 16286438]

[234] Sato T, Hanyu H, Hirao K, Kanetaka H, Sakurai H, Iwamoto T. Efficacy of PPAR-γ agonist pioglitazone in mild Alzheimer disease. Neurobiol Aging 2011; 32(9): 1626-33.
[http://dx.doi.org/10.1016/j.neurobiolaging.2009.10.009] [PMID: 19923038]

[235] Gold M, Alderton C, Zvartau-Hind M, *et al.* Rosiglitazone monotherapy in mild-to-moderate Alzheimer's disease: results from a randomized, double-blind, placebo-controlled phase III study. Dement Geriatr Cogn Disord 2010; 30(2): 131-46.
[http://dx.doi.org/10.1159/000318845] [PMID: 20733306]

[236] Heneka MT, Fink A, Doblhammer G. Effect of pioglitazone medication on the incidence of dementia. Ann Neurol 2015; 78(2): 284-94.
[http://dx.doi.org/10.1002/ana.24439] [PMID: 25974006]

[237] Cheng H, Shang Y, Jiang L, Shi TL, Wang L. The peroxisome proliferators activated receptor-gamma

agonists as therapeutics for the treatment of Alzheimer's disease and mild-to-moderate Alzheimer's disease: a meta-analysis. Int J Neurosci 2016; 126(4): 299-307.
[http://dx.doi.org/10.3109/00207454.2015.1015722] [PMID: 26001206]

[238] Pahnke J, Langer O, Krohn M. Alzheimer's and ABC transporters--new opportunities for diagnostics and treatment. Neurobiol Dis 2014; 72(Pt A): 54-60.
[http://dx.doi.org/10.1016/j.nbd.2014.04.001] [PMID: 24746857]

[239] Zhang Z, Xie M, Ye K. Asparagine endopeptidase is an innovative therapeutic target for neurodegenerative diseases. Expert Opin Ther Targets 2016; 20(10): 1237-45.
[http://dx.doi.org/10.1080/14728222.2016.1182990] [PMID: 27115710]

[240] Wang Z-H, Liu P, Liu X, *et al.* Delta-secretase phosphorylation by SRPK2 enhances its enzymatic activity, provoking pathogenesis in alzheimer's disease. Mol Cell 2017; 67(5): 812-825.e5.
[http://dx.doi.org/10.1016/j.molcel.2017.07.018] [PMID: 28826672]

[241] Basurto-Islas G, Gu J-H, Tung YC, Liu F, Iqbal K. Mechanism of tau hyperphosphorylation involving lysosomal enzyme asparagine endopeptidase in a mouse model of brain ischemia. J Alzheimers Dis 2018; 63(2): 821-33.
[http://dx.doi.org/10.3233/JAD-170715] [PMID: 29689717]

[242] Zhang Z, Obianyo O, Dall E, *et al.* Inhibition of delta-secretase improves cognitive functions in mouse models of Alzheimer's disease. Nat Commun 2017; 8: 14740.
[http://dx.doi.org/10.1038/ncomms14740] [PMID: 28345579]

[243] Wang L, Cheng J, Wang S, Zhang X, Cai X. Screening of inhibitors of *Taenia solium* glycogen synthase Kinase-3β. RSC Advances 2017; 7: 43319.
[http://dx.doi.org/10.1039/C7RA05873J]

[244] Georgievska B, Sandin J, Doherty J, *et al.* AZD1080, a novel GSK3 inhibitor, rescues synaptic plasticity deficits in rodent brain and exhibits peripheral target engagement in humans. J Neurochem 2013; 125(3): 446-56.
[http://dx.doi.org/10.1111/jnc.12203] [PMID: 23410232]

[245] Palomo V, Martinez A. Glycogen synthase kinase 3 (GSK-3) inhibitors: a patent update (2014-2015). Expert Opin Ther Pat 2017; 27(6): 657-66.
[http://dx.doi.org/10.1080/13543776.2017.1259412] [PMID: 27828716]

[246] Lajarín-Cuesta R, Arribas RL, Nanclares C, García-Frutos EM, Gandía L, de Los Ríos C. Design and synthesis of multipotent 3-aminomethylindoles and 7-azaindoles with enhanced protein phosphatase 2A-activating profile and neuroprotection. Eur J Med Chem 2018; 157: 294-309.
[http://dx.doi.org/10.1016/j.ejmech.2018.07.030] [PMID: 30099252]

[247] Sadek B, Stark H. Cherry-picked ligands at histamine receptor subtypes. Neuropharmacology 2016; 106: 56-73.
[http://dx.doi.org/10.1016/j.neuropharm.2015.11.005] [PMID: 26581501]

[248] Estrada M, Pérez C, Soriano E, *et al.* New neurogenic lipoic-based hybrids as innovative Alzheimer's drugs with σ-1 agonism and β-secretase inhibition. Future Med Chem 2016; 8(11): 1191-207.
[http://dx.doi.org/10.4155/fmc-2016-0036] [PMID: 27402296]

[249] Bardgett ME, Davis NN, Schultheis PJ, Griffith MS. Ciproxifan, an H3 receptor antagonist, alleviates hyperactivity and cognitive deficits in the APP Tg2576 mouse model of Alzheimer's disease. Neurobiol Learn Mem 2011; 95(1): 64-72.
[http://dx.doi.org/10.1016/j.nlm.2010.10.008] [PMID: 21073971]

[250] Herrera-Arozamena C, Martí-Marí O, Estrada M, de la Fuente Revenga M, Rodríguez-Franco MI. Recent advances in neurogenic small molecules as innovative treatments for neurodegenerative diseases. Molecules 2016; 21(9): 1165.
[http://dx.doi.org/10.3390/molecules21091165] [PMID: 27598108]

[251] Ramsay RR, Popovic-Nikolic MR, Nikolic K, Uliassi E, Bolognesi ML. A perspective on multi-target

drug discovery and design for complex diseases. Clin Transl Med 2018; 7(1): 3.
[http://dx.doi.org/10.1186/s40169-017-0181-2] [PMID: 29340951]

[252] A Bittar, U Sengupta, R Kayed. Prospects for strain-specific immunotherapy in Alzheimer's disease and tauopathies. npj Vaccines 2018; 3-9.

CHAPTER 3

Could Antibiotics Be Therapeutic Agents in Alzheimer's Disease?

Oscar Gómez-Torres, Cristina Pintado-Losa, María Rodríguez-Pérez and **Emma Burgos-Ramos**[*]

Biochemistry Area, Faculty of Environmental Sciences and Biochemistry, University of Castilla-La Mancha, Toledo, Spain

Abstract: Alzheimer´s disease (AD) is an irreversible neurodegenerative disorder and one of the main aging-dependent maladies of the 21st century. Around 46 million people suffer from AD worldwide and this is projected to double within the next 20 years. Due to the progressive aging of the population and the prediction of an increase in the incidence of this disease, AD constitutes a serious familial and social health problem. Therefore, it is necessary to find new therapeutic strategies which are aimed to prevent, delay the onset, slow the progression and/or improve the symptoms of AD. Currently, the research is focused on finding and identifying new drugs for achieving these goals.

In this chapter of the book, we widely review the neuroprotective role that some antibiotics could play in AD, because these drugs reach the brain quickly and are relatively inexpensive. Likewise, we have found evidence in both *in vitro* and *in vivo* studies and also in some clinical trials. In summary, all the reviewed antibiotics exert neuroprotection because they act on the main pathophysiological features of AD. Nevertheless, it must be taken into account that a long-term treatment with antibiotics could cause adverse effects including antibiotic resistance. Thus, properly clinical trials should be carried out in order to corroborate benefits of these antibiotics in people with AD.

Keywords: Alzheimer´s Disease, Amphotericin B, Amyloid β, Antibiotics, Azithromycin, Clioquinol, Dapsone, Doxycycline, Minocycline, Tetracycline Rapamycin, Rifampicin.

[*] **Corresponding author Emma Burgos-Ramos**: Biochemistry Area, Faculty of Environmental Sciences and Biochemistry, Avenue of Carlos III. University of Castilla-La Mancha, 45071, Toledo, Spain; Tel/Fax: (+34)925268800, Ext: 96813; E-mail: emma.burgos@uclm.es

Atta-ur-Rahman (Ed.)

INTRODUCTION

Alzheimer's disease (AD) is the most common cause of dementia worldwide. Currently, around 46 million people suffer from AD, and these data will be duplicated in 20 years. The aetiology of the disease is multifactorial (genetic, environmental, behavioural…); nevertheless, the greatest risk factor is aging. The global demographic trend indicates that population aging is quickly increasing. The World Health Organization estimates that, by 2040, the proportion of world population aged ≥65 will be 1.3 billion, 14% of the total). Due to the progressive aging of the population and to the prediction of an increase in the incidence of this disease, AD constitutes a serious familial and social health problem. In 2015, direct medical costs, social costs and the cost of informal care added up to a total of $818 billion (US) at the global level. Therefore, it is essential to find therapeutic strategies which are designed to prevent, delay the onset, slow the progression and /or improve the symptoms of AD. Nowadays, the lines of research focus on finding and identifying new drugs to address these issues. In this chapter of the book, we have focused on a thorough review of the neuroprotective role of the antibiotics rifampicin, rapamycin and minocycline as well as other antibiotics, such as azithromycin, erythromycin, clioquinol, amphotericin B and tetracycline, in order to see what their role may be in the treatment of AD as these compounds reach the brain quickly and are relatively inexpensive. Likewise, we have deeply analysed both *in vitro* and *in vivo* studies and clinical trials, in order to explain the possible action mechanisms of these drugs and examining their possible clinical use.

PATHOPHYSIOLOGY OF AD

AD is a neurodegenerative disorder which induces progressive memory loss and cognitive decline, exacerbated by neurotransmitter deficits. The pathophysiology of the disease is complex. The presence of extracellular senile plaques containing amyloid beta peptide (Aβ) and intracellular neurofibrillary tangles (NFTs) of hyperphosphorylated tau protein are neuropathological characteristics in brain of people with AD [1, 2].

The "amyloid theory", which is based on the over expression and aggregation of Aβ, is believed to be one of the main causes of its aetiology [1 - 8].

The amyloid protein precursor (APP) can normally be cleaved by α-secretase and γ-secretase (the non-amyloidogenic pathway) or can aberrantly be processed by β-secretases and γ-secretase (the amyloidogenic pathway), leading to the production of Aβ (Fig. **1**). As a consequence, Aβ peptides spontaneously aggregate into soluble oligomers and combine to form fibrils insoluble beta-sheet conformation and are eventually deposited in diffuse senile plaques resulting in an imbalance between production and clearance of Aβ peptide [2, 3].

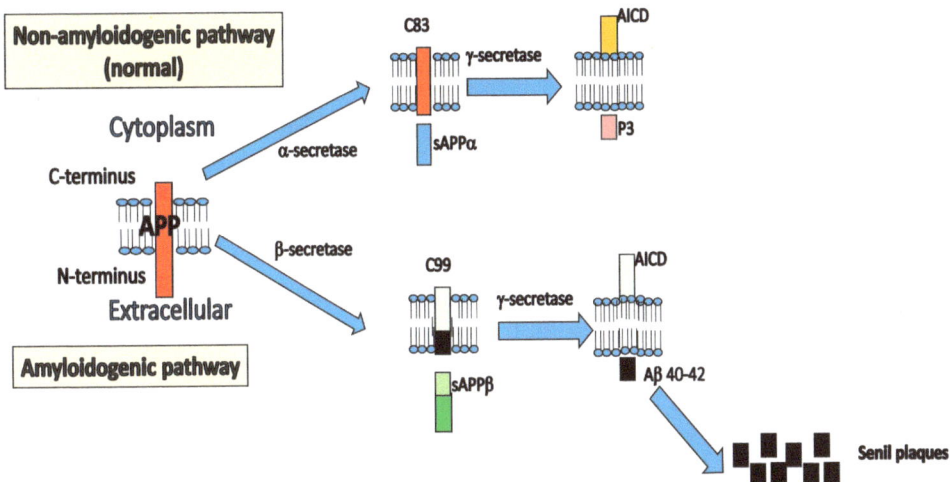

Fig. (1). Scheme of APP processing showing amyloidogenic and non-amyloidogenic pathways. (Adapted from Kim D.& Tsai L. 2009).

APP is a type 1 transmembrane glycoprotein consisting of a long N-terminal extracellular segment (ectodomain), a transmembrane domain and a shorter intracellular C-terminal portion (the cytoplasmic domain). Its expression is mainly localized around the synapse of neuronal tissue. The gene encoding APP is located in chromosome 21. Although its primary role is not fully understood, it is crucial for neuronal plasticity and synapse formation.

The primary proteolytic events on APP, whether pro- or anti-amyloid, occur at or around its transmembrane region.

The APP cleavage site for α-secretase is very close to the cell membrane surface and gives a soluble extracellular APP fragment (sAPPα) and an 83 amino acid, membrane-bound, carboxy terminus fragment (C83). Following intramembrane proteolytic cleavage of C83 by γ-secretase releases a short extracellular p3-peptide (p3) and a cytosolic APP intracellular domain (AICD). The α-APPs fragment has been described having neurotrophic and/or neuroprotective effects and that the AICD has nuclear signalling functions.

Cleavage of APP by β-secretase 1 (also named beta-site amyloid precursor protein cleaving enzyme 1 or BACE1) produces a soluble extracellular fragment (sAPPβ) and a 99 amino acid, membrane-bound, carboxy terminus fragment (C99). Subsequent cleavage of C99 within its transmembrane domain by γ-secretase releases extracellular Aβ and an AICD [4 - 6].

γ-Secretase is a multiprotein complex consisting of Presenilin1 (PS1), nicastrin, Aph-1, and PS2, and all four proteins are necessary for full proteolytic activity [7]. Together with APP gene mutation, several mutations in PS1 and PS2 genes have been described to be involved in familial AD [5].

As previously explained, the level of Aβ in the brain is controlled by its production and clearance. A chronic imbalance between these two processes may result in an accumulation of Aβ in the brain.

The Aβ species are substrates for various proteases. The most studied are neprilysin, that degrades monomeric and oligomeric forms of Aβ, and the insulin-degrading enzyme (IDE), selective for monomers. Other proteases have been implicated in Aβ degradation including endothelin-converting enzyme (ECE-1), which is also selective for monomers, plasmin, that cleaves monomers and fibrils, angiotensin-converting enzyme (ACE) and the cysteine protease cathepsin B [4, 8].

On the other hand, one of the hypotheses is that increased Aβ level is a result of its faulty clearance across the blood-brain barrier (BBB). Aβ clearance across the BBB is mediated by receptor(s) or transporter(s) such as the low density lipoprotein receptor-related protein-1 (LRP1), P-glycoprotein (P-gp) and the multidrug resistance-associated protein 1 (MRP1) [9].

An increasing number of studies suggest that both alteration of expression and functional activity of LRP1 and P-gp contribute to the accumulation of Aβ in the brain and lead to increased risk for developing AD [10]. Moreover, recent evidence indicates a progressive decline in the levels of LRP1 and P-gp at the BBB during normal aging and this decline was positively correlated with accumulation of Aβ in AD [11].

Aβ aggregations are tightly linked to increased oxidative stress, which is accompanied by mitochondrial dysfunction, pronounced inflammation, gliosis, axonal degeneration and impairment of synaptic transmission induced by the deregulated cellular proteostasis [12], which ultimately ends in progressive neuronal loss, predominantly by apoptosis [13]. Even the impaired phagocytic activity of microglia favours the Aβ deposition, exacerbating memory loss [14].

The other major hallmark of AD are the intracellular NFTs [1, 2]. The microtubule associated protein tau, the main constituent of NFTs, is normally synthesized by neuronal cells in order to stabilize the microtubules for proper functioning of the neurons, particularly axonal morphology, growth, and polarity [6]. The activity of the protein tau is mainly regulated by phosphorylation. Tau hyperphosphorylation is thought to result from an imbalance in the function of several protein kinases and phosphatases [15]. Increased phosphorylation of tau, destabilizes tau-microtubule interactions, leading to microtubule instability, transport defects along microtubules, and ultimately neuronal death. Hyperphosphorylated tau detached from microtubules and becomes mislocalized from the axon to the neuronal soma and dendrites and forms this abnormal filaments [1, 2, 6].

Neuroinflammation

As one of its many features, AD is characterized by an Aβ-induced chronic inflammatory state, Aβ peptides and their aggregations can induce an inflammatory reaction that subsequently triggers cognitive decline and the development of neurodegeneration in the brain of people with AD [16]. Microglia and astrocytes are the main cellular players in this state because they release cytokines, interleukins and reactive oxygen species (ROS) after exposure to Aβ, thereby exacerbating a neuroinflammatory reaction [17, 18]. Up-regulation of proinflammatory mediator genes expression was observed in Aβ-treated primary cultures of human microglia [19]. An increased Aβ load in aging TgAPPsw and PSAPP transgenic mice is associated with an increased production of inflammatory cytokines as well as an adjacent neuronal death by releasing reactive oxygen intermediators and proteolytic enzymes [20, 21]. Nevertheless, AD-associated inflammation is named "cytokine vicious cycle", because the generation of Aβ, from amyloidogenic processing of APP, is enhanced by cytokines and interleukins, such as TNF-α, IL-6 and IL-1β. At the same time, Aβ is able to stimulate a nuclear factor kappaB- (NFκB) dependent pathway that is required for cytokine microglial production [22, 23]. Microglia, the guardian phagocyte of CNS, is responsible for initiating an innate immune response through pattern recognition receptors (PRRs) that bind to danger-associated molecular patterns (DAMPs) from pathological triggers. DMAPs can be released from either extracellular or intracellular space after injury. During the course of AD, the most frequent DAMPs is Aβ both in oligomeric form and fibrils or aggregates [24]. Different Aβ forms can bind to cell-surface microglial receptors, including SCARA1, CD36, CD14, CD47 and Toll-like receptors (TLR) [25, 26]. It has been reported that TLR2 and human Aβ-triggered microglial activation is through NFkB-dependent pathway. Moreover, TLR4 can trigger the Aβ-induced activation of murine microglia [27, 28]. So, the microglial binding between Aβ and TLRs actively promotes the neuroinflammation in the brain of people with AD. In the same way, deletion of CD36, TLR4 and TLR6 *in vitro* decreases Aβ-induced cytokine production and prevents amyloid accumulation [29]. In fact, microglia aids in Aβ clearance mechanisms, internalizing phagocytosis-mediated soluble Aβ fragments, which are deleted by neprilysin and IDE. However, in chronic inflammation conditions the microglial phagocytic capacity is impaired by fibrillar Aβ. *In vitro* experiments have shown that microglial phagocytosis is regulated by proinflammatory cytokines [30, 31]. In addition, a study carried out in humans, demonstrated that microglia deteriorate with age, reducing its phagocytic property [32]. In consequence, this insufficient microglial phagocytic capacity induces Aβ pathological accumulation, exacerbating the neurodegeneration in AD.

Astrocytes are also involved in the AD-associated inflammatory response. Like microglial cells, astrocytes produce inflammatory mediators through a limited set of PRRs which bind to Aβ [33]. Additionally, astrocytes degrade Aβ fragments and may promote the Aβ clearance in healthy adult mice [34]. Conversely, under an Aβ-induced chronic inflammation, astrocytes become reactive, expressing high levels of GFAP with ensuing astrocytic hypertrophy and proliferation [35]. Reactive astrocytes have been observed in the hippocampus and cerebral cortex in the majority of transgenic mice models with AD, increasing the pathogenesis of this neurodegenerative disease [36]. Therefore, depending on the levels of inflammatory mediators present in the environment, astrocytes may mediate a beneficial or harmful effect on AD.

The cytokines, including TNF-α, IL-1β, Il-6 and IL-4, are the main inflammatory mediators involved in AD-associated neuroinflammation. Numerous studies have reported high levels of these proinflammatory cytokines in the brains of transgenic mouse model and people with AD reducing the concentration of anti-inflammatory cytokines, such as IL-4 [37, 38]. An *in vitro* study reported that IL-1β can boost Aβ aggregation by favouring amyloidogenic APP processing [38]. IL-1β and IL-6 seem to be important molecular mediators of astrocytic activity [35]. Likewise, TNF-α and Il-1β reduced Aβ phagocytosis and might impair neuronal function by the secretion and production of neurotoxic ROS [30, 39]. Conversely, the transgenic expression of IL-1β in APP/PS1 mice led to a beneficial form of inflammation since Aβ aggregation decreased [40]. Therefore, some proinflammatory cytokines could be considered helpful in reducing AD like pathology in transgenic mouse models.

Finally, the role of tau protein in neuroinflammation should be considered. Tau transgenic mice showed microglial activation, synapse loss and impaired neurologic function [41]. Recently, it has been observed that a potentiated neurodegeneration and neuroinflammation in transgenic mice expressing tau mutant mediated by recruiting proinflammatory monocytes and enhancing sensitivity to neurotoxicity [42].

Autophagy

Proteinopathies are neurodegenerative disorders in which aggregated proteins are abnormally accumulated. AD involves different protein accumulation and based on the evidences many researchers classified AD as a proteinopathy. It is characterized as a dysfunction in the protein homeostasis control processes including autophagy. There is much evidence demonstrating a relationship between autophagic alteration in normal aging and AD [43]. Moreover, it has recently been proposed that the gender differences in the risk of AD are related to differences in autophagy activity throughout the person's life. The lowering levels of estrogen in women during menopause may contribute to an increase AD risk compared to men [44].

Autophagy is a catabolic, physiological, intracellular and ubiquitous process by which old, damaged or misfolded aberrant proteins and/or organelles are degraded in lysosomes, so that cellular components are recycled. Only the proteins with a specific sequence can be degraded in lysosomes [45, 46]. There is overwhelming evidence demonstrating the important neuroprotective role exerted by autophagy and the existing relationship between defective autophagy and the development of different neuropathologies including Parkinson's disease, Huntington's disease and AD [47 - 50]. This process is crucial in maintaining cellular homeostasis and the proper function of cells in many organs, including the brain [43, 51, 52]. It plays an important role in controlling the quality of cellular protein and organelles which could otherwise be harmful to the cell. Autophagy also plays important roles when nutrients are insufficient, in embryonic development, pathogen defense and even, tumor protection [53, 54]. There are three major subtypes of autophagy: microautophagy, chaperone-mediated autophagy and macro-autophagy. While macro-and microautophagy involve the "in bulk" degradation of regions of the cytosol, chaperone-mediated autophagy is a more selective pathway and only proteins with a lysosomal targeting sequence are degraded [55]. Macroautophagy is the most frequently used process for recycling proteins and organelles, being linked most to AD and other neuropathies. This is a multiple step process consisting of sequential steps including sequestration, by which the cytoplasmic components or organelles are firstly surrounded by a simple membrane; phagophore. This structure elongates forming the autophagosome constituted by a double or multi-membrane, forming a cytosolic vesicle; autophagosome. In this process the material is sequestrated, but to be degraded it must be transported to lysosomes. Autophagosome fuses with a lysosome forming an autophagolysosome and then, material is degraded by hydrolases. Finally, amino acids, lipids, nutrients and metabolites are reused [56].

At least 30 proteins are involved in the process of autophagosome formation [57]. Atg1 was the first such protein identified and shown to have intrinsic serine/threonine kinase activity, which is essential for the initiation of autophagy [58]. Autophagy is regulated by phosphatidylinositol 3-kinase (PI3K) type I and type III. PI3K type I is activated by growth factors and suppresses autophagy through the regulation of the mammalian target of rapamycin (mTOR). One of them is Beclin 1 (Atg6), which constitutes a complex in which PI3K III is included and plays an essential role in omegasome formation (lipid bilayer membranes). The expansion of these membranes depends on two pathways; the Atg5–Atg12 pathway and the microtubule-associated protein 1 light chain 3 (LC3) pathway [59]. Soluble LC3 is translocated to autophagic membranes and Phosphatidylethanolamine is ubiquitin-like conjugated to LC3. This mechanism involved in the named conventional autophagy. The levels of Beclin-1 are reduced in both in humans and AD mouse models which are related to a decrease in Aβ clearance [60, 61]. On the contrary, when this protein is up regulated after the injection of a lentivirus, autophagosoma formation and Aβ clearance are increased. Finally, an alternative autophagy model has been proposed, which involves vesicles from trans-Golgi and late endosomes. In this process, Rab 9 attaches to autophagic membranes after Unc51-like (ULK) activation [57] (Fig. **2**).

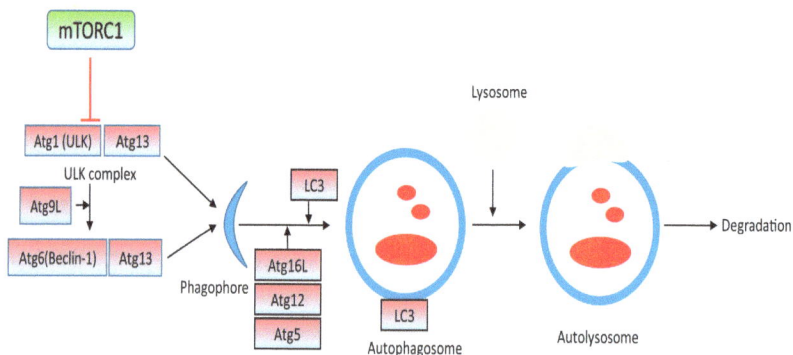

Fig. (2). Molecular mechanisms related to autophagosome formation. (Adapted from Cicchini *et al.*, 2015).

As it has been described, autophagosome biosynthesis impairment is related to Aβ accumulation, but it is also important to highlight the relationship between autophagosome formation and neurofibrillary accumulation. The function of tau protein is related to the normal assembly of microtubules. In AD, the deposits of neurofibrillary tangles alter cytoskeleton activity, which then may alter the autophagosome formation and fusion with lysosomes [62]. Moreover, impairment of autophagolysosomes formation may contribute to tau protein aggregates accumulation and cytotoxicity [63]. This alteration in the autophagolysosoma formation induces cellular dysfunction and death [63, 64]. This was demonstrated by using modified human neuroblastoma cells which express human tau, used as a model of taupathy which were treated with chroriquine [65].

Many other proteins have been involved in this process. It is important to highlight the role of some kinases such Atg1, which exerts serine/threonine kinase activity and initiates autophagy [58]. When PI3K type I is activated, it inhibits autophagy through mTOR that phosphorylate and inhibits ULK. In AD, mTOR overactivation seems to be implicated in the alteration of autophagosome synthesis through the inhibition of kinases such as ULK [66, 67] but also LC3 [68]. Likewise, the role of this kinase is described in more detail in the antibiotic section (rapamycin). Finally, the AMP-activated protein kinase (AMPK) is another kinase which is also phosphorylated and inhibits ULK and blocks autophagy in consequence.

In summary, it has been convincingly demonstrated that autophagy promotes a neuroprotective effect. In fact, there are defects in the autophagic course in many neural diseases, which is the case with AD [47]. Any alteration of this process, formation or function, causes protein aggregation and many studies are focused on the development of new therapeutic drugs which increase autophagosoma formation in AD.

Oxidative Stress

Diverse studies have found that Aβ toxic properties are mediated by different mechanisms and one of them is oxidative stress. However, the exact role of Aβ oligomers in this mechanism is widely debated [69]. The paramagnetic electronic resonance, a high-sensitive method for direct detection of free radicals, has been used to examine the pro-oxidant effect of Aβ [70, 71]. To observe this effect, high doses of the peptide are normally requested but the concrete mechanism is still unknown. Aβ has several metal-binding sites located in its first 15 amino acids constituted by the tyrosine 10 and the histidines 6, 13 and 14 position. These well-known and powerful metal-binding sites have a high affinity for Cu^{2+} as well as for other metallic chelators [72]. Furthermore, Cu^{2+} can be bound by the nitrogen atoms located in the histidines´ imidazole rings, suggesting that the oxygen, which is necessary to enable this binding, can be provided by multiple groups, like the carboxylated lateral chain of the glutamate 5, the hydroxyl group of the tyrosine 10 or the ending amino group as well as from a water molecule [73].

Aβ has the capacity of reducing Cu^{2+} and Fe^{3+} to Cu^+ and Fe^{2+}, respectively. Thus, the O_2 can react with reduced metals generating superoxide anion which combines with two hydrogen atoms to form hydrogen peroxide that could react with another reduced metallic ion and forming the hydroxyl radical by Fenton-Haber Weiss reaction. The radical form of Aβ can extract protons from close lipids all proteins generating lipid peroxides (precursors of 4-hydroxynonenal which affects cellular signal transduction) and carbonyls, respectively [73, 74]. It has also been shown that these metals´ reductions are mediated by a 35 position-located methionine, whose sulfide group has the ability to oxide and donate electrons. Moreover, several studies have confirmed that when this amino acid is substituted, Aβ´s oxidative properties totally disappear [75]. However, the role of this residue is not entirely clear due to the fact that the oxidation of neurotransmitters exposed to Aβ´s 1-16 and 1-12 peptides bound to metal but lacking of Met35, which has previously been confirmed [76]. On the other hand, Aβ can trigger mitochondrial dysfunction owing to the fact that it has been found in different intracellular structures like the mitochondria´s inner membrane or matrix. Aβ can directly inhibit the generation of mitochondrial ATP affecting the correct action of a subunit of ATP synthase. Furthermore, Aβ administration in subtoxic doses and in a chronic manner can inhibit the transportation of some nuclear proteins to the mitochondria. Therefore, Aβ is able to cause changes in mitochondrial permeability with the consequent release of cytochrome C and because of this, the apoptosis is activated [71]. Additionally, an increased expression of the divalent metal transporter 1 has been seen in the animal model APP/SS1 transgenic mouse, in the senile plaques of people with AD and in cellular lines which overexpressing APP, all of them being related to higher levels of iron in human cells exposed to Aβ, suggesting that impairments in iron homeostasis could induce an increment in oxidative stress caused by Aβ [77].

ANTIBIOTICS

Current AD pharmacologic therapy is aimed at treating the cognitive symptoms but do not alter the course of the disease. Cholinesterase inhibitors (rivastigmine, galantamine, donepezil) and memantine (a NMDA antagonist) are the only drugs approved for its treatment. Nonetheless, there is a wide range of components with a different nature and therapeutic purpose, such as antibiotic, antipsychotic and antihypertensive, that exert assorted neuroprotective effects in AD [2, 78]. In the literature there are many studies both pre-clinical and clinical which demonstrate that these candidates may interact with AD-associated pathophysiological mechanisms, inducing beneficial effects. Even recent studies have attributed neuroprotective properties to some foods such as extra virgin olive oil (hydroxytyrosol) [79], grapes (resveratrol) [80], fresh fish (omega 3 fatty acids) [81] and beverages such as green tea [81] and coffee [82].

Here, we review the known neuroprotective effects of some antibiotics on AD development because these inexpensive and interesting candidates are able to cross the BBB and by reaching the brain, target organ of this disorder [83, 84]. Currently, the rifampicin, rapamycin and minocycline are the most common antibiotics used both in preclinical and clinical studies, so in this review, we will bring the neuroprotective role of these antibiotics on AD up to date. Also, we will briefly update the actions of other, lesser studied antibiotics.

Dapsone, Rifampicin and Doxycycline

Dapsone, is a synthetic derivative of diamino-sulfone and was the first antibiotic to be widely used for leprosy. Also, it has shown significant anti-inflammatory activity as an inhibitor of the enzyme myeloperoxidase (MPO) in neutrophils and inhibits the integrin-mediated adherence and chemotaxis of them. Thus, it has been used to treat a number of allergic and autoimmune disorders [85].

Rifampicin, also known as rifampin, is a semisynthetic antibiotic derivative of rifamycin synthesized by the bacterium *Amycolatopsis rifamycinica*. Rifamycins are a subclass of the larger family of ansamycins. The rifamycins' mechanism of action is to selectively inhibit bacterial DNA-dependent RNA polymerase and show no cross-resistance with other antibiotics in clinical use.

Rifampicin is an antibiotic with a very broad spectrum of activity and is used in the treatment of mycobacterium infections, including tuberculosis and leprosy [86, 87]. However, rifampicin and other rifamycins, as well as dapsone, are typically used in combination with other antibacterial drugs to slow or prevent development of resistance. Hepatotoxicity is generally rare alone, but preexisting conditions can be exacerbated. Rifampicin is lipid-soluble and following oral administration, it is rapidly absorbed and diffuses well to most body tissues and fluids, as well as to the brain by crossing the BBB [87].

Apart from their antimicrobial effects, rifampicin and dapsone, between other anti-leprosy drugs, seem to have additional properties that could have an impact on AD [85 - 93]. Thus, an epidemiological study showed that in Japan, a group of patients with leprosy treated with anti-leprosy drugs, had a significantly lower incidence of dementia compared with an untreated group [88]. In addition, histological analyses indicated that elderly non-demented leprosy patients in Japan showed significantly lower levels of amyloid plaques in the brain than non-demented, non-leprosy subjects, whereas the number of NFTs where either unchanged or increased [89, 90]. However, other studies did not confirm these results in the brain samples analyzed [91, 92]. Due to the controversy whether or not anti-leprosy drugs prevent AD, many researches began to study its effect, both *in vitro* and *in vivo* [94, 95].

Tomiyama *et al.* [94], demonstrated that rifampicin inhibited aggregation and fibril formation of synthetic Aβ1–40 and as well as having the strongest activity against the accumulation and toxicity of intracellular Aβ oligomers in PC12 cultured cells, but not dapsone. This protective effect may be achieved by scavenging ROS as well as by inhibiting Aβ aggregation and fibril formation in a dose dependent manner and/or the oligomer–cell membrane interaction. Chemical structure of rifampicin has a naphthohydroquinone ring [96] which is involved in its ROS scavenger role by oxidizing to quinone. Also, it may bind to Aβ by hydrophobic interaction between its lipophilic ansa chain and the hydrophobic region of the peptide blocking association between peptide molecules (Fig. **3**). On the contrary, Mindermann *et al.* [97], also showed that the ansa chain of rifampicin was not essential for the Aβ aggregation inhibitory activities while its lipophilicity contributed significantly in the transportation of the molecule into the brain *in vivo*.

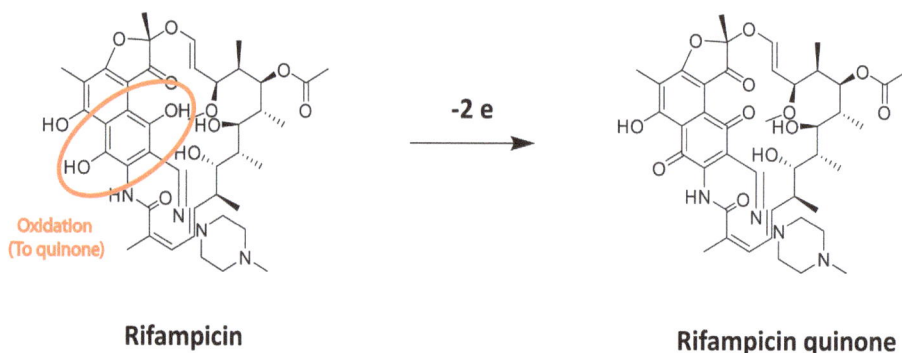

Fig. (3). Rifampicin chemical structure [87] Illustration of the oxidation reaction of rifampicin leading to rifampicin quinone formation. Figure adapted from [Chokkareddy *et al* 2017].

Rifampicin and its analogues, p-benzoquinone and hydroquinone, inhibited the toxicity of preformed aggregates of Aβ polypeptide by binding to peptide fibrils, recognizing a certain conformation, preventing amyloid-cell interaction. Therefore, this antibiotic may mediate the conversion of plaque Aβ from toxic oligomers to non-toxic fibrils *via* monomers [98 - 101]. Similar studies have evaluated the anti-amyloid effect of rifampicin on Aβ aggregation and related toxicity and have revealed that the inhibitory effect was induced more by their binding to peptide fibrils than by their intracellular antioxidant action [102, 103].

Furthermore, rifampicin may promote the efflux of amyloidogenic proteins from the brain into the periphery. Rifampicin has been shown, both *in vitro* and *in vivo* to facilitate Aβ clearance by upregulating the expression of LRP1 and P-gp at the BBB and such clearance may be more efficient for protein monomers than for oligomers [9, 10, 104]. In this line, recently Kaur *et al.* revealed that rifampicin significantly improved memory dysfunction and locomotor impairment in a rat dementia model. They confirmed the activity known to date of rifampicin against Aβ accumulation and neurotoxicity. They found that this antibiotic significantly reduced the oxidative stress, neutrophilic infiltration and amyloid deposition [105].

Moreover, this antibiotic has anti-inflammatory properties by inhibiting microglial activation. It has been described that rifampicin inhibits the LPS-stimulated expression of TLR-2 and TLR-4 and improves neural survival against inflammation and also can inhibit the production of proinflammatory factors through downregulation of NF-κB and MAPKs pathway [106 - 108]. Finally, this antibiotic exerts antiapoptotic properties that provide neuroprotection [108, 109].

Umeda *et al.*, [95], showed that, when orally administered to different mouse models of AD and tauopathy, rifampicin reduces the accumulation of Aβ oligomers inhibiting aggregation and promoting secretion from cells. It also reduced tau protein hyperphosphorylation, synapse loss, microglial activation in a dose-dependent manner, inhibited cytochrome c release from the mitochondria and caspase 3 activation in the hippocampus and improved the memory of the mice, examined using the Morris water maze. Additionally, these authors suggest that this antibiotic restores autophagy-lysosomal function by preventing abnormal protein accumulation beyond the capacity of the protein-degrading system. Rifampicin significantly reduced the levels of p62 in transgenic mice without affecting LC3 conversion.

This results together with its pharmacokinetic properties make rifampicin a suitable medicine to treat neurodegenerative diseases that show extracellular and intracellular protein aggregates in the CNS, such is the case with AD and it has been proposed as being a promising medicine for the prevention of AD and other neurodegenerative diseases [87, 102, 103, 110].

On the other hand, Endoh *et al.* [93], did not find any effect of dapsone or rifampicin on Aβ neurotoxicity using mouse neuronal cultures. Also, an *in vivo* study showed that dapsone does not seem to decrease the production of Aβ in mice [111]. Moreover, a 1-year, randomized, double-blind, placebo-controlled study carried out with dapsone in people with AD was unsuccessful. Dapsone (100 mg/day) and placebo were administered orally, once daily for 52 weeks in 201 people with mild-to-moderate AD. At the end of treatment, there were no significant differences between dapsone and the placebo on either the cognitive or other measures of efficacy. There is, however, a great deal of further analysis yet to be done and specific attention will be paid to the analysis of subgroups of patients [112 - 115].

In contrast to the numerous pre-clinical studies, only a few clinical studies have analyzed the neuroprotective effects of rifampicin in people with AD. None of the trials used rifampicin alone but combined with doxycycline.

Doxycycline, part of the tetracycline antibiotic group, as well as rifampicin, is able to cross the BBB. Doxycycline is used to treat infections caused by bacteria with a broad therapeutic spectrum and there is little evidence of serious adverse events. It is avoided during pregnancy because other tetracyclines have been associated with transient suppression of bone growth and with staining of developing teeth. Doxycycline may also be used for the treatment of malaria, sexually transmitted infection, and to treat or prevent Lyme disease [116].

Related to AD, *in vitro*, doxycycline, disassembles Aβ fibrils [117] and suppresses mutant tau protein production in transgenic mice [118]. It also has an anti-inflammatory effect [119, 120]. In a *Drosophila* model of Aβ toxicity, doxycycline prevents Aβ fibrillation by generating non-toxic structures. These results were also confirmed *in vitro* in a neuroblastoma cell line [121].

Thus, Loeb *et al.* [122], developed a pilot study where oral daily doses of doxycycline 200 mg and rifampin 300 mg were provided for 3 months. This showed to have a therapeutic role in people with mild to moderate AD, improving their cognitive function measured with the Standardized Alzheimer Disease Assessment Scale-Cognitive subscale (SADAScog score). This treatment also reduced blood C-reactive protein (CRP) levels supporting the anti-inflammatory role of these antibiotics.

However, a later clinical trial (ISRCTN15039674) called the DARAD study, designed to confirm or refute these promising pilot results, did not show any beneficial effect on cognition or function in people with AD as a result of taking rifampicin (300mg) or doxycycline (100mg) alone or in combination after twelve months of treatment [123].

The same group registered in 2007 the study NCT00439166. The goal of this multi-centered, randomized, controlled trial, was to determine if the biomarkers Aβ1-40 and Aβ1-42, Phospho-tau and total-tau protein, matrix metalloproteinases (MMP-2, MMP-9), pro-inflammatory cytokines (IL-1β, TNF-α) and anti-inflammatory cytokines (IL-4 and IL-10) present in the cerebrospinal fluid (CSF) of people with AD were affected by treatment with doxycycline and rifampicin at the start and one year after treatment (100 participants). The last register of this phase III trial was in March of 2018. Although it is completed, no result has been published yet. Likewise, the same team registered a pilot study (Phase III) in 2008 with 21 participants (NCT00715858) whose objective was to compare inflammatory biomarkers in blood and CSF in people with AD and age-matched controls and their response to the antibiotics doxycycline and rifampin. The results of this preliminary analysis would be used in defining the direction of further research. On the other hand, the observational clinical trial NCT00692588, was a prospective study started in 2008, aimed at analyzing the changes in brain structure and function using MRI scans in patients who participated in the

DARAD study comparing changes pre *versus* post treatment and normal controls *versus* people with AD, in order to provide more definitive information about the promising use of antibiotics as a treatment. Lastly, in 2010, a clinical trial in phase I (NCT01002079) named the Drug-Drug Interaction Study with Rifampin, was started. The purpose of this study was to determine if the concomitant administration of rifampicin with BMS-708163 in healthy subjects will affect the pharmacokinetics of BMS-708163, a potent, selective and orally bioavailable γ-Secretase Inhibitor [124] and to assess safety and tolerability of co-administration BMS-708163 and rifampicin. In any case the results have yet to be published.

Recently, Izuka *et al*. [125], examined whether rifampicin has a preventive effect on humans. These authors retrospectively reviewed 18F-FDG-PET findings of elderly patients with mycobacterium infection treated with rifampicin. Forty non-demented elderly patients treated with rifampicin for mycobacterium infections who showed AD-type hypometabolism were enrolled. That way, they could evaluate the effect of rifampicin prior to the onset of dementia by using FDG-PET, as rifampicin was administered not for the clinical trial but for the treatment of mycobacterium infections. Having AD-type hypometabolism indicated that the patients were in a state of readiness to decline and this timing seemed optimal to start preventive therapy. Clinical trials of therapies that target Aβ in people with AD have revealed that initiating therapy after the onset of clinical symptoms has little effect on cognitive function. The results showed that the preventive effect of rifampicin depended on the dose and the treatment duration, namely at least 450 mg daily for one year to produce the desired effect.

On the whole, despite the strong evidence for the beneficial effects of rifampicin in cells and animal models of AD (Fig. **4**), there is not agreement about its role in humans. Hence, further studies are necessary to confirm the neuroprotective effect of rifampicin alone or in combination with other antibiotics such as doxycycline and to evaluate their clinical relevance.

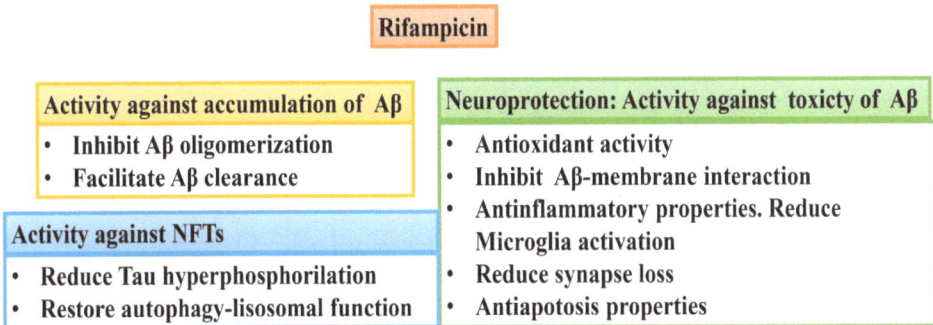

Rifampicin

Activity against accumulation of Aβ	**Neuroprotection: Activity against toxicty of Aβ**
• **Inhibit Aβ oligomerization** • **Facilitate Aβ clearance**	• **Antioxidant activity** • **Inhibit Aβ-membrane interaction** • **Antinflammatory properties. Reduce Microglia activation** • **Reduce synapse loss** • **Antiapotosis properties**
Activity against NFTs	
• **Reduce Tau hyperphosphorilation** • **Restore autophagy-lisosomal function**	

Fig (4). Summary of the main effects of Rifampicin on AD pathophysiology.

Rapamycin

Vezima *et al.* in 1975 obtained *Streptomyces hygroscopicus*, bacteria from the soil of Rapa Nui (Eastern Island) from which rapamycin was isolated [126]. This antibiotic exerted no effect on any bacterial species and was described as an antifungal activity, but the use for this purpose in humans was quickly discarded mainly due to its immunosuppressive effects, as these adverse effects are totally linked to the development of some tumors [127]. However, it was in 1999 when the FDA approved its clinical use for kidney transplant rejection therapy and it was precisely due to its properties as an immunosuppressor [128]. It inhibits the IL-2 induced signal [129] and cell cycle progression in T lymphocytes [130]. On the other hand, surprising results were obtained using this compound on different animal species. It increases life expectancy in nematodes (*Caenorhabditis elegans*), insects (*Drosophila melanogaster)* and laboratory rodents [131 - 133]. In mice, rapamycin exerts a positive effect in lifespan and healthy aging, especially in females, even when the treatment begins in older subjects (19 months of age) [133]. Additionally, many researchers have focused their studies on the effects of this compound and its analogs in different pathologies including coronary stenosis [134], polycystic kidney disease [135] and different cancer types such as the glioblastoma multiform [136] finding, which in general shows a positive effect. One of the pathologies in which more research groups throughout the world are interested in, is AD. This idea is based on the inhibitory effect of rapamycin on mTOR pathway and its connection to AD. This kinase exerts a fundamental role in protein synthesis and proteolysis control which depends on its capacity to phosphorylate, more than 800 different proteins, either directly or indirectly. The overactivation of this pathway has been involved in some of the pathogenic changes related to AD development including neuroinflammation, the formation of neurofibrillary tangles, amyloid beta plaque and the inhibition of autophagosome. The mTOR pathway and the mechanisms through which mTOR participates in the pathophysiology of this disease are explained in more detailed later [137 - 139]. These data are crucial to understanding the importance of rapamycin as an eventual therapeutic drug for Alzheimer's.

In 1993, the protein complex through rapamycin exerted its toxicity in yeast; Target of Rapamycin (TOR) was discovered [140, 141]. It was named mTOR in mammals due to its homology to the yeast TOR/DRR genes [142] and it is downstream activated by different factors such as hormones, growth factors, nutrients, stress, amino acids and it exerts a central role in cellular metabolism, cell motility, growth and survival. The interest in studying mTOR is related to its role in the development of several brain neurodegenerative processes including Huntington's disease, Parkinson's disease, Down syndrome and AD [143, 144]. In the latter case, many studies have demonstrated an mTOR pathway dysregulation in brains of people with AD and in animal models [145 - 147] and these findings suggest that rapamycin or its analogs could be a valid drug for AD treatment. This hypothesis is based on the knowledge about the mTOR structure and function that is reviewed below.

mTOR controls many basic cell functions including mRNA transcription, protein synthesis, proteolysis, lipid biogenesis, mitochondrial metabolism, autophagy, cytoskeleton assembly, cell motility and cell growth [148 - 151]. mTOR constitutes two possible different structural and functional complexes; mTORC1 and mTORC2, which exert different roles in cell activities possibly related to a control of cell function in a precise temporal and spatial manner [152]. Starting with mTORC1, it is composed of mTOR, the regulatory protein associated with mTOR (RAPTOR), the non-core components PRAS40, the DEP domain-containing mTOR interacting protein (Deptor), Tti/Tel2 complex and the mammalian lethal with SEC 13 protein 8 (mLST8) [153]. Once the complex is activated, Raptor binds to factor 4E-binding protein (4EBP1) and ribosomal S6 protein kinase p70 S6 kinase (p70S6K), inducing phosphorylation of 4EBP1 enhancing p70S6K kinase activity. Then, ribosomal biosynthesis, by activating RNA polymerase III dependent transcription and protein synthesis capacity are up-regulated. It also controls energy homeostasis regulating the mitochondrial oxidative activity and biosynthesis of mitochondria and these effects are related to cell cycle progression, size increase and survival. Then, it acts as a regulator of cell growth, exerting different effects on protein homeostasis, nucleotide biosynthesis, lipogenesis, glycogenesis and autophagy [154 - 164]. It is important to highlight the role that mTOR plays in autophagy and protein homeostasis, given its involvement in the development of AD. In general, when a cell has enough nutrients, mTORC1 inhibits autophagy by phosphorylating different kinases including ULK, which are related to the formation of the pre-autophagosome in mammals through mechanisms not completely understood [66, 67].

Finally, it is necessary to mention that part of the activity of mTORC1 depends on PI3K/Akt activation, which in turn activates mTORC1 and participates in the effect of 4EBP1 and p70S6K being that these pathways are deeply related to each other [162, 165]. But, what is the relationship between mTORC1 and rapamycin? Rapamycin inhibits mTORC1 and this effect depends on the intracellular receptor FKBP12 which interacts to rapamycin in turn inhibiting mTORC1 [166]. Then, mTORC1 activity is regulated by rapamycin, but also by hormones such as insulin, growth factors as IGF-I, amino acids, mechanical stimuli and oxidative stress [167].

mTORC2 exerts important actions through F-actin, paxilin, RhoA, essential for actin cytoskeleton function. Moreover, this complex phosphorylates Akt/PKB protein kinase on Ser473, which is related to survival and metabolism [152]. Akt is also phosphorylated on Thr473 by mTORC2 which facilitates the phosphorylation of Akt by PDK1. Then Akt is fully activated [168]. mTORC2 also phosphorylates IGF-1 receptor and insulin receptor, through its tyrosine protein kinase activity [169]. The mTORC2 complex consists of rapamycin-insensitive companion of mTOR (RICTOR), mLSTR8, Deptor, Tti/Tel2 complex, mammalian stress-activated MAP kinase-interacting protein 1 and mTOR. It was previously considered that only mTORC1 was inhibited by rapamycin [170]. However, Sarbassov *et al.* demonstrated that mTORC2 activity is also reduced when cells are treated long term *in vitro* [171] (Fig. **5**).

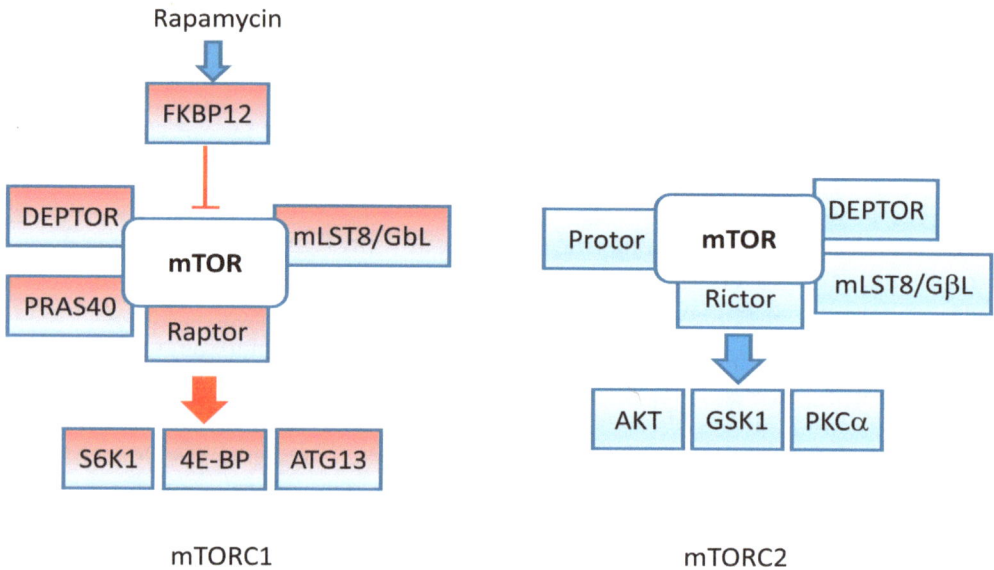

Fig. (5). mTORC1 and mTORC2 constituent proteins and their downstream targets. (Adapted from Lipton and Sahin, 2014).

In the last decade, the relationship between AD development and mTOR complex activity has been investigated in many laboratories around the world, and some of the data obtained are described below. A common early change that occurs in neurons from amnestic mild cognitive impairment and late-stage AD subjects is the increase of mTOR phosphorylation [172]. The hyperphosphorylated residues are Ser2448 and Ser2481, in the temporal lobe of AD human brains, respecting their controls [173]. mTOR phosphorylation itself is altered, but also many other kinases related to this pathway including phosphoinositide 3 kinase, Akt and p70S6K [174, 175]. An *et al.*, studied the temporal lobe from people with AD and demonstrated a positive correlation between neurofibrillary neuron alterations to p70S6K phosphorylation (at Thr421/Ser424) [165]. Thus, the overactivation of mTOR-p70S6K is related to the accumulation of abnormally hyperphosphorylated tau protein and decrease in dephosphorylation [165]. The eIF4E phosphorylated levels are also elevated, especially in late stages, and it is again correlated to hyperphosphorylated tau protein [174]. These data suggest that the changes in eIF4E and p70S6K may be related to neurofibrillary accumulation, being that proteolytic system is unable to compensate for the excess of these proteins. Furthermore, as it has been previously described, the autophagy is one of the impaired mechanisms in AD and some of the factors related to the biosynthesis of autophagosomes has been previously described in this report. Thus, mTOR overactivation seems to be implicated in this process through the inhibition of kinases as ULK [66, 67] and LC3 [68].

Aβ and hyperphosphorylated tau protein produce aggregates which are accumulated in AD leading to neuron death in AD [1, 2]. mTOR inhibits the transcription of genes related to autophagosome induction [68, 176]. Then, mTOR ameliorates Aβ clearance in AD brain [146, 177, 178], Moreover, when mTOR is inhibited by rapamycin or its analogs, autophagy genes are transcribed [179, 180], autophagosome formation is increased and consequently, Aβ accumulations reduced. This has been proved both *in vitro* using 7PA2 cells, as *in vivo* using AD mice models [138]. Finally, it has been demonstrated the role of mTOR in the clearance of other pathological aggregates such huntingtin in Huntington disease mice model [181].

It is well known that learning and memory are affected by AD and some researchers have focused their studies on the relationship between mTOR and memory impairment of this disease. Thus, Lang *et al.*, demonstrated that four weeks of Everolimus (Rapamycin) treatment improved logical memory, mood and life quality in heart transplant patients [182]. On the other hand, Yates *et al.*, using blood lymphocytes samples obtained from people with AD, demonstrated a positive correlation between mTOR pathway dysfunction and cognitive impairment [183]. They proposed that this peripheral modification in the phosphorylating levels of mTOR in these cells could be used as a predictive indicator of AD development. These changes depend on the important role of mTOR in synaptic plasticity which is important for learning and memory. One of the early changes in AD is the reduction in the expression of some of the proteins such as SV2, SNAP-25, synaptophysin and synaptotagmin, which are related to synaptic function and the release of neurotransmitters [184 - 186]. mTOR is related to this effect because rapamycin is able to revert the effect exerted by Aβ, increasing some of these proteins such as SV2 expression [187]. Rapamycin is also able to increase the frequency of miniature excitatory postsynaptic currents after Aβ induced synaptotoxicity in primary hippocampal cultures obtained from mice [187]. Although all these studies demonstrated a negative effect of mTOR on synaptic plasticity, Tischmeyer *et al.* proposed that long term memory consolidation depended on mTOR activation [188]. This is an apparent important discrepancy, but it could be due to differences in the magnitude of the mTOR pathway phosphorylation in both cases.

One important question that should be asked is: What is the mechanism responsible for the overactivation of mTOR in AD? The plausible candidate is Aβ and this hypothesis was elegantly demonstrated by Caccamo *et al.*, using a genetic mice model of AD in which mTOR is not overactivated when Aβ accumulation is prevented by the use of the Aβ antibody (6E10) [189]. On the other hand, Aβ directly injected into wild type mice hippocampus causes mTOR overactivation and this effect is mediated by the proline-rich Akt substrate 40 (PRA40). There is much evidence suggesting that the role of Aβ in mTOR activation. Hence, both Akt and p85 subunit of PI3K are overphosphorylated at Ser473 and Tyr508 in people with AD [172] and the increase of Aβ1-42 overactivates the PI3K/Akt/mTOR axis which in turn affects proteostasis [138]. Furthermore, embryonic cortical neurons from mice treated with Aβ oligomers presented neurite and cell cycle alterations and these effects occurred through PI3K, Akt/mTOR activation [190]. There is an alternative way which increases the activation of these pathways and depends on the insulin receptor. Therefore, it is well known that AD is associated to alterations in the brain's glucose uptake capacity. Some of the studies are focused on the study of the role of the insulin receptor sustrate-1/2 (IRS1/2) because in AD brains, Aβ oligomers increase its phosphorylation at Ser636, 312, 616. These are inhibitory residues and their phosphorylation also mediates the overactivation of Akt and mTOR [191]. However, it is important to highlight that there is a report which has demonstrated a negative effect of Aβ on mTOR signaling. The differences between this and the experiments mentioned above may be related to the dose of Aβ used, being cytotoxic in the latter case [192].

In conclusion, overwhelming results from *in vivo* and *in vitro* experiments demonstrate the important role of mTOR in AD development and its inhibition by rapamycin or its analogs seems to be a potentially good therapeutic strategy. It exerts a positive effect on autophagosome formation, Aβ accumulation, tau protein hyperphosphorylation and cognitive status. However, to our knowledge, no clinical assays have been performed with these drugs for the prevention or treatment of AD.

Minocycline

Minocycline is a semisynthetic second-generation tetracycline, used clinically as a broad-spectrum antibiotic that can easily cross BBB because it is a highly lipophilic molecule. Normally minocycline possesses great popularity as an anti-acne treatment and in rheumatoid arthritis therapy [83]. In comparison to tetracycline structure, minocycline presents a diethyl amino group to the ring D and its lack of functional groups in the ring C, which has been related to its non-antibiotic properties. Its bacteriostatic activity is based on binding the bacterial 30S ribosomal subunit, inhibiting protein synthesis [193]. These chemical modifications enhance tissue absorption of minocycline into the cerebrospinal fluid and CNS and give a longer half-life compared to the original tetracyclines [194, 195] (Fig. **6**).

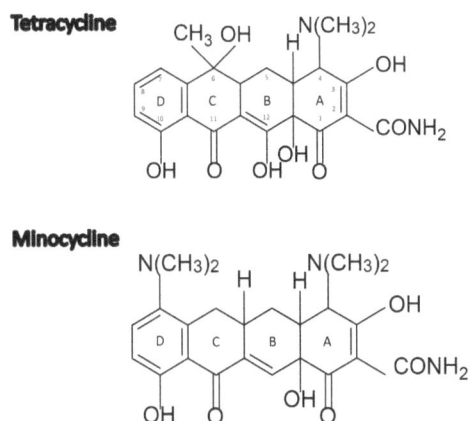

Fig. (6). Chemical structures of tetracycline and minocycline (Obtained from Kim *et al.*, 2009).

There is numerous evidence showing that minocycline exerts neuroprotective effects on experimental models of different neurodegenerative diseases with an inflammatory base such as Huntington´s disease [196], Parkinson´s disease [197], amyotrophic lateral sclerosis [198] and AD, performing different action mechanisms. Nonetheless, we should not forget that a suitable minocycline doses must always be used for carrying out our research. It has been shown that a prolonged minocycline treatment decreased the survival of motor neurons and glial function in the organotypic rat spinal cord cultures [199]. In this same way, a study has reported that high doses of minocycline may induce delayed activation of microglia in aged rats and thus cannot prevent cognitive decline [200]. In humans, intravenous minocycline administration is safe and well tolerated for up to a dose of 10 mg/kg and has been shown to be neuroprotective in experimental trials [201]. Therefore, a concentration lower than 100 M *in vitro* studies and a dose up to 10 mg/kg of body weight *in vivo* studies should be used to analyse the neuroprotective properties of minocycline. However, we should not forget that the long-term minocycline treatment can trigger some rare adverse effects such as hyperpigmentation of the lower extremities, vomit, diarrhea, anorexia, thrombocytopenia and hepatic insufficiency and renal failure in the elderly population [202].

The minocycline anti-inflammatory properties were the first non-antibiotic activity reported due to its capacity to reduce microglial activation and levels of inflammatory mediators. Yrjänheikki´s group was the first to attribute anti-inflammatory properties to minocycline in an ischemia cerebral rat model [203]. AD-associated neuroinflammation is a vicious circle, since Aβ induces microglial production of pro-inflammatory cytokines which favour the Aβ formation and aggregation at the same time [204]. Several studies have found that minocycline can improve cognitive decline by suppressing microglial production of pro-inflammatory cytokines, IL-1β, IL-6, and TNF-α, *in vitro* as well as *in vivo* in models of AD [205, 206]. Moreover, the minocycline-induced decrease in NF-kβ gene expression reduces cytokines production in rhesus monkey brain astrocytes and microglia, because NF-kβ is a cellular stimulator of pro-inflammatory cytokines synthesis [207].

Different *in vitro* studies demonstrated that minocycline blocked LPS-stimulated inflammatory cytokine secretion in the BV2 microglia-derived cell line and on microglia isolated from the brains of mice [208 - 210]. Nevertheless, an *in vitro* and *in vivo* study by Vay *et al.* demonstrated that minocycline not only reduced the activity of proinflammatory cytokines but mitigated the cytokine-induced astrocytic differentiation and microglia activation and recovered the neurogenic and oligodendrogliogenic potential of neural stem cells as well, counteracting the devastating effect of neuroinflammation [211]. Subsequently, many *in vivo* studies have followed corroborating the anti-inflammatory role of minocycline in AD. In 2004, using an experimental model of AD in mice, Hunter described, for the first time, that minocycline reduces cholinergic fibre loss in the hippocampus, ameliorates microglial and astrocytic activation induced by toxins and attenuates pro-inflammatory cytokines secretion as well as cognitive impairment [212]. Also, Biscaro *et al.*, demonstrated that the minocycline reduced microglial cell activation, normalized IL-6 and increased levels of anti-inflammatory IL-10, which is a known inhibitor of TNF-α, improving cognitive deficit in a doubly transgenic (APP and mutant human PS1) mouse model of AD [213]. A recent article has reported that minocycline reduces inflammatory parameters in different brain areas and serum as well as reverses the cognitive decline induced by the administration of Aβ1-42 in mice [214]. Additionally, Clemens *et al.*, have suggested that anti-inflammatory effects of minocycline appears to predominantly be mediated by retinoid signalling, being that the postulated first mode of action for minocycline in human microglial-like cells. They observed that minocycline treatment blocked the retinoic acid degradation in adose-dependent manner, enhancing retinoid acid levels and reducing pro-inflammatory cytokines in microglial cells [215].

Therefore, all the findings obtained so far encourage the scientific community to fully trusts in minocycline's therapeutic capacities which could mitigate the AD-associated neuroinflammation.

Minocycline is also considered an anti-amyloidogenic agent because it reduces the Aβ production and potentiates the Aβ clearance. Additionally, a recent Food and Drug Administration (FDA) approved study demonstrates that minocycline is, among other drugs tested, the most effective for preventing Aβ1-42-induced toxicity and oligomerization in PC12 neurons [216]. Several studies have reported that minocycline administration affects Aβ deposits in APP transgenic mice [217]. In addition, this antibiotic inhibits the formation of Aβ fibrils in the post-mortem brains of people with AD [218]. Furthermore, it has been shown that minocycline decreases Aβ production through the inhibition of β-secretase (BACE1), the main enzyme responsible for the amyloidogenic processing of APP [219]. Regarding Aβ clearance, minocycline increased neprilysin expression, an Aβ-degrading enzyme, in the brains of Aβ-treated rats, preventing the appearance of senile plaques [220]. In this same vein, microglial phagocytic activity plays an important role in Aβ degradation, being reduced during the aging process. Initial studies described that minocycline does not modify the phagocytic capacity of microglial cells [221]. Conversely, a study has demonstrated that minocycline enhances Aβ fibrils phagocytosis in primary microglial cells [222]. Moreover, recent studies have found that autophagy, an important regulator system of protein cellular homeostasis necessary for a health brain, plays a dual role in the metabolism and secretion of Aβ in AD disease. This process is very impaired in AD as shown by the presence of autophagosomes surrounding the senile plaques nucleus within the AD brain [223]. Nilsson *et al.*, observed that autophagy-deficient APP transgenic mouse had a reduced extracellular plaque load and Aβ levels in the brain [224]. Likewise, it has been shown that minocycline is able to induce autophagic cell death in glioma cells, inhibiting the growth of tumour [225] However, the effect of minocycline on AD-associated autophagy has not been studied so far. Previously obtained findings in other pathologies, suggest that this antibiotic could also be a good candidate for boosting autophagy in AD [226]. Hence, further study is necessary in order to describe the effect of minocycline on AD-associated autophagy. Also, a well-functioning autophagy warrants a healthy brain because neurons with complex axonal and dendritic structures are dependent on intense transport and efficient protein turnover in order to allow for a suitable synapse and neuroplasticity [227]. In AD, it is well known that both neuronal function and intercellular communication is completely impaired in humans and murine models of this disease [228]. However, various studies showed that minocycline greatly improves learning and memory by augmenting neuroplasticity and synaptogenesis in the hippocampus of aged mice. The brain-derived neurotrophic factor (BDNF) levels, a regulator factor of synapsis plasticity and memory, are increased by minocycline, as well as the reduction of synapse-associated proteins in hippocampus of aged mice [229]. Furthermore, it has been demonstrated that minocycline promotes neurite growth in PC12 cells after oxygen-glucose deprivation, which supports the beneficial effect of minocycline on the neuroplasticity even more [230]. Along the same line, several studies demonstrated that the Akt- mediated cellular survival pathway is aberrant and is related to cognitive alterations in AD [231]. However, minocycline is able to potentiate cell survival as well as cell damage prevention by PI3K/Akt signalling pathway [232, 233]. Again, all investigations seem to lead to the beneficial effect of minocycline in this devastating neurodegenerative disease.

Aβ accumulation also provokes harmful effects on some neurotransmitters involved in learning and memory such as somatostatin, dopamine and glutamate. The expression levels of somatostatin and dopamine are decreased [234, 235] as well as the glutamate transporter-1 from the human temporal cortex in brains with AD [236]. Minocycline prevents Aβ-induced reduction of somatostatin [220] and protects the somatostatin receptor-effector system from Aβ-induced alterations in an experimental model of AD [237]. Regarding dopamine, several studies have shown that minocycline prevents the dopaminergic neurodegeneration typical of Parkinson´s disease and other closely related conditions of memory loss and mood in people with AD [238 - 240]. Furthermore, the minocycline heightens the spinal excitatory glutamatergic transmission in lamina II rat neurons [241]. Therefore, the impaired neurotransmission systems typical of AD might be improved with minocycline treatment.

Considering the neurofibrillary tangles of hyperphosphorylated tau protein, minocycline decreases the production of abnormal tau protein species in *in vitro* and *in vivo* animal models of AD [242]. At the same time, studies carried out using a transgenic mouse model for AD have shown that minocycline restores hippocampus, cortex and amygdala-dependent learning and memory deficits [243] adding another cognitive effect to this tetracycline.

In summary, all pre-clinical findings indicate that minocycline exerts a great range of neuroprotective effects on AD. However, there are not enough clinical trials to test these properties in people with AD. Currently, there is only one completed clinical trial (NCT01463384), but the results are not conclusive because the number of patients studied is not very high. Also, the National Institute of Health Research (NIHR) is carrying out a clinical trial for AD to evaluate the effects of two years of minocycline treatment on the deterioration of the mental processes in people with early AD (ISRCTN16105064). In the last update (18 August), they reported that all findings have been collected and that they are currently analysing these data.

Therefore, proper designed clinical trials are required to extrapolate minocycline neuroprotective properties from pre-clinical studies in humans. Likewise, more studies need to be carried out in order to describe the possible action mechanisms of minocycline.

Other Antibiotics

Macrolides

There are two main macrolide antibiotics which have been studied in the treatment of AD, azithromycin and erythromycin. On one hand, azithromycin, which acts as antibacterial inhibiting the 50S ribosome subunit during translation [244]. This semisynthetic aminoglycoside antibiotic has been used to treat different pathological stages in various AD models [245] because it is an excellent inhibitor of APP 5′-untranslated region (5′-APP-UTR) of mRNA encoding the APP and it also intervenes at the overlapping signaling pathway to activate α and/or γ-secretase cleavage [246]. Therefore, azithromycin can modulate APP processing because it alters the cleavage of APP and it can also modify the processing of APP in human lens epithelial (B3) cells and in human neuroblastoma (SH-SY5Y) [247]. Recent advancements in drug development highlight this antibiotic as a new possibility in targeting the 5′-APP-UTR in AD. More than 1000 drugs related to this have been preapproved by the FDA, but only 17 have been found to be effective in inhibiting 5′-APP-UTR. However, it is important to mention some possible side effects of azithromycin, like nausea, abdominal pain, vomiting, flatulence, diarrhea and an increase of several liver enzymes [245].

On the other hand, erythromycin, which is an analog of azithromycin, was also investigated in similar studies using an AD mouse model. Both antibiotics markedly change the cleavage of the APP C-terminal fragment in SH-SY5Y cells [248]. Like azithromycin, it alters APP processing resulting in a diminution of cerebral levels of Aβ(1-42) without having any effect on Aβ(1-40) levels. Erythromycin has been postulated as a possible neuroprotective agent because it induces the expression of a 7-kDa APP C-terminal fragment, which may increase the expression of different neuroprotective target genes [249]. This antibiotic also reduced the amount of C99 and C83 C-terminal fragments (both are substrates of γ and α-secretase, respectively, for the generation of Aβ) in the brain cortex of the transgenic mouse model for AD TgCrND8 relative to controls. A postulated explanation for erythromycin action could be that this antibiotic would activate the γ-secretase complex to switch APP cleavage specificity towards the Aβ(1-42) cleavage site way the Aβ(1-40) cleavage site [248].

To summarize, all these pre-clinical studies indicate that the main role of these two macrolide antibiotics in the treatment of AD is due to the fact that they can modify the APP processing.

Clioquinol

Clioquinol (5-chloro-7-iodo-8-hydroxyquinolinol) is a hydroxyquinoline with antifungal and antiprotozoal properties which is prescribed in some topical preparations in order to treat different skin infections [249] and is described as neurotoxic when it is used chronically at high doses [245]. Clioquinol was extensively used in the past as an anti-infective, especially for diarrhea. In 1970, it was withdrawn from oral use after being associated with subacute myelo-optic neuropathy (SMON), a Japanese neurotoxic epidemic [250]. This syndrome is characterized by upper and lower motor neuron lower limb signs and subacute visual changes (optic neuritis). The physical signs of SMON are similar to those of subacute combined degeneration of the cord secondary to vitamin B12 deficiency. Therefore, for people treated with clioquinol, SMON could be avoided by the administration of vitamin B12 [251].

This antibiotic is a hydrophobic metal protein attenuating compound which can cross the BBB. It has a greater affinity for Cu^{2+} and Zn^{2+}, two metal ions critically involved in amyloid-β aggregation and toxicity [252], than for Ca^{2+} and Mg^{2+} ions. In the ´amyloid cascade´, clioquinol might act by either preventing Aβ deposition into amyloid plaques or by promoting Aβ clearance through the mobilization of Aβ from existing deposits [251]. Specifically, clioquinol can prevent the formation of amyloid plaques in transgenic AD mice and dissolve Aβ from postmortem tissue of people with AD [251, 253]. Therefore, clioquinol may act by disaggregating collections of Aβ and Cu, ´dissolving´ accumulations of Aβ [254]. However, an extended clioquinol treatment in a different mouse model led to a reduced survival [255]. Some clinical trials of clioquinol suggested a significant slowing down of the cognitive decline in the most severely affected subgroup of people with AD, with a parallel reduction in plasma Aβ1-42 levels. In addition, clioquinol was also reported to ameliorate disease pathology in a mouse model of Parkinson´s disease and Huntington´s disease [252].

Because of its action mode, different conflicting hypotheses have postulated the ability of clioquinol to interfere with brain metal metabolism. These hypotheses suggest that it might work as a chelating agent. Others studies propose that this antibiotic may act as a ´copper carrier´, behaving as an ionophore that facilitates membrane crossing to complexed copper [256]. It was experimentally proved that an increase in intracellular copper levels raises metalloprotease activities which enhance Aβ clearance [256].

Moreover, a series of selenium-containing clioquinol derivatives were designed, synthesized and evaluated as multifunctional anti-AD agents. *In vitro* examination showed that several target compounds exhibited activities such as the inhibition of metal-induced Aβ aggregation, antioxidative properties, hydrogen peroxide scavenging and the prevention of copper redox cycling. A parallel artificial membrane permeation assay indicated that selenium-containing clioquinol derivatives possessed significant BBB permeability [257].

In TgCRND8, a transgenic AD mouse, this hydroxyquinoline reverses to a large extent, the working memory problems that are characteristic of this mouse model, reducing amyloid plaques in the cortex and hippocampus region of the brain and attenuating astrogliosis. Furthermore, significant effects on the absolute and relative brain concentrations of the three most important biometals (Cu^{2+}, Zn^{2+} and Fe^{2+}) were highlighted following this treatment with clioquinol as well as its distribution within the brain mirrored areas implicated in memory and learning [252, 258]. These observations led to a clinical trial using cloroquinol in which thirty-six subjects with AD (cognitive subscale score of the Alzheimer's Disease Assessment Scale (ADAS-Cog) 20-45; Mini-Mental State Examination score 10-24) were included in this randomized double-bind phase II study. After 9 months, the effect of the treatment was significant in the more clinically severe group (ADAS-Cog ≥25) as a reduction in plasma Aβ(1-42) levels in the clioquinol group was observed and an increased in the pacebo group. While plasma zinc levels also increase in the clioquinol group, copper levels remained unchanged. Overall, clioquinol was well tolerated by subjects at a maximum oral dose of 375 mg twice per day. Furthermore, the planned phase III trial of clioquinol was abandoned and this compound has been withdrawn from development [258].

It has been also demonstrated that a combination of the administration of cloroquinol and inoculation with Aβ1−42 vaccines in the transgenic mouse model APP/PS1 was effective in significantly reducing the deposits of amyloid in the brains of these animals. Furthermore, this study reports that systemic clioquinol induces myelopathies in the dorsal lateral geniculate nucleus, which was almost devoid of amyloid plaques and is the primary site of retinal efferent projections *via* the optic nerve. Inoculation with an Aβ1−42 vaccine was also found to result in a significant increase in plaque-independent astrocytic hyperplasia in the dorsal part of the lateral septal nucleus which was also devoid of plaques, reflecting potential brain inflammatory processes [259].

In conclusion, several clinical trials show clioquinol´s capacity for preventing Aβ deposition and active the mobilization of Aβ from existing deposits, therefore being an interesting candidate to utilize in the treatment of AD. However, more studies are needed in order to check if its behavior as a chelating agent could make this a potential drug for interfering with brain metal metabolism.

Amphotericin B

Amphotericin B is a widely used membrane-active antifungal drug which has several known toxicities including multiorgan failure and potential death that would like to limit its use as a treatment for AD [249]. It has also antiprotozoal, antiviral and indirect antimicrobial activity through immune stimulation. Most intriguing are the reports of the antiprion activity of amphotericin B and its derivatives. In fact, they have been shown to be among the very few agents, which can slow the course of prion disease in animal models. It is currently unclear as to what the mechanism of action could be [260].

This polyene macrolide antibiotic prevents fibrillation in amyloid disease in AD, as it binds specifically to Aβ 25-35 fibrils [261]. Amphotericin B seems to have a complementary face for amyloid fibrils but not the native protein. Moreover, this medication interacts specifically with Congo Red, a very well-known fibril-binding agent. In addition, in kinetic fibril formation studies, amphotericin B was able to significantly kinetically delay the formation of Aβ 25-35 fibrils at pH 7.4 but not insulin fibrils at pH 2 [260]. However, the ability of amphotericin B to affect the conformational changes of neurotoxic soluble oligomers of amyloid peptides, in particular Aβ(1-42) peptide, has not been reported thus far. Notably, this drug was also shown to modulate the aggregation process of prion protein, but no mechanistic details have been provided. In addition, amphotericin B has no measurable impact neither on the secondary structure nor on the time-dependent aggregation profile of the amyloid peptide [261].

In summary, some *in vitro* studies supporting that the importance of this antibiotic in the treatment of AD is based on the prevention of the formation of amyloid fibrils which amphotericin B has.

Tetracycline

Tetracycline is a classical antibiotic used to treat different infections and it has been proposed for AD therapy due to its effects on the aggregation of Aβ protein, particularly its ability to interact *in vivo* with the Aβ oligomers and aggregates in different AD animal models [249]. Specifically, this drug led to the formation of colloidal particles that particularly sequester oligomers, preventing the progression of the amyloid cascade [262].

Furthermore, this antibiotic reduces the resistance of Aβ1-42 to trypsin digestion [117] and it induces an increase in disassembly of preformed fibrils in AD mouse models [249]. These effects were dose dependent and specific to tetracycline as opposed to antibiotics in general. It has been reported that the mechanism of action of this drug is based on the induced changes in the secondary structure of Aβ protein, from soluble form to β-sheet-rich structures corresponding to the presence of oligomers and fibrils, disassembling both Aβ oligomeric and fibrillar β-sheet assemblies, restoring their non-amyloidogenic structures [262, 263]. Moreover, tetracycline-treated transgenic *Caenorhabditis elegans,* a simplified invertebrate model of AD, showed lowered oxidative stress reducing superoxide production and protects from the onset of the paralysis phenotype [262].

In summary, owing to all these findings, tetracycline has been described in several animal models of AD as a possible drug used against this disease because of its capacity to disrupt the amyloid cascade, among other effects.

Finally, Fig. (**7**) recapitulates the principal effects of all these antibiotics.

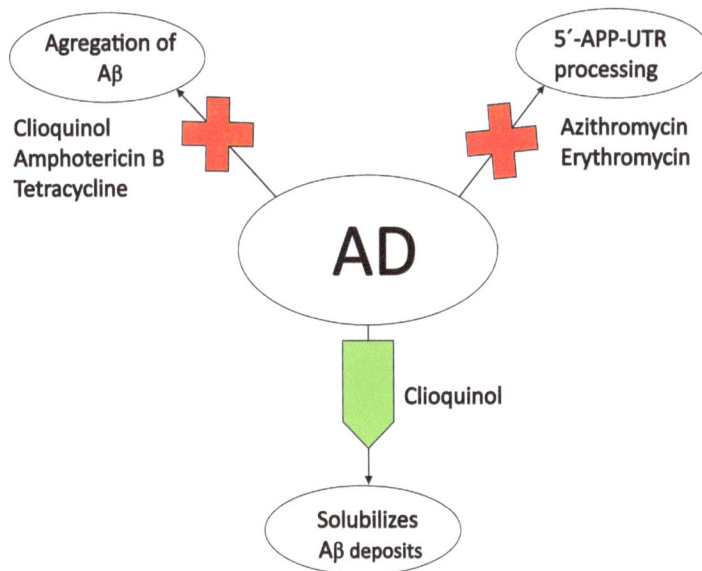

Fig. (7). Scheme of the main effects of other antibiotics.

CONCLUSION

Evidence seems to drive in the same direction, confirming that antibiotics could be promising candidates in the prevention and treatment of AD, exerting beneficial effects and are relatively inexpensive. It is important to emphasize that they can cross the BBB and all of them exert neuroprotective activities. Among them, it is important to highlight the anti-amyloidogenic role exerted by rifampicin, avoiding the Aβ aggregation. We mention again the effect of Rapamycin on autophagy activity which favours the clearance of Aβ and the anti-inflammatory activity of minocycline that reduces the Aβ-induced microglial activation. Likewise, all these cited antibiotics and some others analysed in this chapter exert beneficial effects on many of the pathophysiological features of AD. Nevertheless, it is important to highlight that long-term treatment with antibiotics could cause adverse effects including antibiotic resistance. Therefore, these issues should be considered in current and future clinical trials in order to determine the therapeutic dose range which exerts neuroprotection. In this chapter, we have tried to emphasize the potential neuroprotector role of some antibiotics, mainly used against infections. Notwithstanding, it is still necessary to conduct further, properly executed, clinical trials in order to confirm all the cited properties of these antibiotics in people with AD.

CONSENT FOR PUBLICATION

Not applicable.

CONFLICT OF INTEREST

The authors declare no conflict of interest, financial or otherwise.

ACKNOWLEDGEMENTS

We strongly appreciate Dean Custer, American teacher, who has checked and edited the manuscript.

REFERENCES

[1]   Walsh DM, Selkoe DJ. Deciphering the molecular basis of memory failure in Alzheimer's disease. Neuron 2004; 44(1): 181-93.
[http://dx.doi.org/10.1016/j.neuron.2004.09.010] [PMID: 15450169]

[2]   Kumar A, Singh A, Ekavali . A review on Alzheimer's disease pathophysiology and its management: an update. Pharmacol Rep 2015; 67(2): 195-203.
[http://dx.doi.org/10.1016/j.pharep.2014.09.004] [PMID: 25712639]

[3]   Sambamurti K, Greig NH, Lahiri DK. Advances in the cellular and molecular biology of the beta-amyloid protein in Alzheimer's disease. Neuromolecular Med 2002; 1(1): 1-31.
[http://dx.doi.org/10.1385/NMM:1:1:1] [PMID: 12025813]

[4]   Vardy ER, Catto AJ, Hooper NM. Proteolytic mechanisms in amyloid-beta metabolism: therapeutic implications for Alzheimer's disease. Trends Mol Med 2005; 11(10): 464-72.
[http://dx.doi.org/10.1016/j.molmed.2005.08.004] [PMID: 16153892]

[5]   Salomone S, Caraci F, Leggio GM, Fedotova J, Drago F. New pharmacological strategies for treatment of Alzheimer's disease: focus on disease modifying drugs. Br J Clin Pharmacol 2012; 73(4): 504-17.
[http://dx.doi.org/10.1111/j.1365-2125.2011.04134.x] [PMID: 22035455]

[6]   Kim D, Tsai LH. Bridging physiology and pathology in AD. Cell 2009; 137(6): 997-1000.
       [http://dx.doi.org/10.1016/j.cell.2009.05.042] [PMID: 19524503]

[7] . De Strooper B. Aph-1, Pen-2, and Nicastrin with Presenilin generate an active γ-Secretase complex.
       Neuron 2003; 38(1): 9-12.
       [http://dx.doi.org/10.1016/S0896-6273(03)00205-8] [PMID: 12691659]

[8]   Mueller-Steiner S, Zhou Y, Arai H, *et al.* Antiamyloidogenic and neuroprotective functions of
       cathepsin B: implications for Alzheimer's disease. Neuron 2006; 51(6): 703-14.
       [http://dx.doi.org/10.1016/j.neuron.2006.07.027] [PMID: 16982417]

[9]   Qosa H, Abuznait AH, Hill RA, Kaddoumi A. Enhanced brain amyloid-β clearance by rifampicin and
       caffeine as a possible protective mechanism against Alzheimer's disease. J Alzheimers Dis 2012;
       31(1): 151-65.
       [http://dx.doi.org/10.3233/JAD-2012-120319] [PMID: 22504320]

[10]   Abuznait AH, Cain C, Ingram D, Burk D, Kaddoumi A. Up-regulation of P-glycoprotein reduces
       intracellular accumulation of beta amyloid: investigation of P-glycoprotein as a novel therapeutic
       target for Alzheimer's disease. J Pharm Pharmacol 2011; 63(8): 1111-8.
       [http://dx.doi.org/10.1111/j.2042-7158.2011.01309.x] [PMID: 21718295]

[11]   Silverberg GD, Messier AA, Miller MC, *et al.* Amyloid efflux transporter expression at the blood-
       brain barrier declines in normal aging. J Neuropathol Exp Neurol 2010; 69(10): 1034-43.
       [http://dx.doi.org/10.1097/NEN.0b013e3181f46e25] [PMID: 20838242]

[12]   Pintado C, Macías S, Domínguez-Martín H, Castaño A, Ruano D. Neuroinflammation alters cellular
       proteostasis by producing endoplasmic reticulum stress, autophagy activation and disrupting ERAD
       activation. Sci Rep 2017; 7(1): 8100.
       [http://dx.doi.org/10.1038/s41598-017-08722-3] [PMID: 28808322]

[13]   Freeman LC, Ting JP. The pathogenic role of the inflammasome in neurodegenerative diseases. J
       Neurochem 2016; 136(1) (Suppl. 1): 29-38.
       [http://dx.doi.org/10.1111/jnc.13217] [PMID: 26119245]

[14]   Mandrekar-Colucci S, Landreth GE. Microglia and inflammation in Alzheimer's disease. CNS Neurol
       Disord Drug Targets 2010; 9(2): 156-67.
       [http://dx.doi.org/10.2174/187152710791012071] [PMID: 20205644]

[15]   Alavi Naini SM, Soussi-Yanicostas N. Tau Hyperphosphorylation and Oxidative Stress, a Critical
       Vicious Circle in Neurodegenerative Tauopathies? Oxid Med Cell Longev 2015; 2015: 151979.
       [http://dx.doi.org/10.1155/2015/151979] [PMID: 26576216]

[16]   Apelt J, Schliebs R. Beta-amyloid-induced glial expression of both pro- and anti-inflammatory
       cytokines in cerebral cortex of aged transgenic Tg2576 mice with Alzheimer plaque pathology. Brain
       Res 2001; 894(1): 21-30.
       [http://dx.doi.org/10.1016/S0006-8993(00)03176-0] [PMID: 11245811]

[17]   Krabbe G, Halle A, Matyash V, *et al.* Functional impairment of microglia coincides with Beta-amyloid
       deposition in mice with Alzheimer-like pathology. PLoS One 2013; 8(4): e60921.
       [http://dx.doi.org/10.1371/journal.pone.0060921] [PMID: 23577177]

[18]   Medeiros R, LaFerla FM. Astrocytes: conductors of the Alzheimer disease neuroinflammatory
       symphony. Exp Neurol 2013; 239: 133-8.
       [http://dx.doi.org/10.1016/j.expneurol.2012.10.007] [PMID: 23063604]

[19]   Walker DG, Link J, Lue LF, Dalsing-Hernandez JE, Boyes BE. Gene expression changes by amyloid
       beta peptide-stimulated human postmortem brain microglia identify activation of multiple
       inflammatory processes. J Leukoc Biol 2006; 79(3): 596-610.
       [http://dx.doi.org/10.1189/jlb.0705377] [PMID: 16365156]

[20]   Patel NS, Paris D, Mathura V, Quadros AN, Crawford FC, Mullan MJ. Inflammatory cytokine levels
       correlate with amyloid load in transgenic mouse models of Alzheimer's disease. J Neuroinflammation

2005; 2(1): 9.
[http://dx.doi.org/10.1186/1742-2094-2-9] [PMID: 15762998]

[21]   Halliday G, Robinson SR, Shepherd C, Kril J. Alzheimer's disease and inflammation: a review of cellular and therapeutic mechanisms. Clin Exp Pharmacol Physiol 2000; 27(1-2): 1-8.
[http://dx.doi.org/10.1046/j.1440-1681.2000.03200.x] [PMID: 10696521]

[22]   Griffin WS, Sheng JG, Royston MC, *et al.* Glial-neuronal interactions in Alzheimer's disease: the potential role of a 'cytokine cycle' in disease progression. Brain Pathol 1998; 8(1): 65-72.
[http://dx.doi.org/10.1111/j.1750-3639.1998.tb00136.x] [PMID: 9458167]

[23]   Combs CK, Karlo JC, Kao SC, Landreth GE. β-Amyloid stimulation of microglia and monocytes results in TNFalpha-dependent expression of inducible nitric oxide synthase and neuronal apoptosis. J Neurosci 2001; 21(4): 1179-88.
[http://dx.doi.org/10.1523/JNEUROSCI.21-04-01179.2001] [PMID: 11160388]

[24]   Venegas C, Heneka MT. Danger-associated molecular patterns in Alzheimer's disease. J Leukoc Biol 2017; 101(1): 87-98.
[http://dx.doi.org/10.1189/jlb.3MR0416-204R] [PMID: 28049142]

[25]   Bamberger ME, Harris ME, McDonald DR, Husemann J, Landreth GE. A cell surface receptor complex for fibrillar beta-amyloid mediates microglial activation. J Neurosci 2003; 23(7): 2665-74.
[http://dx.doi.org/10.1523/JNEUROSCI.23-07-02665.2003] [PMID: 12684452]

[26]   Stewart CR, Stuart LM, Wilkinson K, *et al.* CD36 ligands promote sterile inflammation through assembly of a Toll-like receptor 4 and 6 heterodimer. Nat Immunol 2010; 11(2): 155-61.
[http://dx.doi.org/10.1038/ni.1836] [PMID: 20037584]

[27]   Liu S, Liu Y, Hao W, *et al.* TLR2 is a primary receptor for Alzheimer's amyloid β peptide to trigger neuroinflammatory activation. J Immunol 2012; 188(3): 1098-107.
[http://dx.doi.org/10.4049/jimmunol.1101121] [PMID: 22198949]

[28]   Walter S, Letiembre M, Liu Y, *et al.* Role of the toll-like receptor 4 in neuroinflammation in Alzheimer's disease. Cell Physiol Biochem 2007; 20(6): 947-56.
[http://dx.doi.org/10.1159/000110455] [PMID: 17982277]

[29]   Sheedy FJ, Grebe A, Rayner KJ, *et al.* CD36 coordinates NLRP3 inflammasome activation by facilitating intracellular nucleation of soluble ligands into particulate ligands in sterile inflammation. Nat Immunol 2013; 14(8): 812-20.
[http://dx.doi.org/10.1038/ni.2639] [PMID: 23812099]

[30]   Koenigsknecht-Talboo J, Landreth GE. Microglial phagocytosis induced by fibrillar beta-amyloid and IgGs are differentially regulated by proinflammatory cytokines. J Neurosci 2005; 25(36): 8240-9.
[http://dx.doi.org/10.1523/JNEUROSCI.1808-05.2005] [PMID: 16148231]

[31]   von Bernhardi R, Ramírez G, Toro R, Eugenín J. Pro-inflammatory conditions promote neuronal damage mediated by Amyloid Precursor Protein and decrease its phagocytosis and degradation by microglial cells in culture. Neurobiol Dis 2007; 26(1): 153-64.
[http://dx.doi.org/10.1016/j.nbd.2006.12.006] [PMID: 17240154]

[32]   Streit WJ. Microglial senescence: does the brain's immune system have an expiration date? Trends Neurosci 2006; 29(9): 506-10.
[http://dx.doi.org/10.1016/j.tins.2006.07.001] [PMID: 16859761]

[33]   Falsig J, van Beek J, Hermann C, Leist M. Molecular basis for detection of invading pathogens in the brain. J Neurosci Res 2008; 86(7): 1434-47.
[http://dx.doi.org/10.1002/jnr.21590] [PMID: 18061944]

[34]   Wyss-Coray T, Loike JD, Brionne TC, *et al.* Adult mouse astrocytes degrade amyloid-beta *in vitro* and *in situ*. Nat Med 2003; 9(4): 453-7.
[http://dx.doi.org/10.1038/nm838] [PMID: 12612547]

[35]   Hostenbach S, Cambron M, D'haeseleer M, Kooijman R, De Keyser J. Astrocyte loss and astrogliosis

in neuroinflammatory disorders. Neurosci Lett 2014; 565: 39-41.
[http://dx.doi.org/10.1016/j.neulet.2013.10.012] [PMID: 24128880]

[36]　Kulijewicz-Nawrot M, Verkhratsky A, Chvátal A, Syková E, Rodríguez JJ. Astrocytic cytoskeletal atrophy in the medial prefrontal cortex of a triple transgenic mouse model of Alzheimer's disease. J Anat 2012; 221(3): 252-62.
[http://dx.doi.org/10.1111/j.1469-7580.2012.01536.x] [PMID: 22738374]

[37]　Tarkowski E, Andreasen N, Tarkowski A, Blennow K. Intrathecal inflammation precedes development of Alzheimer's disease. J Neurol Neurosurg Psychiatry 2003; 74(9): 1200-5.
[http://dx.doi.org/10.1136/jnnp.74.9.1200] [PMID: 12933918]

[38]　Mrak RE, Sheng JG, Griffin WS. Glial cytokines in Alzheimer's disease: review and pathogenic implications. Hum Pathol 1995; 26(8): 816-23.
[http://dx.doi.org/10.1016/0046-8177(95)90001-2] [PMID: 7635444]

[39]　Tan J, Town T, Crawford F, *et al.* Role of CD40 ligand in amyloidosis in transgenic Alzheimer's mice. Nat Neurosci 2002; 5(12): 1288-93.
[http://dx.doi.org/10.1038/nn968] [PMID: 12402041]

[40]　Ghosh S, Wu MD, Shaftel SS, *et al.* Sustained interleukin-1β overexpression exacerbates tau pathology despite reduced amyloid burden in an Alzheimer's mouse model. J Neurosci 2013; 33(11): 5053-64.
[http://dx.doi.org/10.1523/JNEUROSCI.4361-12.2013] [PMID: 23486975]

[41]　Yoshiyama Y, Higuchi M, Zhang B, *et al.* Synapse loss and microglial activation precede tangles in a P301S tauopathy mouse model. Neuron 2007; 53(3): 337-51.
[http://dx.doi.org/10.1016/j.neuron.2007.01.010] [PMID: 17270732]

[42]　Asai H, Ikezu S, Woodbury ME, Yonemoto GM, Cui L, Ikezu T. Accelerated neurodegeneration and neuroinflammation in transgenic mice expressing P301L tau mutant and tau-tubulin kinase 1. Am J Pathol 2014; 184(3): 808-18.
[http://dx.doi.org/10.1016/j.ajpath.2013.11.026] [PMID: 24418258]

[43]　Cuervo AM, Bergamini E, Brunk UT, Dröge W, Ffrench M, Terman A. Autophagy and aging: the importance of maintaining "clean" cells. Autophagy 2005; 1(3): 131-40.
[http://dx.doi.org/10.4161/auto.1.3.2017] [PMID: 16874025]

[44]　Congdon EE. Sex differences in autophagy contribute to female vulnerability in alzheimer's disease. Front Neurosci 2018; 12: 372.
[http://dx.doi.org/10.3389/fnins.2018.00372] [PMID: 29988365]

[45]　Cuervo AM. Autophagy: in sickness and in health. Trends Cell Biol 2004; 14(2): 70-7.
[http://dx.doi.org/10.1016/j.tcb.2003.12.002] [PMID: 15102438]

[46]　Klionsky DJ, Ohsumi Y. Vacuolar import of proteins and organelles from the cytoplasm. Annu Rev Cell Dev Biol 1999; 15: 1-32.
[http://dx.doi.org/10.1146/annurev.cellbio.15.1.1] [PMID: 10611955]

[47]　Nixon RA. Autophagy in neurodegenerative disease: friend, foe or turncoat? Trends Neurosci 2006; 29(9): 528-35.
[http://dx.doi.org/10.1016/j.tins.2006.07.003] [PMID: 16859759]

[48]　Gomez TA, Clarke SG. Autophagy and insulin/TOR signaling in Caenorhabditis elegans pcm-1 protein repair mutants. Autophagy 2007; 3(4): 357-9.
[http://dx.doi.org/10.4161/auto.4143] [PMID: 17404495]

[49]　Pan T, Kondo S, Zhu W, Xie W, Jankovic J, Le W. Neuroprotection of rapamycin in lactacystin-induced neurodegeneration *via* autophagy enhancement. Neurobiol Dis 2008; 32(1): 16-25.
[http://dx.doi.org/10.1016/j.nbd.2008.06.003] [PMID: 18640276]

[50]　García-Arencibia M, Hochfeld WE, Toh PP, Rubinsztein DC. Autophagy, a guardian against neurodegeneration. Semin Cell Dev Biol 2010; 21(7): 691-8.

[http://dx.doi.org/10.1016/j.semcdb.2010.02.008] [PMID: 20188203]

[51]   Meléndez A, Tallóczy Z, Seaman M, Eskelinen EL, Hall DH, Levine B. Autophagy genes are essential for dauer development and life-span extension in C. elegans. Science 2003; 301(5638): 1387-91.
[http://dx.doi.org/10.1126/science.1087782] [PMID: 12958363]

[52]   Yang Z, Klionsky DJ. Mammalian autophagy: core molecular machinery and signaling regulation. Curr Opin Cell Biol 2010; 22(2): 124-31.
[http://dx.doi.org/10.1016/j.ceb.2009.11.014] [PMID: 20034776]

[53]   Mizushima N. The pleiotropic role of autophagy: from protein metabolism to bactericide. Cell Death Differ 2005; 12 (Suppl. 2): 1535-41.
[http://dx.doi.org/10.1038/sj.cdd.4401728] [PMID: 16247501]

[54]   Mizushima N. A brief history of autophagy from cell biology to physiology and disease Nat Cell Biol 2018; 20: 521-7.
[http://dx.doi.org/10.1038/s41556-018-0092-5]

[55]   Todde V, Veenhuis M, van der Klei IJ. Autophagy: principles and significance in health and disease. Biochim Biophys Acta 2009; 1792(1): 3-13.
[http://dx.doi.org/10.1016/j.bbadis.2008.10.016] [PMID: 19022377]

[56]   Meléndez A, Levine B. Autophagy in C. elegans. WormBook 2009; 24: 1-26.
[http://dx.doi.org/10.1895/wormbook.1.147.1] [PMID: 19705512]

[57]   Fujikake N, Shin M, Shimizu S. Association between autophagy and neurodegenerative diseases. Front Neurosci 2018; 12: 255.
[http://dx.doi.org/10.3389/fnins.2018.00255] [PMID: 29872373]

[58]   Kabeya Y, Kamada Y, Baba M, Takikawa H, Sasaki M, Ohsumi Y. Atg17 functions in cooperation with Atg1 and Atg13 in yeast autophagy. Mol Biol Cell 2005; 16(5): 2544-53.
[http://dx.doi.org/10.1091/mbc.e04-08-0669] [PMID: 15743910]

[59]   Mizushima N, Komatsu M. Autophagy: renovation of cells and tissues. Cell 2011; 147(4): 728-41.
[http://dx.doi.org/10.1016/j.cell.2011.10.026] [PMID: 22078875]

[60]   Jaeger PA, Wyss-Coray T. Beclin 1 complex in autophagy and Alzheimer disease Arch Neurol 2010; 67(): 1181-4.

[61]   Pickford F, Masliah E, Britschgi M, *et al.* The autophagy-related protein beclin 1 shows reduced expression in early Alzheimer disease and regulates amyloid beta accumulation in mice. J Clin Invest 2008; 118(6): 2190-9.
[PMID: 18497889]

[62]   Li L, Zhang X, Le W. Autophagy dysfunction in Alzheimer's disease. Neurodegener Dis 2010; 7(4): 265-71.
[PMID: 20551691]

[63]   Wang Y, Martinez-Vicente M, Krüger U, *et al.* Tau fragmentation, aggregation and clearance: the dual role of lysosomal processing. Hum Mol Genet 2009; 18(21): 4153-70.
[http://dx.doi.org/10.1093/hmg/ddp367] [PMID: 19654187]

[64]   Shacka JJ, Roth KA, Zhang J. The autophagy-lysosomal degradation pathway: role in neurodegenerative disease and therapy. Front Biosci 2008; 13: 718-36.
[http://dx.doi.org/10.2741/2714] [PMID: 17981582]

[65]   Hamano T, Gendron TF, Causevic E, *et al.* Autophagic-lysosomal perturbation enhances tau aggregation in transfectants with induced wild-type tau expression. Eur J Neurosci 2008; 27(5): 1119-30.
[http://dx.doi.org/10.1111/j.1460-9568.2008.06084.x] [PMID: 18294209]

[66]   Mizushima N. The role of the Atg1/ULK1 complex in autophagy regulation. Curr Opin Cell Biol 2010; 22(2): 132-9.

[http://dx.doi.org/10.1016/j.ceb.2009.12.004] [PMID: 20056399]

[67]   Jung CH, Ro SH, Cao J, Otto NM, Kim DH. mTOR regulation of autophagy. FEBS Lett 2010; 584(7): 1287-95.
[http://dx.doi.org/10.1016/j.febslet.2010.01.017] [PMID: 20083114]

[68]   Díaz-Troya S, Pérez-Pérez ME, Florencio FJ, Crespo JL. The role of TOR in autophagy regulation from yeast to plants and mammals. Autophagy 2008; 4(7): 851-65.
[http://dx.doi.org/10.4161/auto.6555] [PMID: 18670193]

[69]   Mitra S, Prasad P. Chakraborty. A unified view of assessing the pro-oxidant *vs* antioxidant nature of amyloid beta conformers. ChemBioChem 2018; 27: 1-11.

[70]   Varadarajan S, Yatin S, Aksenova M, Butterfield DA. Review: Alzheimer's amyloid β-peptid--associated free radical oxidative stress and neurotoxicity. J Struct Biol 2000; 130(2-3): 184-208.
[http://dx.doi.org/10.1006/jsbi.2000.4274] [PMID: 10940225]

[71]   Carrillo-Mora P, Luna R, Colín-Barenque L. Amyloid beta: multiple mechanisms of toxicity and only some protective effects? Oxid Med Cell Longev 2014; 2014: 795375.
[http://dx.doi.org/10.1155/2014/795375] [PMID: 24683437]

[72]   Kontush A, Berndt C, Weber W, *et al.* Amyloid-β is an antioxidant for lipoproteins in cerebrospinal fluid and plasma. Free Radic Biol Med 2001; 30(1): 119-28.
[http://dx.doi.org/10.1016/S0891-5849(00)00458-5] [PMID: 11134902]

[73]   Smith DG, Cappai R, Barnham KJ. The redox chemistry of the Alzheimer's disease amyloid beta peptide. Biochim Biophys Acta 2007; 1768(8): 1976-90.
[http://dx.doi.org/10.1016/j.bbamem.2007.02.002] [PMID: 17433250]

[74]   Butterfield DA, Swomley AM, Sultana R. Amyloid β-peptide (1-42)-induced oxidative stress in Alzheimer disease: importance in disease pathogenesis and progression. Antioxid Redox Signal 2013; 19(8): 823-35.
[http://dx.doi.org/10.1089/ars.2012.5027] [PMID: 23249141]

[75]   Butterfield DA, Boyd-Kimball D. The critical role of methionine 35 in Alzheimer's amyloid beta-peptide (1-42)-induced oxidative stress and neurotoxicity. Biochim Biophys Acta 2005; 1703(2): 149-56.
[http://dx.doi.org/10.1016/j.bbapap.2004.10.014] [PMID: 15680223]

[76]   Dasilva KA, Shaw JE, McLaurin J. Amyloid-beta fibrillogenesis: structural insight and therapeutic intervention. Exp Neurol 2010; 223(2): 311-21.
[http://dx.doi.org/10.1016/j.expneurol.2009.08.032] [PMID: 19744483]

[77]   Wan L, Nie G, Zhang J, *et al.* β-Amyloid peptide increases levels of iron content and oxidative stress in human cell and Caenorhabditis elegans models of Alzheimer disease. Free Radic Biol Med 2011; 50(1): 122-9.
[http://dx.doi.org/10.1016/j.freeradbiomed.2010.10.707] [PMID: 21034809]

[78]   Anand R, Gill KD, Mahdi AA. Therapeutics of Alzheimer's disease: Past, present and future Neuropharmacology 2014 ; 76(Pt A): 27-50.

[79]   Crespo MC, Tomé-Carneiro J, Pintado C, Dávalos A, Visioli F, Burgos-Ramos E. Hydroxytyrosol restores proper insulin signaling in an astrocytic model of Alzheimer's disease. Biofactors 2017; 43(4): 540-8.
[http://dx.doi.org/10.1002/biof.1356] [PMID: 28317262]

[80]   Sarubbo F, Moranta D, Asensio VJ, Miralles A, Esteban S. Effects of Resveratrol and Other Polyphenols on the Most Common Brain Age-Related Diseases. Curr Med Chem 2017; 24(38): 4245-66.
[http://dx.doi.org/10.2174/0929867324666170724102743] [PMID: 28738770]

[81]   Ortiz-López L, Márquez-Valadez B, Gómez-Sánchez A, *et al.* Green tea compound epigallo-catechi--3-gallate (EGCG) increases neuronal survival in adult hippocampal neurogenesis *in vivo* and *in vitro*.

Neuroscience 2016; 322: 208-20.
[http://dx.doi.org/10.1016/j.neuroscience.2016.02.040] [PMID: 26917271]

[82]   Wang Y, Wang Y, Li J, *et al.* Effects of caffeic acid on learning deficits in a model of Alzheimer's disease. Int J Mol Med 2016; 38(3): 869-75.
[http://dx.doi.org/10.3892/ijmm.2016.2683] [PMID: 27430591]

[83]   Aronson AL. Pharmacotherapeutics of newer tetracyclines J Am Vet Med Assoc 1980; 176(10 Spec No): 1061-80.

[84]   Shobo A, Bratkowska D, Baijnath S, *et al.* Visualization of time-dependent distribution of rifampicin in rat brain using MALDI MSI and quantitative LCMS/MS. Assay Drug Dev Technol 2015; 13(5): 277-84.
[http://dx.doi.org/10.1089/adt.2015.634] [PMID: 26070010]

[85]   van Zyl JM, Basson K, Kriegler A, van der Walt BJ. Mechanisms by which clofazimine and dapsone inhibit the myeloperoxidase system. A possible correlation with their anti-inflammatory properties. Biochem Pharmacol 1991; 42(3): 599-608.
[http://dx.doi.org/10.1016/0006-2952(91)90323-W] [PMID: 1650217]

[86]   Bala S, Khanna R, Dadhwal M, *et al.* Reclassification of Amycolatopsis mediterranei DSM 46095 as Amycolatopsis rifamycinica sp. nov. Int J Syst Evol Microbiol 2004; 54(Pt 4): 1145-9.
[http://dx.doi.org/10.1099/ijs.0.02901-0] [PMID: 15280283]

[87]   Rifampin. Tuberculosis (Edinb) 2008; 88(2): 151-4.
[http://dx.doi.org/10.1016/S1472-9792(08)70024-6] [PMID: 18486058]

[88]   McGeer PL, Harada N, Kimura H, McGeer EG, Schulzer M. Prevalence of dementia amongst elderly Japanese with leprosy: apparent effect of chronic drug therapy. Dementia 1992; 3: 146-9.

[89]   Namba Y, Kawatsu K, Izumi S, Ueki A, Ikeda K. Neurofibrillary tangles and senile plaques in brain of elderly leprosy patients. Lancet 1992; 340(8825): 978.
[http://dx.doi.org/10.1016/0140-6736(92)92870-L] [PMID: 1357384]

[90]   Chui DH, Tabira T, Izumi S, Koya G, Ogata J. Decreased beta-amyloid and increased abnormal Tau deposition in the brain of aged patients with leprosy. Am J Pathol 1994; 145(4): 771-5.
[PMID: 7943169]

[91]   Kimura T, Goto M. Existence of senile plaques in the brains of elderly leprosy patients. Lancet 1993; 342(8883): 1364.
[http://dx.doi.org/10.1016/0140-6736(93)92274-W] [PMID: 7901655]

[92]   Goto M, Kimura T, Hagio S, *et al.* Neuropathological analysis of dementia in a Japanese leprosarium. Dementia 1995; 6(3): 157-61.
[PMID: 7620528]

[93]   Endoh M, Kunishita T, Tabira T. No effect of anti-leprosy drugs in the prevention of Alzheimer's disease and beta-amyloid neurotoxicity J Neurol Sci 1999; 1; 165(1): 28-30.

[94]   Tomiyama T, Asano S, Suwa Y, *et al.* Rifampicin prevents the aggregation and neurotoxicity of amyloid β protein *in vitro*. Biochem Biophys Res Commun 1994; 204(1): 76-83.
[http://dx.doi.org/10.1006/bbrc.1994.2428] [PMID: 7945395]

[95]   Umeda T, Ono K, Sakai A, *et al.* Rifampicin is a candidate preventive medicine against amyloid-β and tau oligomers. Brain 2016; 139(Pt 5): 1568-86.
[http://dx.doi.org/10.1093/brain/aww042] [PMID: 27020329]

[96]   Chokkareddy R, Bhajanthri NK, Redhi GG. A novel electrode architecture for monitoring rifampicin in various pharmaceuticals. Int J Electrochem Sci 2017; 12: 9190-203.
[http://dx.doi.org/10.20964/2017.10.13]

[97]   Mindermann T, Landolt H, Zimmerli W, Rajacic Z, Gratzl O. Penetration of rifampicin into the brain tissue and cerebral extracellular space of rats. J Antimicrob Chemother 1993; 31(5): 731-7.

[http://dx.doi.org/10.1093/jac/31.5.731] [PMID: 8335500]

[98] Tomiyama T, Shoji A, Kataoka K, *et al*. Inhibition of amyloid beta protein aggregation and neurotoxicity by rifampicin. Its possible function as a hydroxyl radical scavenger. J Biol Chem 1996; 271(12): 6839-44.
[http://dx.doi.org/10.1074/jbc.271.12.6839] [PMID: 8636108]

[99] Tomiyama T, Kaneko H, Kataoka Ki, Asano S, Endo N. Rifampicin inhibits the toxicity of pre-aggregated amyloid peptides by binding to peptide fibrils and preventing amyloid-cell interaction. Biochem J 1997; 322(Pt 3): 859-65.
[http://dx.doi.org/10.1042/bj3220859] [PMID: 9148761]

[100] Findeis MA. Approaches to discovery and characterization of inhibitors of amyloid beta-peptide polymerization. Biochim Biophys Acta 2000; 1502(1): 76-84.
[http://dx.doi.org/10.1016/S0925-4439(00)00034-X] [PMID: 10899433]

[101] Balali-Mood K, Ashley RH, Hauss T, Bradshaw JP. Neutron diffraction reveals sequence-specific membrane insertion of pre-fibrillar islet amyloid polypeptide and inhibition by rifampicin. FEBS Lett 2005; 579(5): 1143-8.
[http://dx.doi.org/10.1016/j.febslet.2004.12.085] [PMID: 15710403]

[102] Yulug B, Hanoglu L, Kilic E, Schabitz WR. RIFAMPICIN: an antibiotic with brain protective function. Brain Res Bull 2014; 107: 37-42.
[http://dx.doi.org/10.1016/j.brainresbull.2014.05.007] [PMID: 24905548]

[103] Yulug B, Hanoglu L, Ozansoy M, *et al*. Therapeutic role of rifampicin in Alzheimer's disease. Psychiatry Clin Neurosci 2018; 72(3): 152-9.
[http://dx.doi.org/10.1111/pcn.12637] [PMID: 29315976]

[104] Abuznait AH, Patrick SG, Kaddoumi A. Exposure of LS-180 cells to drugs of diverse physicochemical and therapeutic properties up-regulates P-glycoprotein expression and activity. J Pharm Pharm Sci 2011; 14(2): 236-48.
[http://dx.doi.org/10.18433/J36016] [PMID: 21733412]

[105] Kaur P, Sodhi RK. Memory recuperative potential of rifampicin in aluminum chloride-induced dementia: role of pregnane X receptors. Neuroscience 2015; 288: 24-36.
[http://dx.doi.org/10.1016/j.neuroscience.2014.12.033] [PMID: 25545714]

[106] Kim SK, Kim YM, Yeum CE, Jin SH, Chae GT, Lee SB. Rifampicin inhibits the LPS-induced expression of toll-like receptor 2 *via* the suppression of NF-κB DNA-binding activity in RAW 264.7 cells. Korean J Physiol Pharmacol 2009; 13(6): 475-82.
[http://dx.doi.org/10.4196/kjpp.2009.13.6.475] [PMID: 20054495]

[107] Bi W, Zhu L, Wang C, *et al*. Rifampicin inhibits microglial inflammation and improves neuron survival against inflammation. Brain Res 2011; 1395: 12-20.
[http://dx.doi.org/10.1016/j.brainres.2011.04.019] [PMID: 21555117]

[108] Bi W, Zhu L, Jing X, *et al*. Rifampicin improves neuronal apoptosis in LPS-stimulated co☐cultured BV2 cells through inhibition of the TLR-4 pathway. Mol Med Rep 2014; 10(4): 1793-9.
[http://dx.doi.org/10.3892/mmr.2014.2480] [PMID: 25119251]

[109] Stock ML, Fiedler KJ, Acharya S, *et al*. Antibiotics acting as neuroprotectants *via* mechanisms independent of their anti-infective activities. Neuropharmacology 2013; 73: 174-82.
[http://dx.doi.org/10.1016/j.neuropharm.2013.04.059] [PMID: 23748053]

[110] Esposito E, Cuzzocrea S. New therapeutic strategy for Parkinson's and Alzheimer's disease. Curr Med Chem 2010; 17(25): 2764-74.
[http://dx.doi.org/10.2174/092986710791859324] [PMID: 20586718]

[111] Eriksen JL, Sagi SA, Smith TE, *et al*. NSAIDs and enantiomers of flurbiprofen target gamma-secretase and lower Abeta 42 *in vivo*. J Clin Invest 2003; 112(3): 440-9.
[http://dx.doi.org/10.1172/JCI18162] [PMID: 12897211]

[112] Bain A. Alzheimer disease: Dapsone Phase 2 trial results reported Immune Network Ltd press release 2002.

[113] Imbimbo BP. The potential role of non-steroidal anti-inflammatory drugs in treating Alzheimer's disease. Expert Opin Investig Drugs 2004; 13(11): 1469-81.
[http://dx.doi.org/10.1517/13543784.13.11.1469] [PMID: 15500394]

[114] Imbimbo BP. An update on the efficacy of non-steroidal anti-inflammatory drugs in Alzheimer's disease. Expert Opin Investig Drugs 2009; 18(8): 1147-68.
[http://dx.doi.org/10.1517/13543780903066780] [PMID: 19589092]

[115] Walker D, Lue LF. Anti-inflammatory and immune therapy for Alzheimer's disease: current status and future directions. Curr Neuropharmacol 2007; 5(4): 232-43.
[http://dx.doi.org/10.2174/157015907782793667] [PMID: 19305740]

[116] Cross R, Ling C, Day NP, McGready R, Paris DH. Revisiting doxycycline in pregnancy and early childhood--time to rebuild its reputation? Expert Opin Drug Saf 2016; 15(3): 367-82.
[http://dx.doi.org/10.1517/14740338.2016.1133584] [PMID: 26680308]

[117] Forloni G, Colombo L, Girola L, Tagliavini F, Salmona M. Anti-amyloidogenic activity of tetracyclines: studies *in vitro*. FEBS Lett 2001; 487(3): 404-7.
[http://dx.doi.org/10.1016/S0014-5793(00)02380-2] [PMID: 11163366]

[118] Santacruz K, Lewis J, Spires T, *et al*. Tau suppression in a neurodegenerative mouse model improves memory function. Science 2005; 309(5733): 476-81.
[http://dx.doi.org/10.1126/science.1113694] [PMID: 16020737]

[119] Dursun D, Kim MC, Solomon A, Pflugfelder SC. Treatment of recalcitrant recurrent corneal erosions with inhibitors of matrix metalloproteinase-9, doxycycline and corticosteroids. Am J Ophthalmol 2001; 132(1): 8-13.
[http://dx.doi.org/10.1016/S0002-9394(01)00913-8] [PMID: 11438047]

[120] Raza M, Ballering JG, Hayden JM, Robbins RA, Hoyt JC. Doxycycline decreases monocyte chemoattractant protein-1 in human lung epithelial cells. Exp Lung Res 2006; 32(1-2): 15-26.
[http://dx.doi.org/10.1080/01902140600691399] [PMID: 16809218]

[121] Costa R, Speretta E, Crowther DC, Cardoso I. Testing the therapeutic potential of doxycycline in a Drosophila melanogaster model of Alzheimer disease. J Biol Chem 2011; 286(48): 41647-55.
[http://dx.doi.org/10.1074/jbc.M111.274548] [PMID: 21998304]

[122] Loeb MB, Molloy DW, Smieja M, *et al*. A randomized, controlled trial of doxycycline and rifampin for patients with Alzheimer's disease. J Am Geriatr Soc 2004; 52(3): 381-7.
[http://dx.doi.org/10.1111/j.1532-5415.2004.52109.x] [PMID: 14962152]

[123] Molloy DW, Standish TI, Zhou Q, Guyatt G. A multicenter, blinded, randomized, factorial controlled trial of doxycycline and rifampin for treatment of Alzheimer's disease: the DARAD trial. Int J Geriatr Psychiatry 2013; 28(5): 463-70.
[http://dx.doi.org/10.1002/gps.3846] [PMID: 22718435]

[124] Gillman KW, Starrett JE Jr, Parker MF, *et al*. Discovery and Evaluation of BMS-708163, a Potent, Selective and Orally Bioavailable γ-Secretase Inhibitor. ACS Med Chem Lett 2010; 1(3): 120-4.
[http://dx.doi.org/10.1021/ml1000239] [PMID: 24900185]

[125] Iizuka T, Morimoto K, Sasaki Y, *et al*. Preventive Effect of Rifampicin on Alzheimer Disease Needs at Least 450 mg Daily for 1 Year: An FDG-PET Follow-Up Study. Dement Geriatr Cogn Disord Extra 2017; 7(2): 204-14.
[http://dx.doi.org/10.1159/000477343] [PMID: 28690634]

[126] Sehgal SN, Baker H, Vézina C. Rapamycin (AY-22,989), a new antifungal antibiotic. II. Fermentation, isolation and characterization. J Antibiot (Tokyo) 1975; 28(10): 727-32.
[http://dx.doi.org/10.7164/antibiotics.28.727] [PMID: 1102509]

[127] Martel RR, Klicius J, Galet S. Inhibition of the immune response by rapamycin, a new antifungal antibiotic. Can J Physiol Pharmacol 1977; 55(1): 48-51.
[http://dx.doi.org/10.1139/y77-007] [PMID: 843990]

[128] Kahan BD. Efficacy of sirolimus compared with azathioprine for reduction of acute renal allograft rejection: a randomised multicentre study The Rapamune US Study Group Lancet 2000; 356: 1 94-202.
[http://dx.doi.org/10.1016/S0140-6736(00)02480-6]

[129] Bierer BE, Mattila PS, Standaert RF, *et al.* Two distinct signal transmission pathways in T lymphocytes are inhibited by complexes formed between an immunophilin and either FK506 or rapamycin. Proc Natl Acad Sci USA 1990; 87(23): 9231-5.
[http://dx.doi.org/10.1073/pnas.87.23.9231] [PMID: 2123553]

[130] Magnuson B, Ekim B, Fingar DC. Regulation and function of ribosomal protein S6 kinase (S6K) within mTOR signalling networks. Biochem J 2012; 441(1): 1-21.
[http://dx.doi.org/10.1042/BJ20110892] [PMID: 22168436]

[131] Jia K, Levine B. Autophagy is required for dietary restriction-mediated life span extension in C. elegans. Autophagy 2007; 3(6): 597-9.
[http://dx.doi.org/10.4161/auto.4989] [PMID: 17912023]

[132] Kapahi P, Zid BM, Harper T, Koslover D, Sapin V, Benzer S. Regulation of lifespan in Drosophila by modulation of genes in the TOR signaling pathway. Curr Biol 2004; 14(10): 885-90.
[http://dx.doi.org/10.1016/j.cub.2004.03.059] [PMID: 15186745]

[133] Harrison DE, Strong R, Sharp ZD, *et al.* Rapamycin fed late in life extends lifespan in genetically heterogeneous mice. Nature 2009; 460(7253): 392-5.
[http://dx.doi.org/10.1038/nature08221] [PMID: 19587680]

[134] Jakobsen L, Christiansen EH, Maeng M, *et al.* A randomized phase II study of everolimus in combination with chemorradiation in newly diagnosed glioblastoma. Resuls of NRG Oncology RTOG 0913. Am Heart J 2018; 202: 49-53.
[http://dx.doi.org/10.1016/j.ahj.2018.04.019] [PMID: 29807307]

[135] Peces R, Peces C, Pérez-Dueñas V, Cuesta-López E, Azorín S, Selgas R. Rapamycin reduces kidney volume and delays the loss of renal function in a patient with autosomal-dominant polycystic kidney disease. NDT Plus 2009; 2(2): 133-5.
[PMID: 25949309]

[136] Chinnaiyan P, Won M, Wen PY, *et al.* A randomized phase II study of everolimus in combination with chemoradiation in newly diagnosed glioblastoma: results of NRG Oncology RTOG 0913. Neuro-oncol 2018; 20(5): 666-73.
[http://dx.doi.org/10.1093/neuonc/nox209] [PMID: 29126203]

[137] Caccamo A, Branca C, Talboom JS, *et al.* Reducing ribosomal protein S6 kinase 1 expression improves spatial memory and synaptic plasticity in a mouse model of Alzheimer's disease. J Neurosci 2015; 35(41): 14042-56.
[http://dx.doi.org/10.1523/JNEUROSCI.2781-15.2015] [PMID: 26468204]

[138] Caccamo A, Majumder S, Richardson A, Strong R, Oddo S. Molecular interplay between mammalian target of rapamycin (mTOR), amyloid-beta, and Tau: effects on cognitive impairments. J Biol Chem 2010; 285(17): 13107-20.
[http://dx.doi.org/10.1074/jbc.M110.100420] [PMID: 20178983]

[139] Spilman P, Podlutskaya N, Hart MJ, *et al.* Inhibition of mTOR by rapamycin abolishes cognitive deficits and reduces amyloid-beta levels in a mouse model of Alzheimer's disease. PLoS One 2010; 5(4): e9979.
[http://dx.doi.org/10.1371/journal.pone.0009979] [PMID: 20376313]

[140] Kunz J, Henriquez R, Schneider U, Deuter-Reinhard M, Movva NR, Hall MN. Target of rapamycin in

yeast, TOR2, is an essential phosphatidylinositol kinase homolog required for G1 progression. Cell 1993; 73(3): 585-96.
[http://dx.doi.org/10.1016/0092-8674(93)90144-F] [PMID: 8387896]

[141] Cafferkey R, Young PR, McLaughlin MM, *et al.* Dominant missense mutations in a novel yeast protein related to mammalian phosphatidylinositol 3-kinase and VPS34 abrogate rapamycin cytotoxicity. Mol Cell Biol 1993; 13(10): 6012-23.
[http://dx.doi.org/10.1128/MCB.13.10.6012] [PMID: 8413204]

[142] Abraham RT. Mammalian target of rapamycin: immunosuppressive drugs uncover a novel pathway of cytokine receptor signaling. Curr Opin Immunol 1998; 10(3): 330-6.
[http://dx.doi.org/10.1016/S0952-7915(98)80172-6] [PMID: 9638370]

[143] Lee JH, Tecedor L, Chen YH, *et al.* Reinstating aberrant mTORC1 activity in Huntington's disease mice improves disease phenotypes. Neuron 2015; 85(2): 303-15.
[http://dx.doi.org/10.1016/j.neuron.2014.12.019] [PMID: 25556834]

[144] Bové J, Martínez-Vicente M, Vila M. Fighting neurodegeneration with rapamycin: mechanistic insights. Nat Rev Neurosci 2011; 12(8): 437-52.
[http://dx.doi.org/10.1038/nrn3068] [PMID: 21772323]

[145] Cai Z, Zhao B, Li K, *et al.* Mammalian target of rapamycin: a valid therapeutic target through the autophagy pathway for Alzheimer's disease? J Neurosci Res 2012; 90(6): 1105-18.
[http://dx.doi.org/10.1002/jnr.23011] [PMID: 22344941]

[146] Oddo S. The role of mTOR signaling in Alzheimer disease. Front Biosci (Schol Ed) 2012; 4: 941-52.
[http://dx.doi.org/10.2741/s310] [PMID: 22202101]

[147] Wang C, Yu JT, Miao D, Wu ZC, Tan MS, Tan L. Targeting the mTOR signaling network for Alzheimer's disease therapy. Mol Neurobiol 2014; 49(1): 120-35.
[http://dx.doi.org/10.1007/s12035-013-8505-8] [PMID: 23853042]

[148] Stanfel MN, Shamieh LS, Kaeberlein M, Kennedy BK. The TOR pathway comes of age. Biochim Biophys Acta 2009; 1790(10): 1067-74.
[http://dx.doi.org/10.1016/j.bbagen.2009.06.007] [PMID: 19539012]

[149] Kim YC, Guan KL. mTOR: a pharmacologic target for autophagy regulation. J Clin Invest 2015; 125(1): 25-32.
[http://dx.doi.org/10.1172/JCI73939] [PMID: 25654547]

[150] Perluigi M, Di Domenico F, Butterfield DA. mTOR signaling in aging and neurodegeneration: At the crossroad between metabolism dysfunction and impairment of autophagy. Neurobiol Dis 2015; 84: 39-49.
[http://dx.doi.org/10.1016/j.nbd.2015.03.014] [PMID: 25796566]

[151] Wang X, Proud CG. The mTOR pathway in the control of protein synthesis. Physiology (Bethesda) 2006; 21: 362-9.
[http://dx.doi.org/10.1152/physiol.00024.2006] [PMID: 16990457]

[152] Betz C, Hall MN. Where is mTOR and what is it doing there? J Cell Biol 2013; 203(4): 563-74.
[http://dx.doi.org/10.1083/jcb.201306041] [PMID: 24385483]

[153] Kim DH, Sarbassov DD, Ali SM, *et al.* GbetaL, a positive regulator of the rapamycin-sensitive pathway required for the nutrient-sensitive interaction between raptor and mTOR. Mol Cell 2003; 11(4): 895-904.
[http://dx.doi.org/10.1016/S1097-2765(03)00114-X] [PMID: 12718876]

[154] Averous J, Proud CG. When translation meets transformation: the mTOR story. Oncogene 2006; 25(48): 6423-35.
[http://dx.doi.org/10.1038/sj.onc.1209887] [PMID: 17041627]

[155] Ma XM, Blenis J. Molecular mechanisms of mTOR-mediated translational control. Nat Rev Mol Cell Biol 2009; 10(5): 307-18.

[http://dx.doi.org/10.1038/nrm2672] [PMID: 19339977]

[156] Ben-Sahra I, Howell JJ, Asara JM, Manning BD. Stimulation of de novo pyrimidine synthesis by growth signaling through mTOR and S6K1. Science 2013; 339(6125): 1323-8.
[http://dx.doi.org/10.1126/science.1228792] [PMID: 23429703]

[157] Laplante M, Sabatini DM. An emerging role of mTOR in lipid biosynthesis. Curr Biol 2009; 19(22): R1046-52.
[http://dx.doi.org/10.1016/j.cub.2009.09.058] [PMID: 19948145]

[158] Peterson TR, Sengupta SS, Harris TE, *et al.* mTOR complex 1 regulates lipin 1 localization to control the SREBP pathway. Cell 2011; 146(3): 408-20.
[http://dx.doi.org/10.1016/j.cell.2011.06.034] [PMID: 21816276]

[159] Robitaille AM, Christen S, Shimobayashi M, *et al.* Quantitative phosphoproteomics reveal mTORC1 activates de novo pyrimidine synthesis. Science 2013; 339(6125): 1320-3.
[http://dx.doi.org/10.1126/science.1228771] [PMID: 23429704]

[160] Ganley IG, Lam H, Wang J, Ding X, Chen S, Jiang X. ULK1.ATG13.FIP200 complex mediates mTOR signaling and is essential for autophagy. J Biol Chem 2009; 284(18): 12297-305.
[http://dx.doi.org/10.1074/jbc.M900573200] [PMID: 19258318]

[161] Hosokawa N, Hara T, Kaizuka T, *et al.* Nutrient-dependent mTORC1 association with the ULK1-Atg13-FIP200 complex required for autophagy. Mol Biol Cell 2009; 20(7): 1981-91.
[http://dx.doi.org/10.1091/mbc.e08-12-1248] [PMID: 19211835]

[162] Hara K, Maruki Y, Long X, *et al.* Raptor, a binding partner of target of rapamycin (TOR), mediates TOR action. Cell 2002; 110(2): 177-89.
[http://dx.doi.org/10.1016/S0092-8674(02)00833-4] [PMID: 12150926]

[163] Morita M, Gravel SP, Chénard V, *et al.* mTORC1 controls mitochondrial activity and biogenesis through 4E-BP-dependent translational regulation. Cell Metab 2013; 18(5): 698-711.
[http://dx.doi.org/10.1016/j.cmet.2013.10.001] [PMID: 24206664]

[164] Cunningham JT, Rodgers JT, Arlow DH, Vazquez F, Mootha VK, Puigserver P. mTOR controls mitochondrial oxidative function through a YY1-PGC-1α transcriptional complex. Nature 2007; 450(7170): 736-40.
[http://dx.doi.org/10.1038/nature06322] [PMID: 18046414]

[165] An WL, Cowburn RF, Li L, *et al.* Up-regulation of phosphorylated/activated p70 S6 kinase and its relationship to neurofibrillary pathology in Alzheimer's disease. Am J Pathol 2003; 163(2): 591-607.
[http://dx.doi.org/10.1016/S0002-9440(10)63687-5] [PMID: 12875979]

[166] Yang H, Rudge DG, Koos JD, Vaidialingam B, Yang HJ, Pavletich NP. mTOR kinase structure, mechanism and regulation. Nature 2013; 497(7448): 217-23.
[http://dx.doi.org/10.1038/nature12122] [PMID: 23636326]

[167] Kim DH, Sarbassov DD, Ali SM, *et al.* Growing roles for the mTOR pathway Curr Opin Cell Biol 2005; 17: 596-603.

[168] Sarbassov DD, Ali SM, Sabatini DM. Growing roles for the mTOR pathway. Curr Opin Cell Biol 2005; 17(6): 596-603.
[http://dx.doi.org/10.1016/j.ceb.2005.09.009] [PMID: 16226444]

[169] Yin Y, Hua H, Li M, *et al.* mTORC2 promotes type I insulin-like growth factor receptor and insulin receptor activation through the tyrosine kinase activity of mTOR. Cell Res 2016; 26(1): 46-65.
[http://dx.doi.org/10.1038/cr.2015.133] [PMID: 26584640]

[170] Caron E, Ghosh S, Matsuoka Y, *et al.* A comprehensive map of the mTOR signaling network. Mol Syst Biol 2010; 6: 453.
[http://dx.doi.org/10.1038/msb.2010.108] [PMID: 21179025]

[171] Sarbassov DD1. Prolonged rapamycin treatment inhibits mTORC2 assembly and Akt/PKB Mol Cell

2006; 22: 159-68.

[172] Tramutola A, Triplett JC, Di Domenico F, *et al.* Alteration of mTOR signaling occurs early in the progression of Alzheimer disease (AD): analysis of brain from subjects with pre-clinical AD, amnestic mild cognitive impairment and late-stage AD. J Neurochem 2015; 133(5): 739-49.
[http://dx.doi.org/10.1111/jnc.13037] [PMID: 25645581]

[173] Griffin RJ, Moloney A, Kelliher M, *et al.* Activation of Akt/PKB, increased phosphorylation of Akt substrates and loss and altered distribution of Akt and PTEN are features of Alzheimer's disease pathology. J Neurochem 2005; 93(1): 105-17.
[http://dx.doi.org/10.1111/j.1471-4159.2004.02949.x] [PMID: 15773910]

[174] Li X, An WL, Alafuzoff I, Soininen H, Winblad B, Pei JJ. Phosphorylated eukaryotic translation factor 4E is elevated in Alzheimer brain. Neuroreport 2004; 15(14): 2237-40.
[http://dx.doi.org/10.1097/00001756-200410050-00019] [PMID: 15371741]

[175] Chano T, Okabe H, Hulette CM. RB1CC1 insufficiency causes neuronal atrophy through mTOR signaling alteration and involved in the pathology of Alzheimer's diseases. Brain Res 2007; 1168: 97-105.
[http://dx.doi.org/10.1016/j.brainres.2007.06.075] [PMID: 17706618]

[176] Su Y, Lu J, Chen X, *et al.* Rapamycin alleviates hormone imbalance-induced chronic nonbacterial inflammation in rat prostate through activating autophagy *via* the mTOR/ULK1/ATG13 signaling pathway. Inflammation 2018; 41(4): 1384-95.
[http://dx.doi.org/10.1007/s10753-018-0786-7] [PMID: 29675586]

[177] Majumder S, Richardson A, Strong R, Oddo S. Inducing autophagy by rapamycin before, but not after, the formation of plaques and tangles ameliorates cognitive deficits. PLoS One 2011; 6(9): e25416.
[http://dx.doi.org/10.1371/journal.pone.0025416] [PMID: 21980451]

[178] Jiang T, Yu JT, Zhu XC, *et al.* Temsirolimus promotes autophagic clearance of amyloid-β and provides protective effects in cellular and animal models of Alzheimer's disease. Pharmacol Res 2014; 81: 54-63.
[http://dx.doi.org/10.1016/j.phrs.2014.02.008] [PMID: 24602800]

[179] Vakana E, Sassano A, Platanias LC. Induction of autophagy by dual mTORC1-mTORC2 inhibition in BCR-ABL-expressing leukemic cells. Autophagy 2010; 6(7): 966-7.
[http://dx.doi.org/10.4161/auto.6.7.13067] [PMID: 20699667]

[180] Young AR, Narita M, Ferreira M, *et al.* Autophagy mediates the mitotic senescence transition. Genes Dev 2009; 23(7): 798-803.
[http://dx.doi.org/10.1101/gad.519709] [PMID: 19279323]

[181] Ravikumar B, Vacher C, Berger Z, *et al.* Inhibition of mTOR induces autophagy and reduces toxicity of polyglutamine expansions in fly and mouse models of Huntington disease. Nat Genet 2004; 36(6): 585-95.
[http://dx.doi.org/10.1038/ng1362] [PMID: 15146184]

[182] Lang UE, Heger J, Willbring M, Domula M, Matschke K, Tugtekin SM. Immunosuppression using the mammalian target of rapamycin (mTOR) inhibitor everolimus: pilot study shows significant cognitive and affective improvement. Transplant Proc 2009; 41(10): 4285-8.
[http://dx.doi.org/10.1016/j.transproceed.2009.08.050] [PMID: 20005385]

[183] Yates SC, Zafar A, Hubbard P, *et al.* Dysfunction of the mTOR pathway is a risk factor for Alzheimer's disease. Acta Neuropathol Commun 2013; 1: 3.
[http://dx.doi.org/10.1186/2051-5960-1-3] [PMID: 24252508]

[184] Terry RD. Cell death or synaptic loss in Alzheimer disease. J Neuropathol Exp Neurol 2000; 59(12): 1118-9.
[http://dx.doi.org/10.1093/jnen/59.12.1118] [PMID: 11138931]

[185] Parodi J, Sepúlveda FJ, Roa J, Opazo C, Inestrosa NC, Aguayo LG. β-amyloid causes depletion of

synaptic vesicles leading to neurotransmission failure. J Biol Chem 2010; 285(4): 2506-14.
[http://dx.doi.org/10.1074/jbc.M109.030023] [PMID: 19915004]

[186] Reddy PH, Mani G, Park BS, *et al.* Differential loss of synaptic proteins in Alzheimer's disease: implications for synaptic dysfunction. J Alzheimers Dis 2005; 7(2): 103-17.
[http://dx.doi.org/10.3233/JAD-2005-7203] [PMID: 15851848]

[187] Ramírez AE, Pacheco CR, Aguayo LG, Opazo CM. Rapamycin protects against Aβ-induced synaptotoxicity by increasing presynaptic activity in hippocampal neurons. Biochim Biophys Acta 2014; 1842(9): 1495-501.
[http://dx.doi.org/10.1016/j.bbadis.2014.04.019] [PMID: 24794719]

[188] Tischmeyer W, Schicknick H, Kraus M, *et al.* Rapamycin-sensitive signalling in long-term consolidation of auditory cortex-dependent memory. Eur J Neurosci 2003; 18(4): 942-50.
[http://dx.doi.org/10.1046/j.1460-9568.2003.02820.x] [PMID: 12925020]

[189] Caccamo A, Maldonado MA, Majumder S, *et al.* Naturally secreted amyloid-beta increases mammalian target of rapamycin (mTOR) activity *via* a PRAS40-mediated mechanism. J Biol Chem 2011; 286(11): 8924-32.
[http://dx.doi.org/10.1074/jbc.M110.180638] [PMID: 21266573]

[190] Bhaskar K, Miller M, Chludzinski A, Herrup K, Zagorski M, Lamb BT. The PI3K-Akt-mTOR pathway regulates Abeta oligomer induced neuronal cell cycle events. Mol Neurodegener 2009; 4: 14.
[http://dx.doi.org/10.1186/1750-1326-4-14] [PMID: 19291319]

[191] O'Neill C, Kiely AP, Coakley MF, Manning S, Long-Smith CM. Insulin and IGF-1 signalling: longevity, protein homoeostasis and Alzheimer's disease. Biochem Soc Trans 2012; 40(4): 721-7.
[http://dx.doi.org/10.1042/BST20120080] [PMID: 22817723]

[192] Lafay-Chebassier C, Paccalin M, Page G, *et al.* mTOR/p70S6k signalling alteration by Abeta exposure as well as in APP-PS1 transgenic models and in patients with Alzheimer's disease. J Neurochem 2005; 94(1): 215-25.
[http://dx.doi.org/10.1111/j.1471-4159.2005.03187.x] [PMID: 15953364]

[193] Sapadin AN, Fleischmajer R. Tetracyclines: nonantibiotic properties and their clinical implications. J Am Acad Dermatol 2006; 54(2): 258-65.
[http://dx.doi.org/10.1016/j.jaad.2005.10.004] [PMID: 16443056]

[194] Stirling DP, Koochesfahani KM, Steeves JD, Tetzlaff W. Minocycline as a neuroprotective agent. Neuroscientist 2005; 11(4): 308-22.
[http://dx.doi.org/10.1177/1073858405275175] [PMID: 16061518]

[195] Kim HS, Suh YH. Minocycline and neurodegenerative diseases. Behav Brain Res 2009; 196(2): 168-79.
[http://dx.doi.org/10.1016/j.bbr.2008.09.040] [PMID: 18977395]

[196] Chen M, Ona VO, Li M, *et al.* Minocycline inhibits caspase-1 and caspase-3 expression and delays mortality in a transgenic mouse model of Huntington disease. Nat Med 2000; 6(7): 797-801.
[http://dx.doi.org/10.1038/77528] [PMID: 10888929]

[197] Cronin A, Grealy M. Neuroprotective and Neuro-restorative Effects of Minocycline and Rasagiline in a Zebrafish 6-Hydroxydopamine Model of Parkinson's Disease. Neuroscience 2017; 367: 34-46.
[http://dx.doi.org/10.1016/j.neuroscience.2017.10.018] [PMID: 29079063]

[198] Zhu S, Stavrovskaya IG, Drozda M, *et al.* Minocycline inhibits cytochrome c release and delays progression of amyotrophic lateral sclerosis in mice. Nature 2002; 417(6884): 74-8.
[http://dx.doi.org/10.1038/417074a] [PMID: 11986668]

[199] Pinkernelle J, Fansa H, Ebmeyer U, Keilhoff G. Prolonged minocycline treatment impairs motor neuronal survival and glial function in organotypic rat spinal cord cultures. PLoS One 2013; 8(8): e73422.
[http://dx.doi.org/10.1371/journal.pone.0073422] [PMID: 23967343]

[200]   Li W, Chai Q, Zhang H, *et al.* High doses of minocycline may induce delayed activation of microglia in aged rats and thus cannot prevent postoperative cognitive dysfunction. J Int Med Res 2018; 46(4): 1404-13.
[http://dx.doi.org/10.1177/0300060517754032] [PMID: 29458276]

[201]   Fagan SC, Waller JL, Nichols FT, *et al.* Minocycline to improve neurologic outcome in stroke (MINOS): a dose-finding study. Stroke 2010; 41(10): 2283-7.
[http://dx.doi.org/10.1161/STROKEAHA.110.582601] [PMID: 20705929]

[202]   Schadler ED, Cibull TL, Mehlis SL. A severe case of minocycline-induced hyperpigmentation of the lower extremities. Cureus 2018; 10(5): e2672.
[PMID: 30050728]

[203]   Yrjänheikki J, Tikka T, Keinänen R, Goldsteins G, Chan PH, Koistinaho J. A tetracycline derivative, minocycline, reduces inflammation and protects against focal cerebral ischemia with a wide therapeutic window. Proc Natl Acad Sci USA 1999; 96(23): 13496-500.
[http://dx.doi.org/10.1073/pnas.96.23.13496] [PMID: 10557349]

[204]   Schwab C, McGeer PL. Inflammatory aspects of Alzheimer disease and other neurodegenerative disorders. J Alzheimers Dis 2008; 13(4): 359-69.
[http://dx.doi.org/10.3233/JAD-2008-13402] [PMID: 18487845]

[205]   Tikka T, Fiebich BL, Goldsteins G, Keinänen R, Koistinaho J. Minocycline, a tetracycline derivative, is neuroprotective against excitotoxicity by inhibiting activation and proliferation of microglia. J Neurosci 2001; 21(8): 2580-8.
[http://dx.doi.org/10.1523/JNEUROSCI.21-08-02580.2001] [PMID: 11306611]

[206]   Choi Y, Kim HS, Shin KY, *et al.* Minocycline attenuates neuronal cell death and improves cognitive impairment in Alzheimer's disease models. Neuropsychopharmacology 2007; 32(11): 2393-404.
[http://dx.doi.org/10.1038/sj.npp.1301377] [PMID: 17406652]

[207]   Bernardino AL, Kaushal D, Philipp MT. The antibiotics doxycycline and minocycline inhibit the inflammatory responses to the Lyme disease spirochete Borrelia burgdorferi. J Infect Dis 2009; 199(9): 1379-88.
[http://dx.doi.org/10.1086/597807] [PMID: 19301981]

[208]   Wang AL, Yu AC, Lau LT, *et al.* Minocycline inhibits LPS-induced retinal microglia activation. Neurochem Int 2005; 47(1-2): 152-8.
[http://dx.doi.org/10.1016/j.neuint.2005.04.018] [PMID: 15904993]

[209]   Kim SS, Kong PJ, Kim BS, Sheen DH, Nam SY, Chun W. Inhibitory action of minocycline on lipopolysaccharide-induced release of nitric oxide and prostaglandin E2 in BV2 microglial cells. Arch Pharm Res 2004; 27(3): 314-8.
[http://dx.doi.org/10.1007/BF02980066] [PMID: 15089037]

[210]   Fan LW, Pang Y, Lin S, Rhodes PG, Cai Z. Minocycline attenuates lipopolysaccharide-induced white matter injury in the neonatal rat brain. Neuroscience 2005; 133(1): 159-68.
[http://dx.doi.org/10.1016/j.neuroscience.2005.02.016] [PMID: 15893639]

[211]   Vay SU, Blaschke S, Klein R, Fink GR, Schroeter M, Rueger MA. Minocycline mitigates the gliogenic effects of proinflammatory cytokines on neural stem cells. J Neurosci Res 2016; 94(2): 149-60.
[http://dx.doi.org/10.1002/jnr.23686] [PMID: 26525774]

[212]   Hunter CL, Quintero EM, Gilstrap L, Bhat NR, Granholm AC. Minocycline protects basal forebrain cholinergic neurons from mu p75-saporin immunotoxic lesioning. Eur J Neurosci 2004; 19(12): 3305-16.
[http://dx.doi.org/10.1111/j.0953-816X.2004.03439.x] [PMID: 15217386]

[213]   Biscaro B, Lindvall O, Tesco G, Ekdahl CT, Nitsch RM. Inhibition of microglial activation protects hippocampal neurogenesis and improves cognitive deficits in a transgenic mouse model for

Alzheimer's disease. Neurodegener Dis 2012; 9(4): 187-98.
[http://dx.doi.org/10.1159/000330363] [PMID: 22584394]

[214] Garcez ML, Mina F, Bellettini-Santos T, *et al.* Minocycline reduces inflammatory parameters in the brain structures and serum and reverses memory impairment caused by the administration of amyloid β (1-42) in mice. Prog Neuropsychopharmacol Biol Psychiatry 2017; 77: 23-31.
[http://dx.doi.org/10.1016/j.pnpbp.2017.03.010] [PMID: 28336494]

[215] Clemens V, Regen F, Le Bret N, Heuser I, Hellmann-Regen J. Anti-inflammatory effects of minocycline are mediated by retinoid signaling. BMC Neurosci 2018; 19(1): 58.
[http://dx.doi.org/10.1186/s12868-018-0460-x] [PMID: 30241502]

[216] Park SK, Ratia K, Ba M, Valencik M, Liebman SW. Inhibition of Aβ$_{42}$ oligomerization in yeast by a PICALM ortholog and certain FDA approved drugs. Microb Cell 2016; 3(2): 53-64.
[http://dx.doi.org/10.15698/mic2016.02.476] [PMID: 28357335]

[217] Seabrook TJ, Jiang L, Maier M, Lemere CA. Minocycline affects microglia activation, Abeta deposition, and behavior in APP-tg mice. Glia 2006; 53(7): 776-82.
[http://dx.doi.org/10.1002/glia.20338] [PMID: 16534778]

[218] Familian A, Boshuizen RS, Eikelenboom P, Veerhuis R. Inhibitory effect of minocycline on amyloid beta fibril formation and human microglial activation. Glia 2006; 53(3): 233-40.
[http://dx.doi.org/10.1002/glia.20268] [PMID: 16220550]

[219] Ferretti MT, Allard S, Partridge V, Ducatenzeiler A, Cuello AC. Minocycline corrects early, pre-plaque neuroinflammation and inhibits BACE-1 in a transgenic model of Alzheimer's disease-like amyloid pathology. J Neuroinflammation 2012; 9: 62.
[http://dx.doi.org/10.1186/1742-2094-9-62] [PMID: 22472085]

[220] Burgos-Ramos E, Puebla-Jiménez L, Arilla-Ferreiro E. Minocycline prevents Abeta(25-35)-induced reduction of somatostatin and neprilysin content in rat temporal cortex. Life Sci 2009; 84(7-8): 205-10.
[http://dx.doi.org/10.1016/j.lfs.2008.11.019] [PMID: 19101571]

[221] Familian A, Eikelenboom P, Veerhuis R. Minocycline does not affect amyloid beta phagocytosis by human microglial cells. Neurosci Lett 2007; 416(1): 87-91.
[http://dx.doi.org/10.1016/j.neulet.2007.01.052] [PMID: 17317005]

[222] El-Shimy IA, Heikal OA, Hamdi N. Minocycline attenuates Aβ oligomers-induced pro-inflammatory phenotype in primary microglia while enhancing Aβ fibrils phagocytosis. Neurosci Lett 2015; 609: 36-41.
[http://dx.doi.org/10.1016/j.neulet.2015.10.024] [PMID: 26472705]

[223] Nixon RA, Wegiel J, Kumar A, *et al.* Extensive involvement of autophagy in Alzheimer disease: an immuno-electron microscopy study. J Neuropathol Exp Neurol 2005; 64(2): 113-22.
[http://dx.doi.org/10.1093/jnen/64.2.113] [PMID: 15751225]

[224] Nilsson P, Loganathan K, Sekiguchi M, *et al.* Aβ secretion and plaque formation depend on autophagy. Cell Reports 2013; 5(1): 61-9.
[http://dx.doi.org/10.1016/j.celrep.2013.08.042] [PMID: 24095740]

[225] Liu WT, Lin CH, Hsiao M, Gean PW. Minocycline inhibits the growth of glioma by inducing autophagy. Autophagy 2011; 7(2): 166-75.
[http://dx.doi.org/10.4161/auto.7.2.14043] [PMID: 21079420]

[226] Zhang L, Huang P, Chen H, *et al.* The inhibitory effect of minocycline on radiation-induced neuronal apoptosis *via* AMPKα1 signaling-mediated autophagy. Sci Rep 2017; 7(1): 16373.
[http://dx.doi.org/10.1038/s41598-017-16693-8] [PMID: 29180765]

[227] Boland B, Kumar A. Lee s, Platt FM, Weigiel J, Yu WH, Nixon RA. Authophagy induction and autophagosome clearance in neurons: relationshio to autophagic pathology in Alzheimer's disease. J Neurosci 2008; 28(27): 6926-37.
[http://dx.doi.org/10.1523/JNEUROSCI.0800-08.2008] [PMID: 18596167]

[228]   Zott B, Busche MA, Sperling RA, Konnerth A. What happens with the circuit in Alzheimer's disease in mice and humans? Annu Rev Neurosci 2018; 41: 277-97.
[http://dx.doi.org/10.1146/annurev-neuro-080317-061725] [PMID: 29986165]

[229]   Jiang Y, Liu Y, Zhu C, *et al.* Minocycline enhances hippocampal memory, neuroplasticity and synapse-associated proteins in aged C57 BL/6 mice. Neurobiol Learn Mem 2015; 121: 20-9.
[http://dx.doi.org/10.1016/j.nlm.2015.03.003] [PMID: 25838119]

[230]   Tao T, Feng JZ, Xu GH, Fu J, Li XG, Qin XY. Minocycline Promotes Neurite Outgrowth of PC12 Cells Exposed to Oxygen-Glucose Deprivation and Reoxygenation Through Regulation of MLCP/MLC Signaling Pathways. Cell Mol Neurobiol 2017; 37(3): 417-26.
[http://dx.doi.org/10.1007/s10571-016-0374-z] [PMID: 27098315]

[231]   Han X, Yang L, Du H, *et al.* Insulin attenuates beta-amyloid-associated insulin/Akt/EAAT signaling perturbations in human astrocytes. Cell Mol Neurobiol 2016; 36(6): 851-64.
[http://dx.doi.org/10.1007/s10571-015-0268-5] [PMID: 26358886]

[232]   Lu Y, Lei S, Wang N, *et al.* Protective Effect of Minocycline Against Ketamine-Induced Injury in Neural Stem Cell: Involvement of PI3K/Akt and Gsk-3 Beta Pathway. Front Mol Neurosci 2016; 9: 135.
[http://dx.doi.org/10.3389/fnmol.2016.00135] [PMID: 28066173]

[233]   Lu Y, Giri PK, Lei S, *et al.* Pretreatment with minocycline restores neurogenesis in the subventricular zone and subgranular zone of the hippocampus after ketamine exposure in neonatal rats. Neuroscience 2017; 352: 144-54.
[http://dx.doi.org/10.1016/j.neuroscience.2017.03.057] [PMID: 28391017]

[234]   Davies P, Katzman R, Terry RD. Reduced somatostatin-like immunoreactivity in cerebral cortex from cases of Alzheimer disease and Alzheimer senile dementa. Nature 1980; 288(5788): 279-80.
[http://dx.doi.org/10.1038/288279a0] [PMID: 6107862]

[235]   Nordberg A. Neuroreceptor changes in Alzheimer disease. Cerebrovasc Brain Metab Rev 1992; 4(4): 303-28.
[PMID: 1486017]

[236]   Hoshi A, Tsunoda A, Yamamoto T, Tada M, Kakita A, Ugawa Y. Altered expression of glutamate transporter-1 and water channel protein aquaporin-4 in human temporal cortex with Alzheimer's disease. Neuropathol Appl Neurobiol 2018; 44(6): 628-38.
[http://dx.doi.org/10.1111/nan.12475] [PMID: 29405337]

[237]   Burgos-Ramos E, Puebla-Jiménez L, Arilla-Ferreiro E. Minocycline provides protection against beta-amyloid(25-35)-induced alterations of the somatostatin signaling pathway in the rat temporal cortex. Neuroscience 2008; 154(4): 1458-66.
[http://dx.doi.org/10.1016/j.neuroscience.2008.04.036] [PMID: 18555616]

[238]   Zhang L, Shirayama Y, Shimizu E, Iyo M, Hashimoto K. Protective effects of minocycline on 3,4-methylenedioxymethamphetamine-induced neurotoxicity in serotonergic and dopaminergic neurons of mouse brain. Eur J Pharmacol 2006; 544(1-3): 1-9.
[http://dx.doi.org/10.1016/j.ejphar.2006.05.047] [PMID: 16859675]

[239]   Du Y, Ma Z, Lin S, *et al.* Minocycline prevents nigrostriatal dopaminergic neurodegeneration in the MPTP model of Parkinson's disease. Proc Natl Acad Sci USA 2001; 98(25): 14669-74.
[http://dx.doi.org/10.1073/pnas.251341998] [PMID: 11724929]

[240]   Chowdhury R, Guitart-Masip M, Bunzeck N, Dolan RJ, Düzel E. Dopamine modulates episodic memory persistence in old age. J Neurosci 2012; 32(41): 14193-204.
[http://dx.doi.org/10.1523/JNEUROSCI.1278-12.2012] [PMID: 23055489]

[241]   Huang CY, Chen YL, Li AH, Lu JC, Wang HL. Minocycline, a microglial inhibitor, blocks spinal CCL2-induced heat hyperalgesia and augmentation of glutamatergic transmission in substantia gelatinosa neurons. J Neuroinflammation 2014; 11: 7.

[http://dx.doi.org/10.1186/1742-2094-11-7] [PMID: 24405660]

[242] Noble W, Garwood C, Stephenson J, Kinsey AM, Hanger DP, Anderton BH. Minocycline reduces the development of abnormal tau species in models of Alzheimer's disease. FASEB J 2009; 23(3): 739-50.
[http://dx.doi.org/10.1096/fj.08-113795] [PMID: 19001528]

[243] Parachikova A, Vasilevko V, Cribbs DH, LaFerla FM, Green KN. Reductions in amyloid-beta-derived neuroinflammation, with minocycline, restore cognition but do not significantly affect tau hyperphosphorylation. J Alzheimers Dis 2010; 21(2): 527-42.
[http://dx.doi.org/10.3233/JAD-2010-100204] [PMID: 20555131]

[244] Poehlsgaard J, Douthwaite S. The macrolide binding site on the bacterial ribosome. Curr Drug Targets Infect Disord 2002; 2(1): 67-78.
[http://dx.doi.org/10.2174/1568005024605927] [PMID: 12462154]

[245] Sahoo AK, Dandapat J, Dash UC, Kanhar S. Features and outcomes of drugs for combination therapy as multi-targets strategy to combat Alzheimer's disease. J Ethnopharmacol 2018; 215: 42-73.
[http://dx.doi.org/10.1016/j.jep.2017.12.015] [PMID: 29248451]

[246] Morse LJ, Payton SM, Cuny GD, Rogers JT. FDA-preapproved drugs targeted to the translational regulation and processing of the amyloid precursor protein. J Mol Neurosci 2004; 24(1): 129-36.
[http://dx.doi.org/10.1385/JMN:24:1:129] [PMID: 15314261]

[247] Payton S, Cahill CM, Randall JD, Gullans SR, Rogers JT. Drug discovery targeted to the Alzheimer's APP mRNA 5'-untranslated region: the action of paroxetine and dimercaptopropanol. J Mol Neurosci 2003; 20(3): 267-75.
[http://dx.doi.org/10.1385/JMN:20:3:267] [PMID: 14501007]

[248] Tucker S, Ahl M, Cho HH, et al. RNA therapeutics directed to the non coding regions of APP mRNA, in vivo anti-amyloid efficacy of paroxetine, erythromycin, and N-acetyl cysteine. Curr Alzheimer Res 2006; 3(3): 221-7.
[http://dx.doi.org/10.2174/156720506777632835] [PMID: 16842099]

[249] Appleby BS, Nacopoulos D, Milano N, Zhong K, Cummings JLA. A review: treatment of Alzheimer's disease discovered in repurposed agents. Dement Geriatr Cogn Disord 2013; 35(1-2): 1-22.
[http://dx.doi.org/10.1159/000345791] [PMID: 23307039]

[250] Sampson EL, Jenagaratnam L, McShane R. Metal protein attenuating compounds for the treatment of Alzheimer's dementia. Cochrane Database Syst Rev 2014; 2(2): CD005380.
[PMID: 24563468]

[251] Cherny RA, Atwood CS, Xilinas ME, et al. Treatment with a copper-zinc chelator markedly and rapidly inhibits beta-amyloid accumulation in Alzheimer's disease transgenic mice. Neuron 2001; 30(3): 665-76.
[http://dx.doi.org/10.1016/S0896-6273(01)00317-8] [PMID: 11430801]

[252] Grossi C, Francese S, Casini A, et al. Clioquinol decreases amyloid-β burden and reduces working memory impairment in a transgenic mouse model of Alzheimer's disease. J Alzheimers Dis 2009; 17(2): 423-40.
[http://dx.doi.org/10.3233/JAD-2009-1063] [PMID: 19363260]

[253] Cherny RA, Legg JT, McLean CA, et al. Aqueous dissolution of Alzheimer's disease Abeta amyloid deposits by biometal depletion. J Biol Chem 1999; 274(33): 23223-8.
[http://dx.doi.org/10.1074/jbc.274.33.23223] [PMID: 10438495]

[254] Bush AI. The metallobiology of Alzheimer's disease. Trends Neurosci 2003; 26(4): 207-14.
[http://dx.doi.org/10.1016/S0166-2236(03)00067-5] [PMID: 12689772]

[255] Schäfer S, Pajonk FG, Multhaup G, Bayer TA. Copper and clioquinol treatment in young APP transgenic and wild-type mice: effects on life expectancy, body weight, and metal-ion levels. J Mol Med (Berl) 2007; 85(4): 405-13.
[http://dx.doi.org/10.1007/s00109-006-0140-7] [PMID: 17211610]

[256] Filiz G, Price KA, Caragounis A, Du T, Crouch PJ, White AR. The role of metals in modulating metalloprotease activity in the AD brain. Eur Biophys J 2008; 37(3): 315-21.
[http://dx.doi.org/10.1007/s00249-007-0244-1] [PMID: 18270696]

[257] Wang Z, Wang Y, Li W, *et al.* Design, synthesis, and evaluation of multitarget-directed selenium-containing clioquinol derivatives for the treatment of Alzheimer's disease. ACS Chem Neurosci 2014; 5(10): 952-62.
[http://dx.doi.org/10.1021/cn500119g] [PMID: 25121395]

[258] Ritchie CW, Bush AI, Mackninnon A, *et al.* Metal-Protein Attenuation With iodochlorhydroxyquin (Clioquinol) Targeting Aβ Amyloid Deposition and Toxicity in Alzheimer Disease. Arch Neurol 2003; 60: 1685-91.
[http://dx.doi.org/10.1001/archneur.60.12.1685] [PMID: 14676042]

[259] Zhang YH, Raymick J, Sarkar S, *et al.* Efficacy and toxicity of clioquinol treatment and A-beta42 inoculation in the APP/PSI mouse model of Alzheimer's disease. Curr Alzheimer Res 2013; 10(5): 494-506.
[http://dx.doi.org/10.2174/1567205011310050005] [PMID: 23627708]

[260] Hartsel SC, Weiland TR, Amphotericin B. Amphotericin B binds to amyloid fibrils and delays their formation: a therapeutic mechanism? Biochemistry 2003; 42(20): 6228-33.
[http://dx.doi.org/10.1021/bi0270384] [PMID: 12755626]

[261] Smith NW, Annunziata O, Dzyuba SV. Amphotericin B interactions with soluble oligomers of amyloid Abeta1-42 peptide. Bioorg Med Chem 2009; 17(6): 2366-70.
[http://dx.doi.org/10.1016/j.bmc.2009.02.016] [PMID: 19268601]

[262] Diomede L, Cassata G, Fiordaliso F, *et al.* Tetracycline and its analogues protect Caenorhabditis elegans from β amyloid-induced toxicity by targeting oligomers. Neurobiol Dis 2010; 40(2): 424-31.
[http://dx.doi.org/10.1016/j.nbd.2010.07.002] [PMID: 20637283]

[263] Ono K, Hasegawa K, Naiki H, Yamada M. Anti-amyloidogenic activity of tannic acid and its activity to destabilize Alzheimer's β-amyloid fibrils *in vitro*. Biochim Biophys Acta 2004; 1690(3): 193-202.
[http://dx.doi.org/10.1016/j.bbadis.2004.06.008] [PMID: 15511626]

CHAPTER 4

Use of Antipsychotics in Patients with Alzheimer's Disease

Haiyun Xu[a,*], **Jue He**[b, c], **Xiaoyin Zhuang**[a] and **Yuan Shao**[d]

[a] *The Mental Health Center, Shantou University Medical College, Shantou, PR China*

[b] *First Affiliated Hospital, Institute of Neurological Disease, Henan University, Henan, PRChina*

[c] *Xiamen Xian Yue Hospital, Xiamen, Fujian, PR China*

[d] *Shenzhen Mental Health Center, Shenzhen Kangning Hospital, Shenzhen Shi, PR China.*

Abstract: Psychosis is a common and difficult to treat symptom in Alzheimer's disease (AD) patients. For the treatment of psychosis symptom and disruptive behaviors in AD patients, antipsychotics (APs) have been recommended. In this chapter, we reviewed clinical studies with AD patients who were treated by either first- or second-generation antipsychotics (FGAs or SGAs). FGAs showed overall disappointing results in elderly dementia patients, while adverse events occurred frequently, especially motor side effects, sedation, and cognitive impairment. SGAs offer an advance for the management of aggression and psychosis in the context of dementia. Although some of previous studies reported higher mortality in AD patients treated with SGAs, the others described no risks of mortality, even protective effects of them in AD patients. Preclinical studies indicate that early, low-dose and long-term administration of some SGAs may improve behavioral symptoms in various animal models of AD while attenuating pathological changes in the brain. The protective effects are achieved through inhibiting the pathological processes of oxidative stress and neuroinflammation of AD, in addition to the neurotrophic action of SGAs. All these protective actions can be hardly explained by the differential affinities of SGAs to multiple neurotransmitter receptors in the brain. Further studies are required to elucidate the molecular basis of these actions.

Keywords: Alzheimer's Disease, Antipstchotics, Anti-Inflammation, Antioxidatant, Behavioral, Cognitive Impairment, Mortality, Neuroprotection, Psychiatric Disturbances.

---

* **Corresponding author Haiyun Xu:** The Mental Health Center, Shantou University Medical College, Shantou, PR China, Tel: +86-754 -88900728; E-mails: hyxu@stu.edu.cn

Atta-ur-Rahman (Ed.)

INTRODUCTION

Alzheimer's disease (AD) is a neurodegenerative disorder and characterized by progressive decline in cognition and global function thus affecting activities of daily living of patients [1, 2]. As AD progresses, patients may show psychotic symptoms (delusions and hallucinations), aggressive behavior, psychomotor agitation, wandering and depression [3 - 5]. These behavioral and psychiatric disturbances (BPSD) are the source of major problems for families and physicians. Furthermore, they cause immense patient distress and are responsible for caregiver stress, institutionalization, and hospitalization [6 - 8].

Treatment of AD has generally focused on the cognitive and non-cognitive symptoms including BPSD. For cognitive impairment, cholinesterase inhibitors and other cholinergic agents are first line treatment based on the rationale that there is an association between cognitive decline and cholinergic cell loss [9 - 11]. The treatment of non-cognitive symptoms includes pharmacologic and environmental interventions. For BPSD, if non- pharmacological interventions are not suitable, a treatment with antipsychotics (APs) is recommended [12, 13].

Despite the potential negative effects, APs are frequently prescribed for AD patients, especially for those in long-term care settings [14, 15]. However, little is known about the effects of APs on the natural history of AD. Moreover, significant concerns exist among physicians and public health authorities about the reported increased mortality in elderly dementia patients exposed to APs. In this chapter, we reviewed clinical studies with AD patients who were treated by either first- or second-generation antipsychotics (FGAs or SGAs). The outcomes of the treatments were commented in the context of therapeutic efficacy and side effects, as well as mortality of the patients. In addition, we included basic research literature with animal and cell culture experiments by summarizing effects of SGAs on behavioral alterations in animals of commonly used AD models and discussing cellular and molecular mechanisms underlying the therapeutic effects of them.

BPSD AND THEIR PROGRESSIONS IN AD PATIENTS

BPSD in patients with dementia vary considerably in their prevalence and trajectory [16]. Published prevalence estimates of psychosis in patients with AD range from 10 to 73% with an overall median of 34% within clinic populations, and from 7 to 20% in community and clinical trials populations depending on definitions used [17]. In a population-based sample of incident AD cases [18], 50% of the participants (328 individuals with incident AD) experienced neuro-psychiatric symptoms (NPS) at baseline, including depression (26%), irritability or apathy (17% each). In a more recent study using the data from the Cache County Dementia Progression Study, Peters *et al.* [19] reported that 50.9% of the sample had at least one symptom of NPS. At the baseline, psychosis cluster and affective cluster were seen in 18.1% and 38.8% of the patients, respectively. The apathy/ indifference was seen in 16.9% of individuals at baseline and agitation/aggression was seen in 10% of individuals. At baseline, 25.9% of individuals had at least one mild NPS and 25.0% of individuals had at least one clinically significant NPS. In a most recent study [20], psychotic symptoms affected half of the patients (N = 445) over a three-year period. Both delusions and hallucinations were associated with greater cognitive and functional impairment, dementia severity, and caregiver burden. Furthermore, the presence of both delusions and hallucinations was associated with worse outcomes than the presence of only one of these symptoms.

Patients with AD have heterogeneous rates of NPS progression. In a study by Tschanz *et al.* [18], the authors described results from the Cache County Dementia Progression Study, an ongoing population-based study of AD that characterizes the course of symptoms in the domains of cognition, function and NPS from a point near the onset of dementia. They reported that most NPI (neuropsychiatric inventory, a systematic way to measure multiple NPS domains) symptoms increased over time, and 89% of survivors were experiencing symptoms by the 7th year post onset. However, for hallucinations, anxiety, and irritability, the percent of individuals affected declined at the final visit, possibly reflecting the fluctuating nature of NPS. The pattern of NPS also shifted over time because apathy became the most commonly reported symptom by the 5th year post onset.

To investigate whether NPS predict cognitive and functional decline progression in AD, Palmer *et al* [21] longitudinally examined 177 memory-clinic AD outpatients and followed up for two years. After multiple adjustment, patients with the affective syndrome had an increased risk of functional decline, whereas the risk of cognitive decline was associated with the manic syndrome. In addition, previous studies have linked NPS in cognitive impairment, no dementia (CIND) patients to dementia conversion [22 - 25]. In a recent study, NPS even at low severity increased risk of transition from CIND to dementia with a conversion rate of 12% per year, similar to the rates observed in a prior study [24]. These results highlight the importance of incorporating a thorough psychiatric examination in the evaluation of AD patients.

USE OF ANTIPSYCHOTIC DRUGS IN AD PATIENTS

Antipsychotic medications for treating BPSD include both FGAs and SGAs [26,

27]. FGAs have been extensively studied in elderly dementia patients with overall disappointing results [28 - 31]. Furthermore, adverse events occurred frequently, especially motor side effects, sedation, cognitive impairment, orthostatic hypotension, constipation, and urinary hesitancy. These have led to increasing reluctance and some statutory obstacles to the use of FGAs in the elderly, particularly in institutional settings [31].

SGAs have potential advantages over FGAs such as a better side effect profile. They may offer an advance for the management of aggression and psychosis in the context of dementia. In the first multicenter, double-bland, randomized, placebo-controlled studies of an SGA in treatment of psychosis and behavioral disturbances in an elderly dementia population, significantly greater reductions in Mini-Mental State Examination (MMSE) score were seen on the Behavioral Pathology in AD Rating Scale (BEHAVE-AD) total score and psychosis and aggressiveness subscale scores in patients who took risperidone (1 or 2 mg/day) than in those taking placebo [32]. Similarly, in an international, randomized, double-bland, flexible-dosing trial of risperidone, haloperidol, or placebo in AD patients, those treated with risperidone showed significantly greater reductions than the placebo group in BEHAVE-AD aggressiveness scores [33]. Moreover, risperidone was associated with significantly fewer side effects than FGAs, particularly extrapyramidal symptoms (EPS). In addition, this drug lacks anti-cholinergic properties and thus may be useful in treating the elderly with dementia complicated with aggression, agitation, or psychosis [34 - 37]. In line with this expectation, risperidone was reported to significantly improve symptoms of aggression in institutionalized elderly patients with dementia complicated by behavioral disturbances. The EPS profile in these risperidone-treated patients was similar to that of those treated with placebo, while the drug induced fewer EPS than haloperidol at clinically effective doses [32, 38]. In another randomized placebo-controlled trial, treatment with low-dose (mean = 0.95 mg/day) risperidone resulted in significant improvement in aggression, agitation, and psychosis in elderly patients with dementia including AD [37].

In a systemic review aiming at determining whether evidence supports the use of SGAs for the treatment of aggression, agitation and psychosis in people with AD, nine of sixteen placebo controlled trials that had been completed with SGAs were analyzed. The included trials led to the results that there were significant improvements in aggression and psychosis with risperidone and olanzapine treatment compared to placebo, while the risperidone and olanzapine treated patients had a significantly higher incidence of serious adverse cerebrovascular events (including stroke), EPS and other important adverse outcomes including urinary tract and upper respiratory tract infection, as well as peripheral oedema (in the case of risperidone), and somnolence (in the case of olanzapine) [39]. In a recent systemic review [40], 15 studies of SGAs involving risperidone, olanzapine, quetiapine, and aripiprazole were analyzed. Statistically significant results on change in NPS scores compared with placebo were noted in two studies of risperidone [32, 37], two studies of olanzapine [41, 42], and one study of aripiprazole [43]. Olanzapine was associated with reductions in agitation and NPS [44]. Risperidone was associated with greater reductions in agitation when compared with rivastigmine in one study [45]. Quetiapine at 200 mg daily was found to be associated with a higher proportion of individuals with significant global improvement than placebo, while 100 mg was not associated with significant benefit [46]. Another systemic review, which included nine trials with 606 randomized participants to measure the effects of withdrawal of APs on behaviors in older people with dementia in community or nursing home settings, concluded that older people with dementia and NPS using long-term APs can be withdrawn without detrimental effects on their behaviors [47]. However, in a study of risperidone treatment for AD patients with psychosis and agitation, symptom relapse occurred when the antipsychotic was discontinued in a randomly assigned, double-blind fashion [48].

RISK OF MORTALITY ASSOCIATED WITH ANTIPSYCHOTIC USE IN AD PATIENTS

Previous studies have examined the association between APs and risk of mortality in AD patients, but the results are inconsistent. Some of them reported higher mortality in AD patients treated with APs. For example, in a population-based, retrospective cohort study, the use of FGAs and SGAs was associated with higher risk of mortality [49]. Indeed, a previous meta-analysis of randomized controlled trials with SGAs found that these agents were associated with a small increase in risk for death in AD patients [50], leading the Food and Drug Administration to issue a black box warning. Following that, in a placebo-controlled withdrawal trial, 439 AD patients who were randomized to continue APs use for 12 months had increased risk of mortality at 12 months compared with those who received placebo, and the difference became more pronounced after the first year [51]. Similarly, in a recent cohort study, persons with dementia using APs had twice as high risk of mortality from the first dispensing of an AP compared with those using solely other psychotropics and the risk remained consistently higher for over 6 years [52]. In a more recent study with the data from a nationwide register-based study that included all 70,718 community-dwellers newly diagnosed with AD during 2005–2011 in Finland, AP use was associated with an increased risk of mortality. The risk of mortality was increased from the first days of use and attenuated gradually but remained increased even after two years of use [53].

Some of other studies described no risks of mortality [54, 55], even protective effects among AD patients [56, 57]. For example, in a study involving institutional geriatric patients with dementia from Finland using APs, the use of SGAs (risperidone, olanzapine) led to lower two-year mortality risk compared with that in nonusers [56]. Another study [51] found a non-significant increase in mortality in people who continued AP treatment indicated by cumulative probability of survival (70% in the AP treatment group compared with 77% in the withdrawal group) during 12 months. Moreover, AD patients taking either FGAs or SGAs had a lower risk of mortality compared to those not treated with APs in a retrospective case-control cohort study using the Taiwan National Health Insurance Research Database [58]. Similarly, neither FGAs nor SGAs increased the risk of death in AD patients in a large longitudinal observational study [14]. These results are in line with previous studies showing a lower risk of mortality in patients with dementia taking SGAs (mainly quetiapine) [59, 60].

The protective effects of APs in AD patients mentioned above do not support the association between increased mortality and AP use reported in a nationwide register-based study [53]. The discrepancy may be attributed to multiple factors such as patients' characteristics. Indeed, age, education level, male gender, MMSE scores, EPS, and psychosis, but not AP use, were associated with risk of death in AD patients in a large longitudinal observational study [14]. Furthermore, interactions among multiple factors that have yet to be identified may be associated with AP-associated death in nursing home patients. For example, in a study conducted in Finnish institutionalized patients with dementia, while neither FGAs nor SGAs were associated with death, the use of restraints doubled the risk of mortality [56]. In addition, most of studies which reported the association of APs with mortality were short term studies [50, 59 - 62]. Some of these studies found the risk of mortality being strongest shortly after the start of treatment. This risk of mortality reduced thereafter, indicating that the associated risk could be due to an uncontrolled BPSD at the start days of treatment rather than the AP itself [59, 63, 64].

Comparing and contrasting specific SGA in AD patients is beyond the scope of this article. For relevant information, readers are referred to a systematic review which evaluated the evidence in a balanced way from a number of trials of SGAs to treat aggression, agitation or psychosis in AD patients [39], in addition to an updated review on the current evidence supporting or negating the use of psychotropic medications in AD patients, along with safety concerns, monitoring, regulations, and recommendations [65]. For the current guidance on antipsychotic usage in AD patients, Creese *et al.* suggested that the use of APs should always follow an assessment of the patient to determine underlying causes, followed by non- pharmacological management such as psychosocial interventions or environmental modifications. If used, the best evidence base remains for risperidone over the short term only (<12 weeks). For agitation, results from dextromethorphan-quinidine trials show the greatest promise while there remain safety concerns around citalopram [66]. Similarly, an expert panel of 11 international members with clinical and research expertise in BPSD management agreed to a preference for an escalating approach to the management of BPSD in AD commencing with the identification of underlying causes. For BPSD overall and for agitation, caregiver training, environmental adaptations, person-centered care, and tailored activities were identified as first-line approaches prior to any pharmacologic approaches. If pharmacologic strategies were needed, citalopram and analgesia were prioritized ahead of APs. For psychosis, pharmacologic options, and in particular, risperidone, were prioritized following the assessment of underlying causes. Two tailored non-drug approaches (DICE and music therapy) were agreed upon as the most promising non-pharmacologic treatment approaches for BPSD overall and agitation, with dextro- methorphan/quinidine as a promising potential pharmacologic candidate for agitation [67].

BEHAVIORAL SYMPTOMS IN ANIMAL MODELS OF AD

The pathological hallmarks of AD include senile plaques, neurofibrillary tangles with tau protein, and neuronal loss in the brain. Neuronal loss is preceded by synaptic loss and dendrite retraction, which occur early in the disease [68]. β-amyloid peptide (Aβ) is a major component of senile plaques and is derived through processing of the larger transmembrane amyloid precursor protein (APP) [69 - 71]. Two critical enzymes in the processing are β-secretase and a presenilin (PS)-dependent γ-secretase complex [72, 73]. β-secretase mediates the initial step of Aβ production by β-cleavage of APP demonstrated by the finding that β-secretase inhibitor reduces Aβ production in APP transgenic mice, an established animal model of AD [74]. In addition to the APP transgenic mice, animal models used in preclinical studies include single or multi transgenic animals of APP, PS and/or Tau mutations, and non-transgenic animal models of aging resulted from Aβ or tau injection into the brain [75].

A single injection of Aβ (3 nmol/3 µl/per site, intracerebroventricular) was shown to induce abnormal behaviors including memory impairment and anxiety in a mouse model of AD [76]. Transgenic mice over-expressing genes such as APP [77] and PS1 [78], whose mutations are associated with familial AD, have been shown to mimic many traits of AD, including memory impairment, amyloid plaques, tauopathy, neuronal loss, and gliosis. Transgenic mouse models of AD that carry APP and/or PS1 mutated genes showed AD-like pathology and memory impairment [78 - 81]. Double transgenic mice carrying both APP and PS1 mutations showed accelerated AD phenotype compared with APP single transgenic mice [80]. The APP/PS1 double transgenic mouse model of AD showed an impaired anxiety-like behavior in an open field test [82] and in an elevated plus maze [83], and a sensorimotor gating deficit which has been identified in neuropsychiatric diseases in a prepulse inhibition task [84]. The triple transgenic mouse model of AD (APPswe, Tau_{P301L}, $PS1_{M146V}$) with both Aβ- and tau-related lesions has shown memory deficit and an early apathy-like behavior [85]. In addition, behavioral changes associated with non-cognitive, emotional domains appeared before the onset of definitive cognitive deficits in an APP^{NL-} $^{-F/NL-G-F}$ mice in which three familial AD-associated mutations were introduced into the endogenous mouse APP locus to recapitulate Aβ pathology observed in AD: the Swedish (NL) mutation elevating total Aβ production, the Beyreuther/Iberian (F) mutation increasing the Aβ42/Aβ40 ratio; and the Arctic (G) mutation promoting Aβ aggregation [86].

EFFECTS OF SGAS ON NON-COGNITIVE BEHAVIORAL ALTERATIONS IN ANIMAL STUDIES

Non-cognitive behavioral symptoms of AD including aggression, agitation and anxiety are associated with poor outcomes for individuals with dementia and considered to be the earlier cause of the institutionalization in AD patients [87, 88]. SGAs have been widely used in the treatment of schizophrenia, bipolar disorder and psychotic depression [89, 90], and have shown their efficacy in treating psychosis in AD as well as cognition in Parkinson's disease [91, 92]. Although it is a common clinical practice of controlling behavioral symptoms using APs in the late stage of AD, few well-proven treatment options are current available because of side effects of APs due to their high dose application for acute behavioral symptom control.

Recent animal studies have tested the potential beneficial effects of SGAs in the early stages of AD mimicked in animal models. For example, early chronic administration of quetiapine, an SGA, attenuated the anxiety level of the APP/PS1 transgenic mouse in an open field test [82, 93]. Risperidone, another SGA, ameliorated the alterations in locomotor activity and exploratory behavior in an $Aβ_{1-42}$-induced mouse model of AD [94], and improved the enhanced marble burying behavior which is considered to model the spectrum of anxiety, psychotic and obsessive-compulsive like symptoms in the triple transgenic AD mice [95].

EFFECTS OF SGAS ON COGNITIVE IMPAIRMENTS IN ANIMAL STUDIES

Effects of SGAs on the memory impairments in animals have been evaluated in recent years. In our previous studies, quetiapine was shown to attenuate the methamphetamine-induced memory impairment and neurotoxicity [96, 97] and counteract the phencyclidine-induced reference memory impairment while decreasing Bcl-X_L/Bax ratio in the cortex of rats [98]. Olanzapine attenuated the okadaic acid-induced spatial memory impairment and hippocampal cell death in rats [99], and risperidone improved cognitive deficits but enhanced Notch signaling in a MK-801-induced mouse model of schizophrenia [100]. All these previous animal studies suggest that SGAs may have beneficial effects on cognitive impairments in AD.

In line with the above suggestion, chronic quetiapine administration was shown to attenuate spatial memory impairment in water maze and Y-maze tests [82, 101] and improve the impairment of short- and long-term (1 and 24 hr) retention in an object recognition task in the APP/PS1 transgenic AD mice [102]. Risperidone significantly reversed the $A\beta_{1-42}$-induced dysfunction in learning and memory in the Morris water maze and step-through passive avoidance tests in the $A\beta_{1-42}$-induced mouse model of AD [94]. Long-term clozapine treatment attenuated memory impairment and reduced $A\beta$ level in the transgenic APPswe/PS1dE9 mouse model of AD [103].

Clinical studies argued that adverse effects of SGAs may offset its advantages in their schedule of treatment. For example, quetiapine effectively alleviated psychoses in AD, but its possible beneficial effects on cognition in AD were uncertain [104, 105]. However, AD model studies have shown that the pre-treatment and early treatment of low dose chronic quetiapine (2.5 or 5 mg/kg/day for more than 1 month) improved memory impairment in the APP/PS1 AD mice whose cerebral amyloid plaques were detectable at the age of 3 month [102, 103]. Possible reasons for the uncertain effects of SGAs on cognition in AD patients may be associated with the facts of 1) SGAs were most often (51.5%) given to severely dementia patients [106] and 2) the likelihood of SGA prescribing was higher for patients aged 70 years and older than for those < 70 years, as well as 3) SGA prescribing was higher for patients visiting secondary care institutions than for those visiting primary care institutions [107].

MECHANISMS UNDERLYING BENEFICIAL EFFECTS OF SGAS IN AD

Anti-Oxidative Action Of SGAs

In vitro studies have shown that SGAs are effective in reducing PC12 cell death induced by addition of hydrogen peroxide, β-amyloid peptide, or MPP^+. All these treatments share a common capacity of leading to oxidative stress [108 - 110], indicating an anti-oxidative action of SGAs. In support of this indication, quetiapine was shown to block the hydroxyl radical-induced Aβ25-35 aggregation and scavenge the hydroxyl radical from the Fenton system or in the Aβ25-35 solution [111]. Olanzapine and quetiapine prevented the Aβ-induced apoptosis and the overproduction of intracellular reactive oxygen species (ROS), and attenuated Aβ-induced activity changes in the antioxidant enzymes in PC12 cells [112].

The anti-oxidative actions of SGAs may account for the beneficial effects of them shown in some of AD patients. In line with this assumption, there is substantial evidence supporting that increased oxidative stress is one of the mechanisms by which Aβ induces neuronal cell death [113 - 115]. Indeed, oxidative damage to proteins, lipids and DNA has been demonstrated in post-mortem tissue of AD patients [116, 117]. Furthermore, Aβ, being a source of free radicals [118], induces the production of ROS [119]. On the other hand, oxidative stress has been shown to potentiate the generation of Aβ [120], and thereby cause further rises in ROS. Furthermore, the toxicity of Aβ was attenuated by antioxidants [121] and exacerbated by radical oxidative stress [114]. More significantly, quetiapine improved object recognition memory impairment in 12-month old AD mice, and the memory improvement was parallel to the protective effect of quetiapine on the hippocampal oxidative stress in AD mice, suggesting an association between the suppression of oxidative stress and the memory improving effect exerted by quetiapine in AD [103]. Risperidone significantly reversed the $Aβ_{1-42}$-induced dysfunction in learning, memory, locomotor activity and exploratory behavior while reversed the $Aβ_{1-42}$-induced decrease of cell viability and mitochondrial membrane potential in cultured cortical neurons [95]. Relevantly, amentoflavone, a biflavonoid compound, attenuated memory and learning deficits in an Aβ-induced AD model by decreasing apoptotic death of neuronal cells, which might be attributed to its ability to reduce excessive oxidative stress by modulating Nrf2 signaling [122]. Taken together, *via* the anti-oxidative actions, some of SGAs ameliorate ROS and inhibit Aβ aggregation in brain neurons thus delaying neuropathological changes and improving behavioral impairment in animal models of AD (Fig. **1**). This mechanism may help to interpret the protective effects of some SGAs in AD patients.

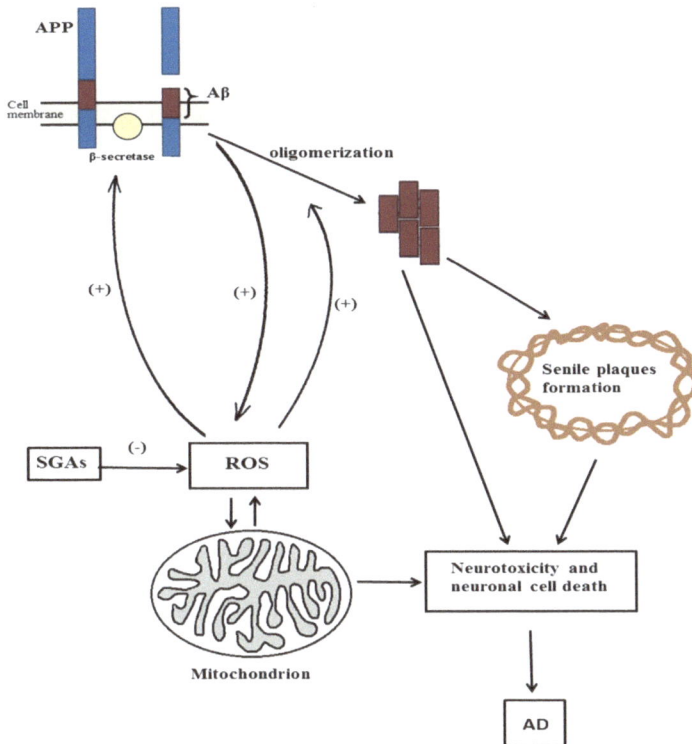

Fig. (1). The anti-oxidative action of SGAs. During the oligomerization process of Aβ, ROS is produced. ROS, in turn, accelerates the production of Aβ and oligomerization process. The accumulated ROS may cause damages to mitochondriion and large molecules, eventually lead to neuronal apoptosis. SGAs, *via* scavenging free radicals (for example hydroxyl radical), decrease levels of ROS and protect neurons against the toxicities of Aβ.

Anti-Inflammatory Action Of SGAs

Numerous studies show the presence of a number of markers of inflammation in the AD brain: accumulation of activated microglia occurring mainly around amyloid plaques accompanied by excessive or dysregulated release of proinflammatory cytokines and chemokines, which contributes to neuronal death and degeneration [123]. Moreover, it has been well known that higher inflammatory levels are related to higher risk of cognitive impairment in AD patients [124]. On the other hand, some of SGAs show anti-inflammatory actions in experimental and clinical studies [125 - 128]. Therefore, it is plausible to propose that the anti-inflammatory action of SGAs may contribute to their beneficial effects shown in AD patients. In line with this assumption, a recent human study showed that SGAs (risperidone and quetiapine) decreased the expression of CD68, an inflammatory molecule participating in the macrophage functions, in psychotic AD patients. And the expression of CD68 was positively correlated with pain, dementia and mental disorder symptoms in the patients [129]. Moreover, this anti-inflammatory action has been demonstrated in a recent experimental study [94]. It was shown that quetiapine improved behavioral performance of APP/PS1 mice while the treatment attenuated microglial activation, reduced proinflammatory cytokine levels, and inhibited activation of astrocytes in the APP/PS1 mice. These anti-inflammatory effects were associated with the suppressing action of quetiapine on the NF-κB p65 pathway which has been implicated in microglial activation and neuroinflammation, given the finding that the protein level of p65 was significantly increased in the cortex and hippocampus of APP/PS1 transgenic mice compared with that in nontransgenic mice, but this increase was significantly attenuated in the mice treated with quetiapine. In primary microglia, quetiapine significantly attenuated the p65 translocation from the cytoplasm to the nucleus induced by $A\beta_{1-42}$ while the drug significantly reduced the Aβ-induced secretion of IL-1β and TNFα (Fig. **2**) [94].

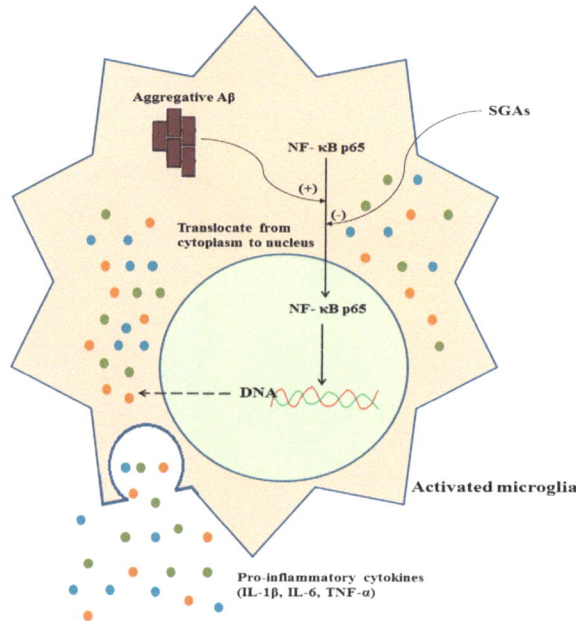

Fig. (2). The anti-inflammatory action of SGAs. The oligomerized Aβ in the cytoplasm facilitates the translocation of NF-κB65 from the cytoplasm to nucleus thus induces inflammatory actions such as the production and release of pro-inflammatory cytokines, which may exacerbate the ROS-induced damage to the cell. SGAs inhibit the translocation of NF-κB65 from the cytoplasm to nucleus thus suppress the Aβ-induced inflammatory actions.

8.3. Neurotrophic Action Of SGAs

In addition to the anti-oxidant and anti-inflammatory actions, SGAs have potential neurotrophic and neuroprotective effects which may contribute to their beneficial effeccts on behavioral impairment and the observed changes in β-secretase and Aβ levels in AD animals (Fig. **3**). In line with this interpretation, neurotrophins have been found to rescue Aβ-induced deficits in hippocampal synaptic plasticity [130]. In a previous animal study, quetiapine attenuated the immobilization stress-induced decrease of brain-derived neurotrophic factor (BDNF), an important neurotrophin mainly expressed and distributed in brain neurons, in rat hippocampus [131]. Quetiapine also modulated the level of BDNF in the basolateral amygdala and hippocampus of AD mice [132]. Furthermore, some of SGAs up-regulated the cerebral level of Bcl-2 in AD transgenic mice and rats [83, 133] thus endow the drug with its neuroprotective capacities, given that Bcl-2 may inhibit apoptosis by sequestering preforms of death-driving caspases and preventing the release of mitochondrial apoptotic factors into the cytoplasm [134, 135]. Also Bcl-2 has been reported to protect against the generation of ROS, to increase antioxidant defenses, and to decrease levels of ROS and oxidative damage [136]. These results associated the neuroprotective effects of quetiapine with its anti-oxidative action. Indeed, quetiapine suppressed the up-regulation of nitrotyrosine, a protein marker of oxidative stress, in the APP/PS1 transgenic mice [103, 137]. Therefore, quetiapine exerts its beneficial effects on the pathogenesis of AD by a complex pharmacological profile, confirming the conception that quetiapine is a multitarget drug.

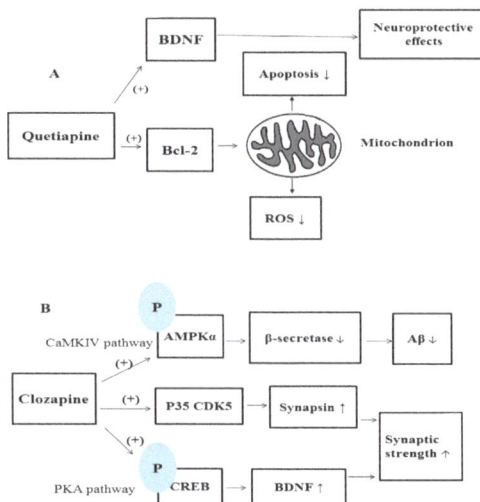

Fig. (3). The neurotrophic action of SGAs. (A) Quetiapine up-regulates the expression of BDNF and Bcl-2 thus exerts neuroprotective effects. **(B)** Clozapine increases the expression of p35 and CDK65, and facilitates the phosphorylation of AMPKα and CREB thus exerts cytoprotective effects.

Similar to quetiapine, clozapine, with strong affinity for a number of serotonin, dopamine, muscarinic, adrenergic, and other biogenic amine receptors [138], may act on multiple receptors thus exert a complex pharmacological actions. In line with this assumption, clozapine adminstration improved memeory impairments and inhibited amyloidogenic processing of APP deposition and Aβ generation in APP/PS1 mice. In the meanwhile, the treatment increased the expression of Trk and BDNF, CDK5, and p35, and AMPK phosphorylation [139], which have been reported as the pathways involved in regulation of Aβ levels and neuroprotection [140 - 143]. In addition, chronic adminstration of quetiapine induced synapsin expression in the hippocampus and phosphorylated synapsin 1 at Ser9 and Ser549 sites in the hippocampus and cortex of AD mice [139].

CONCLUDING REMARKS

Treatment of AD has generally focused on the cognitive and non-cognitive symptoms including BPSD. For BPSD, if non-pharmacological interventions are not suitable, a treatment with APs is recommended. Of the APs, FGAs have been extensively studied in elderly dementia patients with overall disappointing results. They show some adverse effects, especially motor side effects, sedation, cognitive impairment, orthostatic hypotension, constipation, and urinary hesitancy. On the other hand, SGAs have potential advantages over FGAs such as a better side effect profile. They may offer an advance for the management of aggression and psychosis in the context of dementia. However, some of previous studies reported higher mortality in AD patients treated with APs while the others described no risks of mortality, even protective effects among AD patients. The discrepancy may be attributed to multiple factors including age, education level, male gender, MMSE scores, EPS, and psychosis, but not AP use.

Recent studies have tested the potential beneficial effects of SGAs in the early stages of AD mimicked in animal models. The results indicate that early, low-dose and long-term administration of some SGAs including clozapine, quetiapine and risperidone, improves behavioral changes, which are relevant to the cognitive and non-cognitive symptoms seen in AD patients, in various animal models of AD while attenuating pathological changes in the brain. The protective effects are achieved through inhibiting the pathological processes of oxidative stress and neuroinflammation of AD, in addition to the neurotrophic action of SGAs. All these protective actions can be hardly explained by the differential affinities of SGAs to multiple neurotransmitter receptors in the brain. Further studies are required to elucidate the molecular basis of these actions.

CONSENT FOR PUBLICATION

Not applicable.

CONFLIT OF INTEREST

The authors declare no conflict of interest, financial or otherwise

ACKNOWLEDGEMENTS

Declared none.

REFERENCES

[1]   Birks JS, Chong LY, Grimley Evans J. Rivastigmine for Alzheimer's disease. Cochrane Database Syst Rev 2015; 9: CD001191.
[PMID: 26393402]

[2]   Scheltens P, Blennow K, Breteler MM, *et al.* Alzheimer's disease. Lancet 2016; 388(10043): 505-17.
[http://dx.doi.org/10.1016/S0140-6736(15)01124-1] [PMID: 26921134]

[3]   Becker JT, Boller F, Lopez OL, Saxton J, McGonigle KL. The natural history of Alzheimer's disease. Description of study cohort and accuracy of diagnosis. Arch Neurol 1994; 51(6): 585-94.
[http://dx.doi.org/10.1001/archneur.1994.00540180063015] [PMID: 8198470]

[4]   Devanand DP, Jacobs DM, Tang MX, *et al.* The course of psychopathologic features in mild to moderate Alzheimer disease. Arch Gen Psychiatry 1997; 54(3): 257-63.
[http://dx.doi.org/10.1001/archpsyc.1997.01830150083012] [PMID: 9075466]

[5]   Wragg RE, Jeste DV. Overview of depression and psychosis in Alzheimer's disease. Am J Psychiatry 1989; 146(5): 577-87.
[http://dx.doi.org/10.1176/ajp.146.5.577] [PMID: 2653053]

[6]   Coen RF, Swanwick GR, O'Boyle CA, Coakley D. Behaviour disturbance and other predictors of carer burden in Alzheimer's disease. Int J Geriatr Psychiatry 1997; 12(3): 331-6.
[http://dx.doi.org/10.1002/(SICI)1099-1166(199703)12:3<331::AID-GPS495>3.0.CO;2-J] [PMID: 9152717]

[7]   Rymer S, Salloway S, Norton L, Malloy P, Correia S, Monast D. Impaired awareness, behavior disturbance, and caregiver burden in Alzheimer disease. Alzheimer Dis Assoc Disord 2002; 16(4): 248-53.
[http://dx.doi.org/10.1097/00002093-200210000-00006] [PMID: 12468899]

[8]   Robert P. Understanding and managing behavioural symptoms in Alzheimer's disease and related dementias: focus on rivastigmine. Curr Med Res Opin 2002; 18(3): 156-71.
[http://dx.doi.org/10.1185/030079902125000561] [PMID: 12094826]

[9]   Richards SS, Hendrie HC. Diagnosis, management, and treatment of Alzheimer disease: a guide for the internist. Arch Intern Med 1999; 159(8): 789-98.
[http://dx.doi.org/10.1001/archinte.159.8.789] [PMID: 10219924]

[10]   Schneider LS. New therapeutic approaches to Alzheimer's disease. J Clin Psychiatry 1996; 57 (Suppl. 14): 30-6.
[PMID: 9024334]

[11]   Krall WJ, Sramek JJ, Cutler NR. Cholinesterase inhibitors: a therapeutic strategy for Alzheimer disease. Ann Pharmacother 1999; 33(4): 441-50.
[http://dx.doi.org/10.1345/aph.18211] [PMID: 10332536]

[12]   Highlights from the annual scientific assembly: managing the stages of Alzheimer's disease--new management options. South Med J 2002; 95(1): 102-6.
[PMID: 11827239]

[13]   Raskind MA, Peskind ER. Alzheimer's disease and related disorders. Med Clin North Am 2001; 85(3): 803-17.
[http://dx.doi.org/10.1016/S0025-7125(05)70341-2] [PMID: 11349485]

[14]   Lopez OL, Becker JT, Chang YF, *et al.* The long-term effects of conventional and atypical antipsychotics in patients with probable Alzheimer's disease. Am J Psychiatry 2013; 170(9): 1051-8.
[http://dx.doi.org/10.1176/appi.ajp.2013.12081046] [PMID: 23896958]

[15]   Nielsen RE, Valentin JB, Lolk A, Andersen K. Effects of antipsychotics on secular mortality trends in patients with Alzheimer's disease. J Clin Psychiatry 2018; 79(3): 17m11595.
[http://dx.doi.org/10.4088/JCP.17m11595] [PMID: 29659206]

[16]   Cerejeira J, Lagarto L, Mukaetova-Ladinska EB. Behavioral and psychological symptoms of dementia. Front Neurol 2012; 3: 73.
[http://dx.doi.org/10.3389/fneur.2012.00073] [PMID: 22586419]

[17]   Schneider LS, Dagerman KS. Psychosis of Alzheimer's disease: clinical characteristics and history. J Psychiatr Res 2004; 38(1): 105-11.
[http://dx.doi.org/10.1016/S0022-3956(03)00092-X] [PMID: 14690773]

[18]   Tschanz JT, Corcoran CD, Schwartz S, *et al.* Progression of cognitive, functional, and neuropsychiatric symptom domains in a population cohort with Alzheimer dementia: the Cache County Dementia Progression study. Am J Geriatr Psychiatry 2011; 19(6): 532-42.
[http://dx.doi.org/10.1097/JGP.0b013e3181faec23] [PMID: 21606896]

[19]   Peters ME, Schwartz S, Han D, *et al.* Neuropsychiatric symptoms as predictors of progression to severe Alzheimer's dementia and death: the Cache County Dementia Progression Study. Am J Psychiatry 2015; 172(5): 460-5.
[http://dx.doi.org/10.1176/appi.ajp.2014.14040480] [PMID: 25585033]

[20]   Connors MH, Ames D, Woodward M, Brodaty H. Psychosis and clinical outcomes in Alzheimer disease: a longitudinal study. Am J Geriatr Psychiatry 2018; 26(3): 304-13.
[http://dx.doi.org/10.1016/j.jagp.2017.10.011] [PMID: 29174998]

[21]   Palmer K, Lupo F, Perri R, *et al.* Predicting disease progression in Alzheimer's disease: the role of neuropsychiatric syndromes on functional and cognitive decline. J Alzheimers Dis 2011; 24(1): 35-45.
[http://dx.doi.org/10.3233/JAD-2010-101836] [PMID: 21157019]

[22]   Tschanz JT, Welsh-Bohmer KA, Lyketsos CG, *et al.* Conversion to dementia from mild cognitive disorder: the Cache County Study. Neurology 2006; 67(2): 229-34.
[http://dx.doi.org/10.1212/01.wnl.0000224748.48011.84] [PMID: 16864813]

[23]   Morris JC, Storandt M, Miller JP, *et al.* Mild cognitive impairment represents early-stage Alzheimer disease. Arch Neurol 2001; 58(3): 397-405.
[http://dx.doi.org/10.1001/archneur.58.3.397] [PMID: 11255443]

[24]   Tuokko H, Frerichs R, Graham J, *et al.* Five-year follow-up of cognitive impairment with no dementia. Arch Neurol 2003; 60(4): 577-82.
[http://dx.doi.org/10.1001/archneur.60.4.577] [PMID: 12707072]

[25]   Bennett DA, Wilson RS, Schneider JA, *et al.* Natural history of mild cognitive impairment in older persons. Neurology 2002; 59(2): 198-205.
[http://dx.doi.org/10.1212/WNL.59.2.198] [PMID: 12136057]

[26]   Alexopoulos GS, Jeste DV, Chung H, Carpenter D, Ross R, Docherty JP. The expert consensus guideline series. Treatment of dementia and its behavioral disturbances. Introduction: methods, commentary, and summary. Postgrad Med 2005; (Spec No): 6-22.
[PMID: 17203561]

[27]   Reus VI, Fochtmann LJ, Eyler AE, *et al.* The American psychiatric association practice guideline on the use of antipsychotics to treat agitation or psychosis in patients with dementia. Am J Psychiatry 2016; 173(5): 543-6.
[http://dx.doi.org/10.1176/appi.ajp.2015.173501] [PMID: 27133416]

[28]   Schneider LS, Pollock VE, Lyness SA. A metaanalysis of controlled trials of neuroleptic treatment in dementia. J Am Geriatr Soc 1990; 38(5): 553-63.
[http://dx.doi.org/10.1111/j.1532-5415.1990.tb02407.x] [PMID: 1970586]

[29]   Sunderland T. Treatment of the elderly suffering from psychosis and dementia. J Clin Psychiatry 1996; 57 (Suppl. 9): 53-6.
[PMID: 8823351]

[30]   Practice guideline for the treatment of patients with Alzheimer's disease and other dementias of late life. Am J Psychiatry 1997; 154(5) (Suppl.): 1-39.
[PMID: 9140238]

[31]   Daniel DG. Antipsychotic treatment of psychosis and agitation in the elderly. J Clin Psychiatry 2000; 61 (Suppl. 14): 49-52.
[PMID: 11154017]

[32]   Katz IR, Jeste DV, Mintzer JE, Clyde C, Napolitano J, Brecher M. Comparison of risperidone and placebo for psychosis and behavioral disturbances associated with dementia: a randomized, double-blind trial. J Clin Psychiatry 1999; 60(2): 107-15.
[http://dx.doi.org/10.4088/JCP.v60n0207] [PMID: 10084637]

[33]   De Deyn P, Lemmens P, De Smedt G. Risperidone in the treatment of behavioral disturbances in dementia [abstract]. 36th annual meeting of the American College of Neuropsychopharmacology. Dec 8-12, 1997; Kamuela, Hawaii.

[34]   Zaudig M. A risk-benefit assessment of risperidone for the treatment of behavioural and psychological symptoms in dementia. Drug Saf 2000; 23(3): 183-95.
[http://dx.doi.org/10.2165/00002018-200023030-00002] [PMID: 11005702]

[35]   Jeste DV, Lacro JP, Bailey A, Rockwell E, Harris MJ, Caligiuri MP. Lower incidence of tardive dyskinesia with risperidone compared with haloperidol in older patients. J Am Geriatr Soc 1999; 47(6): 716-9.
[http://dx.doi.org/10.1111/j.1532-5415.1999.tb01595.x] [PMID: 10366172]

[36]   Jeste DV, Rockwell E, Harris MJ, Lohr JB, Lacro J. Conventional *vs.* newer antipsychotics in elderly patients. Am J Geriatr Psychiatry 1999; 7(1): 70-6.
[http://dx.doi.org/10.1097/00019442-199902000-00010] [PMID: 9919323]

[37]   Brodaty H, Ames D, Snowdon J, *et al.* A randomized placebo-controlled trial of risperidone for the treatment of aggression, agitation, and psychosis of dementia. J Clin Psychiatry 2003; 64(2): 134-43.
[http://dx.doi.org/10.4088/JCP.v64n0205] [PMID: 12633121]

[38]   De Deyn PP, Rabheru K, Rasmussen A, *et al.* A randomized trial of risperidone, placebo, and haloperidol for behavioral symptoms of dementia. Neurology 1999; 53(5): 946-55.
[http://dx.doi.org/10.1212/WNL.53.5.946] [PMID: 10496251]

[39]   Ballard C, Waite J, Birks J. The effectiveness of atypical antipsychotics for the treatment of aggression and psychosis in Alzheimer's disease. Cochrane Database Syst Rev 2006; 1(1): CD003476.
[PMID: 16437455]

[40]   Seitz DP, Gill SS, Herrmann N, *et al.* Pharmacological treatments for neuropsychiatric symptoms of dementia in long-term care: a systematic review. Int Psychogeriatr 2013; 25(2): 185-203.
[http://dx.doi.org/10.1017/S1041610212001627] [PMID: 23083438]

[41]   De Deyn PP, Carrasco MM, Deberdt W, *et al.* Olanzapine *versus* placebo in the treatment of psychosis with or without associated behavioral disturbances in patients with Alzheimer's disease. Int J Geriatr Psychiatry 2004; 19(2): 115-26.

[http://dx.doi.org/10.1002/gps.1032] [PMID: 14758577]

[42]   Street JS, Clark WS, Gannon KS, *et al.* Olanzapine treatment of psychotic and behavioral symptoms in patients with Alzheimer disease in nursing care facilities: a double-blind, randomized, placebo-controlled trial. Arch Gen Psychiatry 2000; 57(10): 968-76.
[http://dx.doi.org/10.1001/archpsyc.57.10.968] [PMID: 11015815]

[43]   Mintzer JE, Tune LE, Breder CD, *et al.* Aripiprazole for the treatment of psychoses in institutionalized patients with Alzheimer dementia: a multicenter, randomized, double-blind, placebo-controlled assessment of three fixed doses. Am J Geriatr Psychiatry 2007; 15(11): 918-31.
[http://dx.doi.org/10.1097/JGP.0b013e3181557b47] [PMID: 17974864]

[44]   Verhey FRJ, Verkaaik M, Lousberg R. Olanzapine *versus* haloperidol in the treatment of agitation in elderly patients with dementia: results of a randomized controlled double-blind trial. Dement Geriatr Cogn Disord 2006; 21(1): 1-8.
[http://dx.doi.org/10.1159/000089136] [PMID: 16244481]

[45]   Holmes C, Wilkinson D, Dean C, *et al.* Risperidone and rivastigmine and agitated behaviour in severe Alzheimer's disease: a randomised double blind placebo controlled study. Int J Geriatr Psychiatry 2007; 22(4): 380-1.
[http://dx.doi.org/10.1002/gps.1667] [PMID: 17380475]

[46]   Zhong KX, Tariot PN, Mintzer J, Minkwitz MC, Devine NA. Quetiapine to treat agitation in dementia: a randomized, double-blind, placebo-controlled study. Curr Alzheimer Res 2007; 4(1): 81-93.
[http://dx.doi.org/10.2174/156720507779939805] [PMID: 17316169]

[47]   Declercq T, Petrovic M, Azermai M, *et al.* Withdrawal *versus* continuation of chronic antipsychotic drugs for behavioural and psychological symptoms in older people with dementia. Cochrane Database Syst Rev 2013; 3(3): CD007726.
[PMID: 23543555]

[48]   Devanand DP, Schultz SK, Sultzer DL. Discontinuation of risperidone in Alzheimer's disease. N Engl J Med 2013; 368(2): 187-8.
[PMID: 23301737]

[49]   Gill SS, Bronskill SE, Normand SL, *et al.* Antipsychotic drug use and mortality in older adults with dementia. Ann Intern Med 2007; 146(11): 775-86.
[http://dx.doi.org/10.7326/0003-4819-146-11-200706050-00006] [PMID: 17548409]

[50]   Schneider LS, Dagerman KS, Insel P. Risk of death with atypical antipsychotic drug treatment for dementia: meta-analysis of randomized placebo-controlled trials. JAMA 2005; 294(15): 1934-43.
[http://dx.doi.org/10.1001/jama.294.15.1934] [PMID: 16234500]

[51]   Ballard C, Hanney ML, Theodoulou M, *et al.* The dementia antipsychotic withdrawal trial (DART-AD): long-term follow-up of a randomised placebo-controlled trial. Lancet Neurol 2009; 8(2): 151-7.
[http://dx.doi.org/10.1016/S1474-4422(08)70295-3] [PMID: 19138567]

[52]   Langballe EM, Engdahl B, Nordeng H, Ballard C, Aarsland D, Selbæk G. Short- and long-term mortality risk associated with the use of antipsychotics among 26,940 dementia outpatients: a population-based study. Am J Geriatr Psychiatry 2014; 22(4): 321-31.
[http://dx.doi.org/10.1016/j.jagp.2013.06.007] [PMID: 24016844]

[53]   Koponen M, Taipale H, Lavikainen P, *et al.* Risk of mortality associated with antipsychotic monotherapy and polypharmacy among community-dwelling persons with Alzheimer's disease. J Alzheimers Dis 2017; 56(1): 107-18.
[http://dx.doi.org/10.3233/JAD-160671] [PMID: 27935554]

[54]   Barak Y, Baruch Y, Mazeh D, Paleacu D, Aizenberg D. Cardiac and cerebrovascular morbidity and mortality associated with antipsychotic medications in elderly psychiatric inpatients. Am J Geriatr Psychiatry 2007; 15(4): 354-6.
[http://dx.doi.org/10.1097/JGP.0b13e318030253a] [PMID: 17384319]

[55]   Nonino F, De Girolamo G, Gamberini L, Goldoni CA. Survival among elderly Italian patients with dementia treated with atypical antipsychotics: observational study. Neurol Sci 2006; 27(6): 375-80.
[http://dx.doi.org/10.1007/s10072-006-0716-6] [PMID: 17205222]

[56]   Raivio MM, Laurila JV, Strandberg TE, Tilvis RS, Pitkälä KH. Neither atypical nor conventional antipsychotics increase mortality or hospital admissions among elderly patients with dementia: a two-year prospective study. Am J Geriatr Psychiatry 2007; 15(5): 416-24.
[http://dx.doi.org/10.1097/JGP.0b013e31802d0b00] [PMID: 17463191]

[57]   Suh GH, Shah A. Effect of antipsychotics on mortality in elderly patients with dementia: a 1-year prospective study in a nursing home. Int Psychogeriatr 2005; 17(3): 429-41.
[http://dx.doi.org/10.1017/S1041610205002243] [PMID: 16252375]

[58]   Chu CS, Li WR, Huang KL, Su PY, Lin CH, Lan TH. The use of antipsychotics is associated with lower mortality in patients with Alzheimer's disease: A nationwide population-based nested case-control study in Taiwan. J Psychopharmacol (Oxford) 2018; 32(11): 1182-90. Epub ahead of print
[http://dx.doi.org/10.1177/0269881118780016] [PMID: 29926765]

[59]   Kales HC, Kim HM, Zivin K, et al. Risk of mortality among individual antipsychotics in patients with dementia. Am J Psychiatry 2012; 169(1): 71-9.
[http://dx.doi.org/10.1176/appi.ajp.2011.11030347] [PMID: 22193526]

[60]   Rossom RC, Rector TS, Lederle FA, Dysken MW. Are all commonly prescribed antipsychotics associated with greater mortality in elderly male veterans with dementia? J Am Geriatr Soc 2010; 58(6): 1027-34.
[http://dx.doi.org/10.1111/j.1532-5415.2010.02873.x] [PMID: 20487081]

[61]   Rochon PA, Normand SL, Gomes T, et al. Antipsychotic therapy and short-term serious events in older adults with dementia. Arch Intern Med 2008; 168(10): 1090-6.
[http://dx.doi.org/10.1001/archinte.168.10.1090] [PMID: 18504337]

[62]   Wang PS, Schneeweiss S, Avorn J, et al. Risk of death in elderly users of conventional vs. atypical antipsychotic medications. N Engl J Med 2005; 353(22): 2335-41.
[http://dx.doi.org/10.1056/NEJMoa052827] [PMID: 16319382]

[63]   Wang LJ, Lee SY, Yuan SS, et al. Risk of mortality among patients treated with antipsychotic medications: A nationwide population-based study in Taiwan. J Clin Psychopharmacol 2016; 36(1): 9-17.
[http://dx.doi.org/10.1097/JCP.0000000000000451] [PMID: 26658260]

[64]   Wei YJ, Simoni-Wastila L, Zuckerman IH, et al. Quality of psychopharmacological medication prescribing and mortality in Medicare beneficiaries in nursing homes. J Am Geriatr Soc 2014; 62(8): 1490-504.
[http://dx.doi.org/10.1111/jgs.12939] [PMID: 25041166]

[65]   Madhusoodanan S, Ting MB. Pharmacological management of behavioral symptoms associated with dementia. World J Psychiatry 2014; 4(4): 72-9.
[http://dx.doi.org/10.5498/wjp.v4.i4.72] [PMID: 25540722]

[66]   Creese B, Da Silva MV, Johar I, Ballard C. The modern role of antipsychotics for the treatment of agitation and psychosis in Alzheimer's disease. Expert Rev Neurother 2018; 18(6): 461-7.
[http://dx.doi.org/10.1080/14737175.2018.1476140] [PMID: 29764230]

[67]   Kales HC, Lyketsos CG, Miller EM, Ballard C. Management of behavioral and psychological symptoms in people with Alzheimer's disease: an international Delphi consensus. Int Psychogeriatr 2018; 1-8. Epub ahead of print
[http://dx.doi.org/10.1017/S1041610218000534] [PMID: 30068400]

[68]   Levin EC, Acharya NK, Sedeyn JC, et al. Neuronal expression of vimentin in the Alzheimer's disease brain may be part of a generalized dendritic damage-response mechanism. Brain Res 2009; 1298: 194-207.

[http://dx.doi.org/10.1016/j.brainres.2009.08.072] [PMID: 19728994]

[69] Storey E, Cappai R. The amyloid precursor protein of Alzheimer's disease and the Abeta peptide. Neuropathol Appl Neurobiol 1999; 25(2): 81-97.
[http://dx.doi.org/10.1046/j.1365-2990.1999.00164.x] [PMID: 10215996]

[70] Suh YH, Checler F. Amyloid precursor protein, presenilins, and alpha-synuclein: molecular pathogenesis and pharmacological applications in Alzheimer's disease. Pharmacol Rev 2002; 54(3): 469-525.
[http://dx.doi.org/10.1124/pr.54.3.469] [PMID: 12223532]

[71] Van Gassen G, Annaert W. Amyloid, presenilins, and Alzheimer's disease. Neuroscientist 2003; 9(2): 117-26.
[http://dx.doi.org/10.1177/1073858403252227] [PMID: 12708616]

[72] Fukumoto H, Rosene DL, Moss MB, Raju S, Hyman BT, Irizarry MC. Beta-secretase activity increases with aging in human, monkey, and mouse brain. Am J Pathol 2004; 164(2): 719-25.
[http://dx.doi.org/10.1016/S0002-9440(10)63159-8] [PMID: 14742275]

[73] Lazarov O, Robinson J, Tang YP, et al. Environmental enrichment reduces Abeta levels and amyloid deposition in transgenic mice. Cell 2005; 120(5): 701-13.
[http://dx.doi.org/10.1016/j.cell.2005.01.015] [PMID: 15766532]

[74] Wilquet V, De Strooper B. Amyloid-beta precursor protein processing in neurodegeneration. Curr Opin Neurobiol 2004; 14(5): 582-8.
[http://dx.doi.org/10.1016/j.conb.2004.08.001] [PMID: 15464891]

[75] Asai M, Hattori C, Iwata N, et al. The novel beta-secretase inhibitor KMI-429 reduces amyloid beta peptide production in amyloid precursor protein transgenic and wild-type mice. J Neurochem 2006; 96(2): 533-40.
[http://dx.doi.org/10.1111/j.1471-4159.2005.03576.x] [PMID: 16336629]

[76] Puzzo D, Gulisano W, Palmeri A, Arancio O. Rodent models for Alzheimer's disease drug discovery. Expert Opin Drug Discov 2015; 10(7): 703-11.
[http://dx.doi.org/10.1517/17460441.2015.1041913] [PMID: 25927677]

[77] Pinz MP, Dos Reis AS, Vogt AG, et al. Current advances of pharmacological properties of 7-chloro-4-(phenylselanyl) quinoline: Prevention of cognitive deficit and anxiety in Alzheimer's disease model. Biomed Pharmacother 2018; 105: 1006-14.
[http://dx.doi.org/10.1016/j.biopha.2018.06.049] [PMID: 30021335]

[78] Hsiao K, Chapman P, Nilsen S, et al. Correlative memory deficits, Abeta elevation, and amyloid plaques in transgenic mice. Science 1996; 274(5284): 99-102.
[http://dx.doi.org/10.1126/science.274.5284.99] [PMID: 8810256]

[79] Duff K, Eckman C, Zehr C, et al. Increased amyloid-beta42(43) in brains of mice expressing mutant presenilin 1. Nature 1996; 383(6602): 710-3.
[http://dx.doi.org/10.1038/383710a0] [PMID: 8878479]

[80] Games D, Adams D, Alessandrini R, et al. Alzheimer-type neuropathology in transgenic mice overexpressing V717F beta-amyloid precursor protein. Nature 1995; 373(6514): 523-7.
[http://dx.doi.org/10.1038/373523a0] [PMID: 7845465]

[81] Holcomb L, Gordon MN, McGowan E, et al. Accelerated Alzheimer-type phenotype in transgenic mice carrying both mutant amyloid precursor protein and presenilin 1 transgenes. Nat Med 1998; 4(1): 97-100.
[http://dx.doi.org/10.1038/nm0198-097] [PMID: 9427614]

[82] Sturchler-Pierrat C, Abramowski D, Duke M, et al. Two amyloid precursor protein transgenic mouse models with Alzheimer disease-like pathology. Proc Natl Acad Sci USA 1997; 94(24): 13287-92.
[http://dx.doi.org/10.1073/pnas.94.24.13287] [PMID: 9371838]

[83] He J, Luo H, Yan B, et al. Beneficial effects of quetiapine in a transgenic mouse model of Alzheimer's

disease. Neurobiol Aging 2009; 30(8): 1205-16.
[http://dx.doi.org/10.1016/j.neurobiolaging.2007.11.001] [PMID: 18079026]

[84]   Pugh PL, Richardson JC, Bate ST, Upton N, Sunter D. Non-cognitive behaviours in an APP/PS1
        transgenic model of Alzheimer's disease. Behav Brain Res 2007; 178(1): 18-28.
        [http://dx.doi.org/10.1016/j.bbr.2006.11.044] [PMID: 17229472]

[85]   Wang H, He J, Zhang R, *et al.* Sensorimotor gating and memory deficits in an APP/PS1 double
        transgenic mouse model of Alzheimer's disease. Behav Brain Res 2012; 233(1): 237-43.
        [http://dx.doi.org/10.1016/j.bbr.2012.05.007] [PMID: 22595040]

[86]   Bourgeois A, Lauritzen I, Lorivel T, Bauer C, Checler F, Pardossi-Piquard R. Intraneuronal
        accumulation of C99 contributes to synaptic alterations, apathy-like behavior, and spatial learning
        deficits in 3×TgAD and 2×TgAD mice. Neurobiol Aging 2018; 71: 21-31.
        [http://dx.doi.org/10.1016/j.neurobiolaging.2018.06.038] [PMID: 30071370]

[87]   Sakakibara Y, Sekiya M, Saito T, Saido TC, Iijima KM. Cognitive and emotional alterations in App
        knock-in mouse models of Aβ amyloidosis. BMC Neurosci 2018; 19(1): 46.
        [http://dx.doi.org/10.1186/s12868-018-0446-8] [PMID: 30055565]

[88]   Kales HC, Gitlin LN, Lyketsos CG. Management of neuropsychiatric symptoms of dementia in
        clinical settings: recommendations from a multidisciplinary expert panel. J Am Geriatr Soc 2014;
        62(4): 762-9.
        [http://dx.doi.org/10.1111/jgs.12730] [PMID: 24635665]

[89]   Lyketsos CG, Carrillo MC, Ryan JM, *et al.* Neuropsychiatric symptoms in Alzheimer's disease.
        Alzheimers Dement 2011; 7(5): 532-9.
        [http://dx.doi.org/10.1016/j.jalz.2011.05.2410] [PMID: 21889116]

[90]   Purdon SE, Malla A, Labelle A, Lit W. Neuropsychological change in patients with schizophrenia
        after treatment with quetiapine or haloperidol. J Psychiatry Neurosci 2001; 26(2): 137-49.
        [PMID: 11291531]

[91]   Jeste DV, Dolder CR. Treatment of non-schizophrenic disorders: focus on atypical antipsychotics. J
        Psychiatr Res 2004; 38(1): 73-103.
        [http://dx.doi.org/10.1016/S0022-3956(03)00094-3] [PMID: 14690772]

[92]   Juncos JL, Roberts VJ, Evatt ML, *et al.* Quetiapine improves psychotic symptoms and cognition in
        Parkinson's disease. Mov Disord 2004; 19(1): 29-35.
        [http://dx.doi.org/10.1002/mds.10620] [PMID: 14743357]

[93]   Madhusoodanan S, Shah P, Brenner R, Gupta S. Pharmacological treatment of the psychosis of
        Alzheimer's disease: what is the best approach? CNS Drugs 2007; 21(2): 101-15.
        [http://dx.doi.org/10.2165/00023210-200721020-00002] [PMID: 17284093]

[94]   Zhu S, Shi R, Li V, *et al.* Quetiapine attenuates glial activation and proinflammatory cytokines in
        APP/PS1 transgenic mice *via* inhibition of nuclear factor-κB pathway. Int J Neuropsychopharmacol
        2014; 18(3): pyu022.
        [http://dx.doi.org/10.1093/ijnp/pyu022] [PMID: 25618401]

[95]   Wu L, Feng X, Li T, Sun B, Khan MZ, He L. Risperidone ameliorated $A\beta_{1-42}$-induced cognitive and
        hippocampal synaptic impairments in mice. Behav Brain Res 2017; 322(Pt A): 145-56.
        [http://dx.doi.org/10.1016/j.bbr.2017.01.020] [PMID: 28093254]

[96]   Torres-Lista V, López-Pousa S, Giménez-Llort L. Marble-burying is enhanced in 3xTg-AD mice, can
        be reversed by risperidone and it is modulable by handling. Behav Processes 2015; 116: 69-74.
        [http://dx.doi.org/10.1016/j.beproc.2015.05.001] [PMID: 25957954]

[97]   He J, Yang Y, Yu Y, Li X, Li XM. The effects of chronic administration of quetiapine on the
        methamphetamine-induced recognition memory impairment and dopaminergic terminal deficit in rats.
        Behav Brain Res 2006; 172(1): 39-45.
        [http://dx.doi.org/10.1016/j.bbr.2006.04.009] [PMID: 16712969]

[98]   He J, Xu H, Yang Y, Zhang X, Li XM. Chronic administration of quetiapine alleviates the anxiety-like behavioural changes induced by a neurotoxic regimen of dl-amphetamine in rats. Behav Brain Res 2005; 160(1): 178-87.
[http://dx.doi.org/10.1016/j.bbr.2004.11.028] [PMID: 15836913]

[99]   He J, Xu H, Yang Y, Rajakumar D, Li X, Li XM. The effects of chronic administration of quetiapine on the phencyclidine-induced reference memory impairment and decrease of Bcl-XL/Bax ratio in the posterior cingulate cortex in rats. Behav Brain Res 2006; 168(2): 236-42.
[http://dx.doi.org/10.1016/j.bbr.2005.11.014] [PMID: 16360889]

[100]   He J, Yang Y, Xu H, Zhang X, Li XM. Olanzapine attenuates the okadaic acid-induced spatial memory impairment and hippocampal cell death in rats. Neuropsychopharmacology 2005; 30(8): 1511-20.
[http://dx.doi.org/10.1038/sj.npp.1300757] [PMID: 15886720]

[101]   Xue F, Chen YC, Zhou CH, et al. Risperidone ameliorates cognitive deficits, promotes hippocampal proliferation, and enhances Notch signaling in a murine model of schizophrenia. Pharmacol Biochem Behav 2017; 163: 101-9.
[http://dx.doi.org/10.1016/j.pbb.2017.09.010] [PMID: 29037878]

[102]   Zhu S, He J, Zhang R, et al. Therapeutic effects of quetiapine on memory deficit and brain β-amyloid plaque pathology in a transgenic mouse model of Alzheimer's disease. Curr Alzheimer Res 2013; 10(3): 270-8.
[http://dx.doi.org/10.2174/1567205011310030006] [PMID: 22950911]

[103]   Luo G, Liu M, He J, et al. Quetiapine attenuates recognition memory impairment and hippocampal oxidative stress in a transgenic mouse model of Alzheimer's disease. Neuroreport 2014; 25(9): 647-50.
[http://dx.doi.org/10.1097/WNR.0000000000000150] [PMID: 24642954]

[104]   Caballero J, Hitchcock M, Scharre D, Beversdorf D, Nahata MC. Cognitive effects of atypical antipsychotics in patients with Alzheimer's disease and comorbid psychiatric or behavioral problems: a retrospective study. Clin Ther 2006; 28(10): 1695-700.
[http://dx.doi.org/10.1016/j.clinthera.2006.10.008] [PMID: 17157125]

[105]   Scharre DW, Chang SI. Cognitive and behavioral effects of quetiapine in Alzheimer disease patients. Alzheimer Dis Assoc Disord 2002; 16(2): 128-30.
[http://dx.doi.org/10.1097/00002093-200204000-00011] [PMID: 12040309]

[106]   Hessmann P, Dodel R, Baum E, et al. Antipsychotic treatment of community-dwelling and institutionalised patients with dementia in Germany. Int J Psychiatry Clin Pract 2018; 22(3): 232-9.
[http://dx.doi.org/10.1080/13651501.2017.1414269] [PMID: 29235398]

[107]   Seo N, Song I, Park H, Ha D, Shin JY. Trends in the prescribing of atypical antipsychotics in elderly patients with dementia in Korea□. Int J Clin Pharmacol Ther 2017; 55(7): 581-7.
[http://dx.doi.org/10.5414/CP202951] [PMID: 28372635]

[108]   Qing H, Xu H, Wei Z, Gibson K, Li XM. The ability of atypical antipsychotic drugs vs. haloperidol to protect PC12 cells against MPP+-induced apoptosis. Eur J Neurosci 2003; 17(8): 1563-70.
[http://dx.doi.org/10.1046/j.1460-9568.2003.02590.x] [PMID: 12752374]

[109]   Wei Z, Bai O, Richardson JS, Mousseau DD, Li XM. Olanzapine protects PC12 cells from oxidative stress induced by hydrogen peroxide. J Neurosci Res 2003; 73(3): 364-8.
[http://dx.doi.org/10.1002/jnr.10668] [PMID: 12868070]

[110]   Wei Z, Mousseau DD, Richardson JS, Dyck LE, Li XM. Atypical antipsychotics attenuate neurotoxicity of beta-amyloid(25-35) by modulating Bax and Bcl-X(l/s) expression and localization. J Neurosci Res 2003; 74(6): 942-7.
[http://dx.doi.org/10.1002/jnr.10832] [PMID: 14648600]

[111]   Xu H, Wang H, Zhuang L, et al. Demonstration of an anti-oxidative stress mechanism of quetiapine: implications for the treatment of Alzheimer's disease. FEBS J 2008; 275(14): 3718-28.

[http://dx.doi.org/10.1111/j.1742-4658.2008.06519.x] [PMID: 18554300]

[112] Wang H, Xu H, Dyck LE, Li XM. Olanzapine and quetiapine protect PC12 cells from beta-amyloid peptide(25-35)-induced oxidative stress and the ensuing apoptosis. J Neurosci Res 2005; 81(4): 572-80.
[http://dx.doi.org/10.1002/jnr.20570] [PMID: 15948179]

[113] Behl C, Davis JB, Lesley R, Schubert D. Hydrogen peroxide mediates amyloid beta protein toxicity. Cell 1994; 77(6): 817-27.
[http://dx.doi.org/10.1016/0092-8674(94)90131-7] [PMID: 8004671]

[114] Ho PI, Collins SC, Dhitavat S, *et al.* Homocysteine potentiates beta-amyloid neurotoxicity: role of oxidative stress. J Neurochem 2001; 78(2): 249-53.
[http://dx.doi.org/10.1046/j.1471-4159.2001.00384.x] [PMID: 11461960]

[115] Drake J, Link CD, Butterfield DA. Oxidative stress precedes fibrillar deposition of Alzheimer's disease amyloid beta-peptide (1-42) in a transgenic Caenorhabditis elegans model. Neurobiol Aging 2003; 24(3): 415-20.
[http://dx.doi.org/10.1016/S0197-4580(02)00225-7] [PMID: 12600717]

[116] Mecocci P, MacGarvey U, Beal MF. Oxidative damage to mitochondrial DNA is increased in Alzheimer's disease. Ann Neurol 1994; 36(5): 747-51.
[http://dx.doi.org/10.1002/ana.410360510] [PMID: 7979220]

[117] Hensley K, Carney JM, Mattson MP, *et al.* A model for beta-amyloid aggregation and neurotoxicity based on free radical generation by the peptide: relevance to Alzheimer disease. Proc Natl Acad Sci USA 1994; 91(8): 3270-4.
[http://dx.doi.org/10.1073/pnas.91.8.3270] [PMID: 8159737]

[118] Butterfield DA, Yatin SM, Varadarajan S, Koppal T. Amyloid beta-peptide-associated free radical oxidative stress, neurotoxicity, and Alzheimer's disease. Methods Enzymol 1999; 309: 746-68.
[http://dx.doi.org/10.1016/S0076-6879(99)09050-3] [PMID: 10507060]

[119] Frederikse PH, Garland D, Zigler JS Jr, Piatigorsky J. Oxidative stress increases production of beta-amyloid precursor protein and beta-amyloid (Abeta) in mammalian lenses, and Abeta has toxic effects on lens epithelial cells. J Biol Chem 1996; 271(17): 10169-74.
[http://dx.doi.org/10.1074/jbc.271.17.10169] [PMID: 8626578]

[120] Tong Y, Zhou W, Fung V, *et al.* Oxidative stress potentiates BACE1 gene expression and Abeta generation. J Neural Transm (Vienna) 2005; 112(3): 455-69.
[http://dx.doi.org/10.1007/s00702-004-0255-3] [PMID: 15614428]

[121] Prasad KN, Hovland AR, Cole WC, *et al.* Multiple antioxidants in the prevention and treatment of Alzheimer disease: analysis of biologic rationale. Clin Neuropharmacol 2000; 23(1): 2-13.
[http://dx.doi.org/10.1097/00002826-200001000-00002] [PMID: 10682224]

[122] Chen C, Li B, Cheng G, Yang X, Zhao N, Shi R. Amentoflavone ameliorates Aβ$_{1-42}$-induced memory deficits and oxidative stress in cellular and rat model. Neurochem Res 2018; 43(4): 857-68.
[http://dx.doi.org/10.1007/s11064-018-2489-8] [PMID: 29411261]

[123] Lucin KM, Wyss-Coray T. Immune activation in brain aging and neurodegeneration: too much or too little? Neuron 2009; 64(1): 110-22.
[http://dx.doi.org/10.1016/j.neuron.2009.08.039] [PMID: 19840553]

[124] Rosano C, Marsland AL, Gianaros PJ. Maintaining brain health by monitoring inflammatory processes: a mechanism to promote successful aging. Aging Dis 2012; 3(1): 16-33.
[PMID: 22500269]

[125] Mei F, Guo S, He Y, *et al.* Quetiapine, an atypical antipsychotic, is protective against autoimmune-mediated demyelination by inhibiting effector T cell proliferation. PLoS One 2012; 7(8): e42746.
[http://dx.doi.org/10.1371/journal.pone.0042746] [PMID: 22912731]

[126] Shao Y, Peng H, Huang Q, Kong J, Xu H. Quetiapine mitigates the neuroinflammation and

oligodendrocyte loss in the brain of C57BL/6 mouse following cuprizone exposure for one week. Eur J Pharmacol 2015; 765: 249-57.
[http://dx.doi.org/10.1016/j.ejphar.2015.08.046] [PMID: 26321148]

[127] Maes M, Meltzer HY, Bosmans E. Immune-inflammatory markers in schizophrenia: comparison to normal controls and effects of clozapine. Acta Psychiatr Scand 1994; 89(5): 346-51.
[http://dx.doi.org/10.1111/j.1600-0447.1994.tb01527.x] [PMID: 8067274]

[128] Tourjman V, Kouassi É, Koué MÈ, et al. Antipsychotics' effects on blood levels of cytokines in schizophrenia: a meta-analysis. Schizophr Res 2013; 151(1-3): 43-7.
[http://dx.doi.org/10.1016/j.schres.2013.10.011] [PMID: 24200418]

[129] Bahramabadi R, Samadi M, Vakilian A, Jafari E, Fathollahi MS, Arababadi MK. Evaluation of the effects of anti-psychotic drugs on the expression of CD68 on the peripheral blood monocytes of Alzheimer patients with psychotic symptoms. Life Sci 2017; 179: 73-9.
[http://dx.doi.org/10.1016/j.lfs.2017.04.024] [PMID: 28465247]

[130] Zeng Y, Zhao D, Xie CW. Neurotrophins enhance CaMKII activity and rescue amyloid-β-induced deficits in hippocampal synaptic plasticity. J Alzheimers Dis 2010; 21(3): 823-31.
[http://dx.doi.org/10.3233/JAD-2010-100264] [PMID: 20634586]

[131] Xu H, Qing H, Lu W, et al. Quetiapine attenuates the immobilization stress-induced decrease of brain-derived neurotrophic factor expression in rat hippocampus. Neurosci Lett 2002; 321(1-2): 65-8.
[http://dx.doi.org/10.1016/S0304-3940(02)00034-4] [PMID: 11872258]

[132] Tempier A, He J, Zhu S, et al. Quetiapine modulates conditioned anxiety and alternation behavior in Alzheimer's transgenic mice. Curr Alzheimer Res 2013; 10(2): 199-206.
[http://dx.doi.org/10.2174/1567205011310020010] [PMID: 22950914]

[133] Bai O, Zhang H, Li XM. Antipsychotic drugs clozapine and olanzapine upregulate bcl-2 mRNA and protein in rat frontal cortex and hippocampus. Brain Res 2004; 1010(1-2): 81-6.
[http://dx.doi.org/10.1016/j.brainres.2004.02.064] [PMID: 15126120]

[134] Adams JM, Cory S. The Bcl-2 protein family: arbiters of cell survival. Science 1998; 281(5381): 1322-6.
[http://dx.doi.org/10.1126/science.281.5381.1322] [PMID: 9735050]

[135] Bruckheimer EM, Cho SH, Sarkiss M, Herrmann J, McDonnell TJ. The Bcl-2 gene family and apoptosis. Adv Biochem Eng Biotechnol 1998; 62: 75-105.
[http://dx.doi.org/10.1007/BFb0102306] [PMID: 9755641]

[136] Howard S, Bottino C, Brooke S, Cheng E, Giffard RG, Sapolsky R. Neuroprotective effects of bcl-2 overexpression in hippocampal cultures: interactions with pathways of oxidative damage. J Neurochem 2002; 83(4): 914-23.
[http://dx.doi.org/10.1046/j.1471-4159.2002.01198.x] [PMID: 12421364]

[137] He J, Liu F, Zu Q, et al. Chronic administration of quetiapine attenuates the phencyclidine-induced recognition memory impairment and hippocampal oxidative stress in rats. Neuroreport 2018; 29(13): 1099-103.
[http://dx.doi.org/10.1097/WNR.0000000000001078] [PMID: 30036204]

[138] Roth BL, Sheffler DJ, Kroeze WK. Magic shotguns *versus* magic bullets: selectively non-selective drugs for mood disorders and schizophrenia. Nat Rev Drug Discov 2004; 3(4): 353-9.
[http://dx.doi.org/10.1038/nrd1346] [PMID: 15060530]

[139] Choi Y, Jeong HJ, Liu QF, et al. Clozapine improves memory impairment and reduces Aβ level in the Tg-APPswe/PS1dE9 mouse model of Alzheimer's disease. Mol Neurobiol 2017; 54(1): 450-60.
[http://dx.doi.org/10.1007/s12035-015-9636-x] [PMID: 26742522]

[140] Allen SJ, Watson JJ, Shoemark DK, Barua NU, Patel NK. GDNF, NGF and BDNF as therapeutic options for neurodegeneration. Pharmacol Ther 2013; 138(2): 155-75.
[http://dx.doi.org/10.1016/j.pharmthera.2013.01.004] [PMID: 23348013]

[141] Cai Z, Yan LJ, Li K, Quazi SH, Zhao B. Roles of AMP-activated protein kinase in Alzheimer's disease. Neuromolecular Med 2012; 14(1): 1-14.
[http://dx.doi.org/10.1007/s12017-012-8173-2] [PMID: 22367557]

[142] Autry AE, Monteggia LM. Brain-derived neurotrophic factor and neuropsychiatric disorders. Pharmacol Rev 2012; 64(2): 238-58.
[http://dx.doi.org/10.1124/pr.111.005108] [PMID: 22407616]

[143] Pedrós I, Petrov D, Allgaier M, *et al.* Early alterations in energy metabolism in the hippocampus of APPswe/PS1dE9 mouse model of Alzheimer's disease. Biochim Biophys Acta 2014; 1842(9): 1556-66.
[http://dx.doi.org/10.1016/j.bbadis.2014.05.025] [PMID: 24887203]

CHAPTER 5

Approaches Based on Cholinergic Hypothesis and Cholinesterase Inhibitors in the Treatment of Alzheimer's Disease

Zeynep ÖZDEMİR[1,*], Azime-Berna ÖZÇELİK and **Mehtap UYSAL[2]**

[1] *İnönü University, Faculty of Pharmacy, Department of Pharmaceutical Chemistry, Malatya, Turkey*

[2] *Gazi University, Faculty of Pharmacy, Department of Pharmaceutical Chemistry, Ankara, Turkey*

Abstract: Alzheimer's disease (AD) is a progressive and fatal neurodegenerative disease which deteriorates a person's ability to perform daily activities, leading to various neuropsychiatric symptoms and behavioral disorders in later stages of the disease. The pathophysiology of AD is complex. Biochemical studies indicate that some neuromediator levels, especially acetylcholine, are reduced in the brain cortex. Loss of neurons and axons associated with the onset of the disease causes lower levels of acetylcholine release and it is more difficult to maintain the continuity of nerve conduction in low concentration neurotransmitter levels. This has led to the introduction of a "cholinergic hypothesis" that emphasizes the importance of acetylcholine deficiency in the development of the symptoms of the disease. As the (AChE) and (BChE) enzymes from the serine hydrolases which perform the hydrolysis process of acetylcholine (ACh), the inhibition of these enzymes is an important method of raising the level of acetylcholine. Studies indicate that increases in acetylcholine levels due to acetylcholine esterase inhibition may improve cognitive impairment in the early stages of AD.

Keywords: AChE Inhibitors, Acetylcholine, Alzheimer's Disease, BChE Inhibitors, Cholinergic Hypothesis, Donepezile, Galantamine, Huperzin A, Metrifonate, Physostigmine, Rivastigmine, Tacrine.

INTRODUCTION

Alzheimer's disease (AD) is a fatal neurodegenerative disease characterized by the individual's inability to perform daily activities of the person and progressive cognitive dysfunctions and various neuropsychiatric symptoms in the later stages

---

* **Corresponding author Zeynep ÖZDEMİR:** İnönü University, Faculty of Pharmacy, Department of Pharmaceutical Chemistry, Malatya, Turkey; E-mail: zeynep.bulut@inonu.edu.tr

Atta-ur-Rahman (Ed.)

of the disease [1 - 3]. While disturbed memory is the first clinical symptom of the disease, distant memory is relatively conserved during illness. The ability to use visual tests, objects and equipment as well as the ability to make calculations decreases as the disease progresses. The levels of alertness and motor strength are maintained until late in the illness. Muscle contractions are common in advanced stages of the disease, but motor impairment does not occur even in these stages [3, 4]. Most Alzheimer's patients need constant care after a few years from the onset of the disease [1]. Death usually occurs 6 to 12 years after the onset of the disease, and the reasons of death are complications arising from immobility problems such as pneumonia and pulmonary embolism [5].

The prevalence of AD, one of the types of dementia seen in the elderly, doubles every five years, reaching 30% at the age of 85 while it is %1 at the age of 60 [6]. According to the results of the epidemiological studies on AD, it is estimated that the number of patients is approximately 5 million in the United States and 24 million in the world. Estimates suggest that this number will be doubled after 20 years and it can be attained 42 million in 2020 and 81 million in 2040 [1, 3, 7].

A definite diagnosis of AD can only be made with post-mortem brain examinations, as well as a careful clinical evaluation of the patient, differential diagnosis of other similar diseases, and appropriate laboratory tests (Fig. **1**) [3, 8].

Although the physiopathology of AD is not yet fully understood, it is thought that the disease is caused by the undetected disappearance of brain cells and three main hypotheses on the formation of the disease are suggested. According to the cholinergic hypothesis, the oldest hypothesis, it is thought that AD is caused by the decrease of acetylcholine (Ach) which is an important neurotransmitter [9 - 11]. There is a significant decrease in cholinergic receptors and choline acetyltransferase (ChAT) levels in the cerebral cortex of Alzheimer's patients. While many of the previously maintained treatment approaches are based on this hypothesis, clinical trials have shown that treatment strategies to elevate acetylcholine levels provide symptomatic relief [12]. Recent studies on the cholinergic hypothesis indicate that the use of cholinesterase inhibitors may affect the formation of amyloid fibrillation [13 - 15]. Another hypothesis, Tau hypothesis, suggests that hyperphosphorylation of Tau proteins, which are a member of the microtubule-associated protein family and provide tubule stability, results in impaired intraneural fibrin plexus in AD. These formations are reported to accelerate the development of AD and lead to neuronal death by damaging the intracellular transport system and signal transmission [16]. The third hypothesis, known as the Amyloid hypothesis, suggests that β-amyloid peptides which accumulate in certain regions of the brain resulting in insoluble fibrils, destroy the nerve cells by forming senile plaques and reduce the amount of neurotransmitter by breaking down neuronal connections in people who suffer from Alzheimer's disease [17].

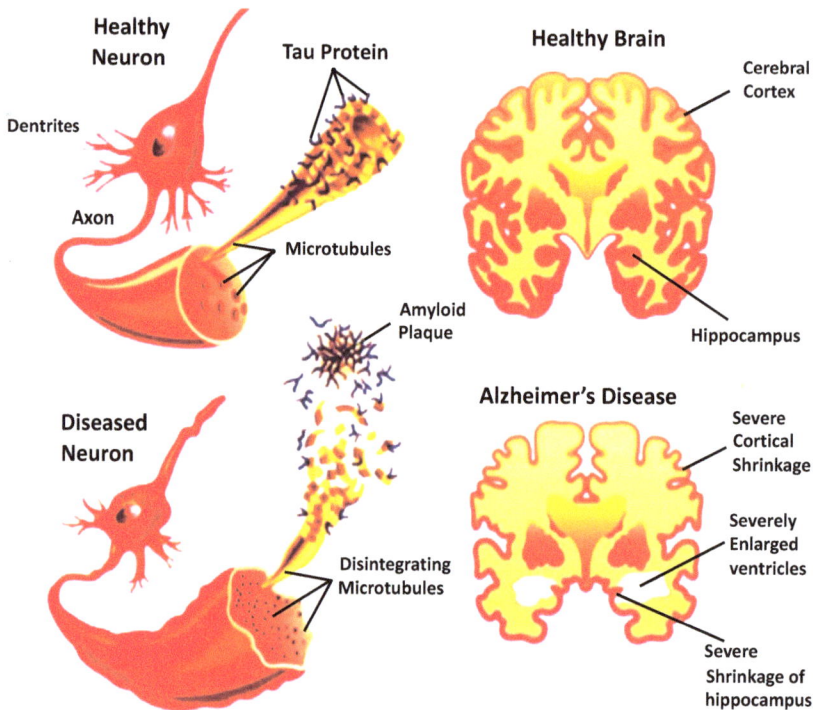

Fig. (1). Comparison of healthy brain with AD brain.

One significant finding in the diagnosis of Alzheimer's disease is the decrease in the number of neurons in the region of the limbic system that manages memory processes [1]. Neurochemical complexity in AD has been extensively researched. It was shown that many neurotransmitters, most commonly acetylcholine, are reduced in parallel to neuronal loss, when the neurotransmitter content of the cerebral cortex is measured directly as a result of biochemical studies in patients with AD [18]. There is a relationship between AD and changes in many neurotransmitters, including cholinergic, noncholinergic, serotonergic, dopaminergic, aminoacidergic, and neuropeptidergic; but the relationship between changes in the cholinergic system and AD is the most well-known. According to this relationship which is described as cholinergic hypothesis, the decrease in acetylcholine and choline acetyl transferase levels is associated with AD. The use of atropine-like central acting cholinergic antagonists is also associated with dementia-like disorders. Selective lack of acetylcholine in AD leads to the introduction of "cholinergic hypothesis" that emphasizes the importance of acetylcholine deficiency in the development of AD's symptoms [19, 20]. The cholinergic hypothesis was first introduced about 30 years ago and contributed significantly to explain the function disorder of the neurons including acetylcholine in the brain and cognitive reduction in Alzheimer's patients. This idea has underpinned the techniques of AD treatment and majority of drug development studies so far. Cholinergic deficiency is the result of atrophy and degeneration in subcortical cholinergic neurons, especially in the basal forebrain which provides cholinergic innervation to the entire cerebral cortex. Current studies on the brains of the sufferers with mild cognitive impairment in which the activity of choline acetyltransferase and / or (AChE) is not affected or even up-regulated or on patients with early stage AD have demonstrated the effectiveness of the cholinergic hypothesis, also the justification for the use of cholinomimetics, particularly in the treatment of impairment in early stages. According to this hypothesis, cognitive impairments can also be observed following degeneration of cholinergic neurons in the medial frontal lobe in Alzheimer's patients. Therefore, there is a direct relationship between cholinergic abnormalities and AD. Post-mortem brain studies and altered cholinergic symptoms such as degeneration of neurons controlling neocortical regions, loss of cholinergic terminals in the cerebral cortex, loss of choline acetyl transferase activity and acetyl choline synthesis, decreased cholinergic receptivity and increased (AChE) activity demonstrate this in many Alzheimer's patients [21, 22]. For this reason, in the treatment of AD, increasing acetylcholine level and suppressing (AChE) enzyme is one of the frequently used treatment modalities. Cholinergic abnormalities can also cause noncognitive behavioral abnormalities and accumulation of toxic neuritic plaques in AD. Besides, the methods which are cholinergic-based will continue as an approach to progress of rational drugs in the medication of AD and another types of mental disorder.

Although cholinesterase inhibitors which are given in AD do not affect mortality, they are used in therapy because of their effectiveness in the quality of life and stabilization of the disease stage [23]. Increases in acetylcholine levels due to (AChE) inhibition may improve cognitive impairment in early stages of AD [3, 18, 24]. In a study investigating the association of cholinesterase inhibitors with mortality in patients with clinically diagnosed AD, of the 2464 patients with AD, 1296 were prescribed AChEIs and were found to be strongly associated with lower mortality in patients using AChEI (risk ratio = 0.57; 95% CI 0.51-0.64) [25]. Additionally, in a meta-analysis of the effects of cholinesterase inhibitors on the risks and benefits of Alzheimer's disease, cholinesterase inhibitors and placebo were compared. In this study, 42 randomized placebo-controlled clinical study groups covering 16106 patients were included. The results of the study showed that cholinesterase inhibitors improve symptoms and decrease mortality compared to placebo [26].

It has also been found that treatment with combined memantine, which is N-methyl-D-aspartate (NMDA) receptor antagonist and cholinesterase inhibitor (ChEI) enhances the activity [27].

Research which focuses on the brain of individuals having mild cognitive impairment (MCI) or patients who are in initial stages of AD are being more and more crucial as a diagnostic method. These kind of examinations may support the evolution and/or recognition of the neuroprotective methods as well as new approaches in the diagnosis and treatment of more specific diseases. The brains of patients in the final stage of the disease were analyzed in most of the previous studies for providing the correlation between the level of cognitive decline and neuropathology of the disease. This may not be particularly useful for new research which aims to change the course of the disease if it is recognized at initial stages.

The consequences of a few published studies in which brains of MCI and / or mild AD patients were analyzed. Some of these researches has led to begin challenging the effectiveness of the cholinergic hypothesis. For instance, *Davis et al.* indicated that (AChE) and choline acetyltransferase (ChAT) activity did not decrease in post-mortem neocortical tissues of mild AD [28]. As a consequence of their work, the authors conclude that a cholinergic marker is unlikely to be used as an early indicator of AD; a cholinergic deficit cannot be detected before the patient becomes symptomatic. Cholinergic treatment is exclusively suitable for patients with excessive disease. Furthermore, *DeKosky et al.* did not define decrease in ChAT activity in a number of cortical areas studied in MCI or mild Alzheimer's patients, and even the activity was literally up-regulated in the frontal cortex and hippocampus [29]. In another study, it was shown that neurons containing ChAT and protein which carries vesical acetylcholine and conserved in the nucleus basalis of MCI and / or mild Alzheimer's patients [30].

The more *in vivo* imaging techniques become widespread, the uncertainties about cholinergic function in central nervous system of MCI sufferers or patients at early AD will probably appear more clear. Nevertheless, different *in vivo* imaging researches performed so far in Alzheimer's patients seem to promote the cholinergic hypothesis. For instance, PET studies using [11C] N-methylpiperidi--4-yl-propionate show that cortical AChE (AChE) activity is actually decreased in individuals who suffer from Alzheimer's disease [31]. [11C] Nicotine-based PET researches show that lack of nicotinic receptors is indeed early markers in AD, and these reports also indicate that cortical nicotinic receptor deficits are considerably associated with cognitive impairment [32]. Other PET studies using non-selective muscarinic ligands, [123I] quinuclidinylbenzylate and [11C] N-methyl-4-piperidyl benzylate show decreases in binding with neocortical regions not only age- but also AD-related [33]. In addition, single photon emission computed tomography (SPECT) studies using [123I] benzoveseamacol binding show that vesicular acetylcholine transporter decreases throughout the entire cerebral cortex and hippocampus in early Alzheimer's patients [34]. It appears appropriate to examine if the function of central cholinergic neurons deteriorates in the elderly or not, because age is now considered to be the strongest risk factor known for AD. Challenges in the cholinergic hypothesis of AD obscure the existence of evidence in help of the relationship between aging, cholinergic dysfunction and cognitive decline. In the early 1980s, when selective and sensitive chemical enzymatic techniques were improved to measure the dynamic aspects of neurotransmitter function, a research on the impact of age on brain cholinergic function has begun. Methods for fast balanced of brain acetylcholine and choline levels have also been introduced with rapid freezing or focused microwave assisted for regular use. For instance, *Gibson et al.* have investigated the entire acetylcholine synthesis in brain of rats aged 3 to 30 months and they have stated that biosynthesis of acetylcholine, measured by injection of a radiolabeled precursor, decreased to 75% in 30 months old animals. Mild hypoxia has reduced acetylcholine synthesis by 90%. Furthermore, following potassium stimulation, the acetylcholine releasing abilities of aged cholinergic neurons are damaged more than their capability to synthesize transmittance [35]. According to the experiments using several techniques to raise acetylcholine output or to damage cholinergic neurons, it has been supported that elderly brain cholinergic neurons function is comparatively normal until they are forced [36, 37]. Thus, it seems reasonable to arrive at the conclusion that any continuous damage to the forebrain cholinergic neurons may inhibit these cells to be able to release adequate amount of transmitter for routine functioning. *Sarter et al.* have analyzed this probability directly on a group of tests in which chemical lesions of basal forebrain cholinergic neurons were formed in rats in early ages in order to reduce basal forebrain cholinergic cells to a limited extent [38]. The rats used in the study were previously well trained in performing the task which requires constant attention. As a result of the observations, it was determined that the task performance of both groups is similar in the beginning, and when the lesioned group has displayed a crucial dysfunction, there was no significant difference in task prolificacy between the lesioned and control rats at 31 months of age. It has been found that the most of the differences that are related to the ages of individuals and especially with regard to the dynamic aspects of cholinergic neurons have been observed in the results of these researches which were applied in elderly rodents [38].

Acetylcholine (Ach)

Acetylcholine is synthesized from choline and acetylcoenzymes by a reaction catalyzed by the choline transferase enzyme. This reaction involves the transfer of an acetyl group from the acetylcoenzyme A to the choline (Fig. **2**). The resulting neurotransmitter is then stored in the vesicles of the nerve endings and released in response to neuronal depolarization [39].

The released neurotransmitter is rapidly cleaved in choline and acetic acid by AChE (AChE) (Fig. **2**) in the synapse cavity or ensures that the effect occurs by binding to receptors in the postsynaptic membrane and Fig. (**3**) [40].

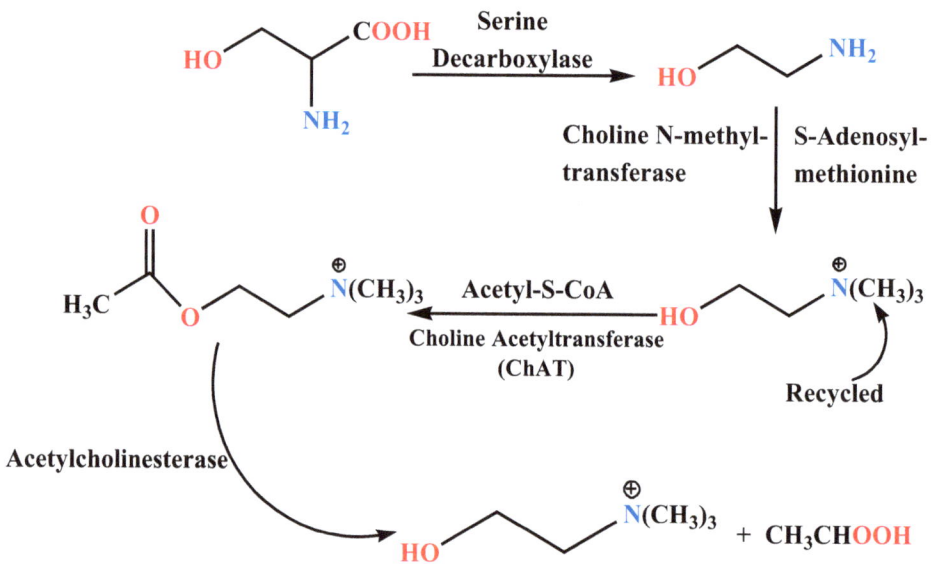

Fig. (2). Biosynthesis, storage, release and hydrolysis of acetylcholine.

There are cholinergic synapses everywhere in the human central nervous system. Specific cholinesterases are localized in presynaptic and postsynaptic membranes. The choline resulting from fragmentation is taken up into the axon by active transport to form acetylcholine molecules, while acetic acid is removed from the synaptic region by the blood stream. In addition to specific acetylcholine esterases in the body, there are also nonspecific cholinesterases such as pseudo-cholinesterase and (BChE). These are usually present in blood and liver and also hydrolyze other choline esters besides the acetylcholine [39, 40].

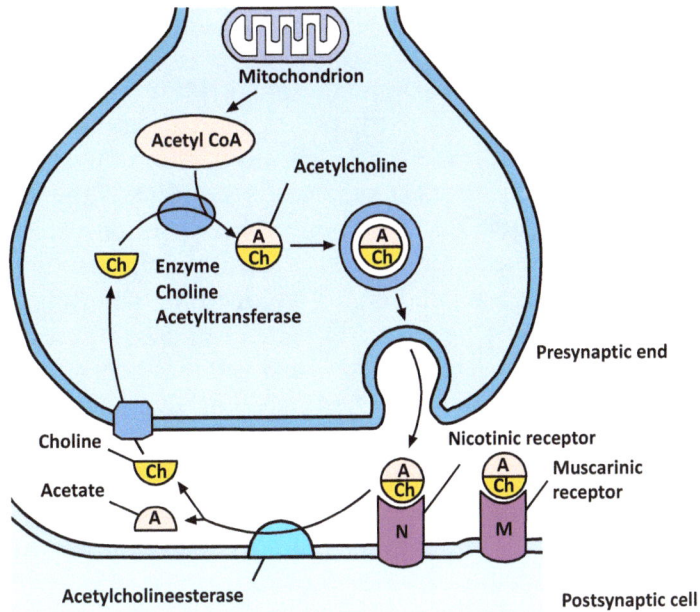

Fig. (3). Biosynthesis, storage, release and hydrolysis of acetylcholine.

There are two acetylcholine receptors different from each other in structure, location, pharmacological activity. These receptors which have specific agonists and antagonists are defined as nicotinic and muscarinic receptors based on the ability to selectively bind nicotine and muscarine, the natural alkaloids. Nicotinic acetylcholine receptors in the neuromuscular motor terminals and ganglia are the first neurotransmitter receptors isolated and purified in their active form [41, 42]. It is thought that the receptor consists of two polypeptide chain monomers linked to one another by a disulfide bond and has five subunits. Acetylcholine allows passage of small cations such as Ca^{++}, Na^+, and K^+ by leading to an increase in membrane permeability when they are bound to these receptors. The physiological effect of this condition is the formation of depolarization at the motor terminals, and therefore the muscular contraction or nerve stimulation in the neuromuscular junction. Muscarinic receptors have a significant impact on the regulation of the functions of the organs that are stimulated by the autonomic nervous system [43]. The effect of acetylcholine on these receptors in parasympathetic synapses may be stimulatory or inhibitory. Acetylcholine stimulates the secretion of salivary and secretion glands and leads to constriction of respiratory way. The compound also inhibits heart contractions and relaxes smooth muscles in blood vessels. Five subtypes of muscarinic receptors (M_1, M_2, M_3, M_4 and M_5) have been shown. M_1 is localized in neuronal structures such as the central nervous system and ganglia, M_2 in the heart, M_3 in the smooth muscle and secretory glands, and M_4 in the striatum and lungs [44].

Cholinergic synapses are presented in the thalamus, the striatum, the limbic system and neocortex at high density, and this situation indicated that cholinergic transmission is crucial for memory, learning, attention and other high brain activities [45]. Many different investigations indicate the excessive roles of cholinergic systems in general brain homeostasis and plasticity. This causes the cholinergic system of the brain to play a central role in ongoing research into normal cognition and age-related cognitive decline, including dementia, such as AD [21]. Cholinergic hypothesis in AD focuses on loss of progressive limbic and neocortical cholinergic innervation. It is believed that neurofibrillary degeneration of the basal forebrain is the primary cause of dysfunction and death of the forebrain cholinergic neurons, and this leads to a common presynaptic cholinergic denervation. Cholinesterase inhibitors increase the use of acetylcholine in synapses in the brain and are one of several drug therapies proven to be clinically useful in the treatment of AD's dementia and this confirms that the cholinergic system is an important therapeutic target of the disease [46].

Cholinesterase Enzymes

The (AChE) and (BChE) enzymes of the serine hydrolase enzyme group are generally known as cholinesterases (Fig. **4**) [47]. Of these two types of cholinesterase enzymes in the human brain, AChE is encoded by chromosome 7 and BChE is encoded by chromosome 3 [48].

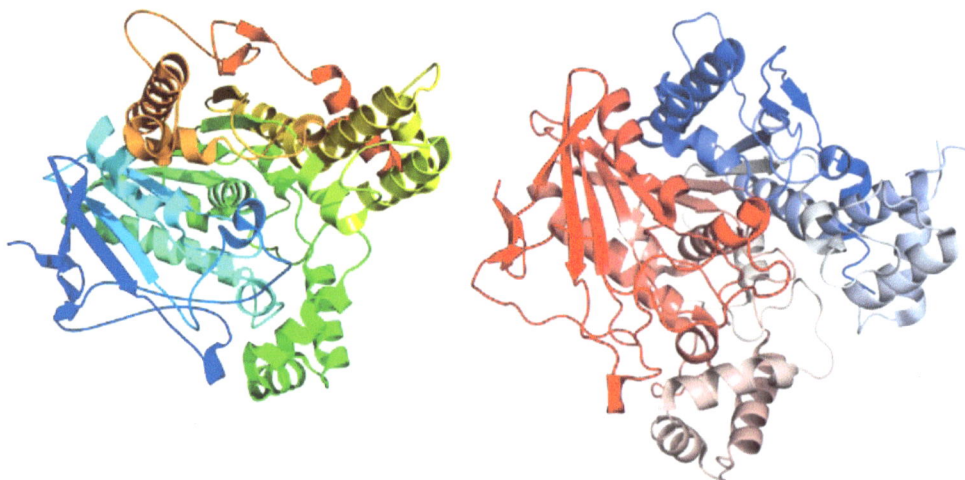

Fig. (4). Structure of AChE and BChE.

Although the effects of AChE in cholinergic transmission are well known, the effects of BChE have not been elucidated sufficiently [47]. The synaptic acetylcholine hydrolysis in healthy brain cells is mainly performed by AChE, and the contribution of BChE to this hydrolysis is considered to be minimal [49].

Although the cholinesterase enzymes are encoded in different chromosomes, the amino acid sequences of both enzymes are similar to each other by 65%. AChE is commonly existed in the normal adult brain, BChE is present in limited quantities, and it is thought that 80% of the cholinesterase activity in the brain is responsible for AChE and the remaining 20% is responsible for BChE [50, 51]. BChE enzyme levels are elevated in the early stages of nervous system development, while this level is decreased in later stages. AChE is located in the cell body of cholinergic neurons, axons and proximals of dendritic extensions, while BChE is existed in cell body and dendrites [48].

Although BChE has been shown to induce cholinergic transmission in smooth muscle, its function in the nervous system is poorly known [52, 53]. Recent studies for better understanding of BChE function suggested that the mice which have no AChE genes may develop to adulthood despite the occurrence of peripheral or central cholinergic dysfunction such as tremor, ileus and weakness. It was also shown that these mice were extremely sensitive to the inhibition of (BChE) and organophosphate effects [54]. This suggests that both AChE and BCHE play an active role in neuronal development and cholinergic transmission.

Three hypothesis have been proposed for the enzyme binding sites of inhibitor molecules in AChE inhibitory activity. One of these hypotheses is that the heterocyclic ring in the structure of the AChEI compounds acts as a peripheral anionic site of the enzyme. The second is that the nitrogen atom in the structure of many strong AChE inhibitors acts as a positive charge center. It was determined by X-ray crystallography researches of AChE / donepezil and AChE / galantamin complexes that this group could interact with the catalytic center of AChE. The third hypothesis is also that the hydrophobic groups, such as the phenyl ring attached to the nitrogen atom, are existed as the choline binding site [55].

The current crystallographic studies on the structural properties of AChE and BChE active sites have been well elucidated by studying on interactions with key binding site residues of inhibitors used in the treatment of AD, such as tacrin, donepezil and galantamine [56 - 59]. Donepezil interacts with the residues of the peripheral anionic region (PAS), including the rim and the cationic active site (CAS) and the acyl binding pocket at the active gorge, whereas galantamine and tacrine do not interact. (Figs. **5** and **6**) [60 - 63].

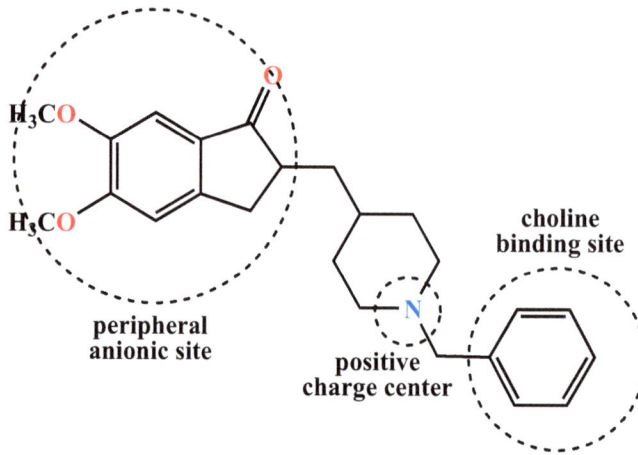

Fig. (5). Structural hypothesis for AChEIs (donepezil) [64].

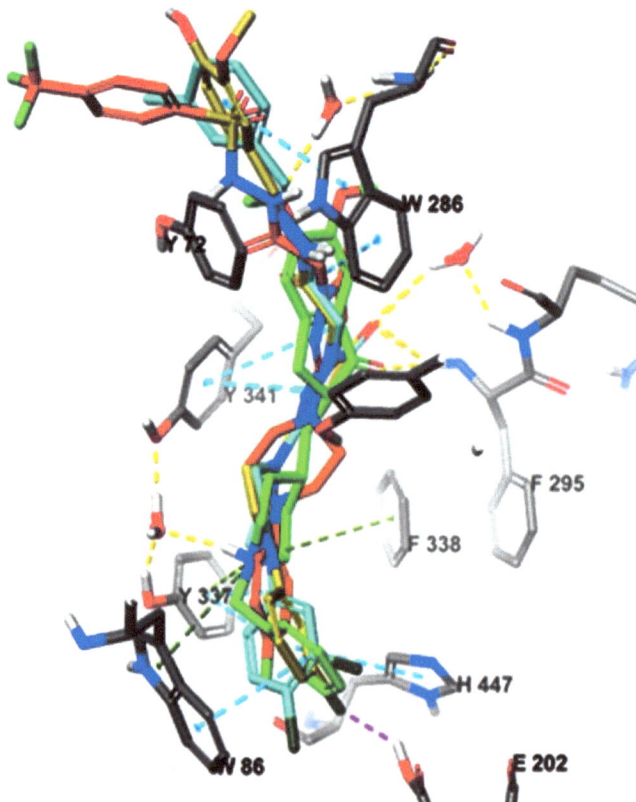

Fig. (6). Binding modes of donepezil with the AChE active site gorge [65].

Current Cholinesterases Inhibitor Drugs

Acetylcholine is released from the ganglion, postganglionic nerve and motor terminals. It is then rapidly hydrolyzed by acetylcholine esterase. The inhibition of acetylcholine esterase prolongs the life of the neurotransmitter in the junction and thus produces pharmacological effects similar to those observed when acetylcholine is administered [39, 40]. Although (AChE) is the main target of cholinesterase inhibitor drugs, (BChE) is also inhibited by these drugs. Anticholinesterases are used in the treatment of diseases such as myasthenia gravis, atonies in the gastrointestinal tract, glaucoma, alzheimer. These compounds are also used as nerve gases and insecticides [21, 66, 67]. According to their mechanism of action, anti-AChE agents can be divided into three groups as competitive agonists, short acting inhibitors (carbamates) and long acting inhibitors (organophosphorus). The first group including edrophonium is composed of quaternary alcohols. These agents can be reversibly bound to the active site of the enzyme both electrostatically and with hydrogen bonds, thereby inhibiting the enzymatic reach of acetylcholine. The enzyme-inhibitor complex does not have a covalent bond; thus, life of this complex is from 2 to 10 minutes. The next one includes carbamate esters such as neostigmine and physostigmine. These compounds are subjected to a hydrolysis process which consists of two steps as described for acetylcholine and it takes time from 30 minutes to 6 hours. The last group includes organophosphates. These agents create a phosphorylated active site by being subjected to primary binding and hydrolysis by the enzyme. Since the phosphorylated enzyme is extremely stable, it can be hydrolyzed in the water during almost hundreds of hours. However, strong nucleophiles such as pralidoxime can break the phosphorus enzyme bond. Therefore, these agents are also known as irreversible cholinesterase inhibitors [40].

Several cholinesterase inhibitors such as tacrine, donepezil, rivastigmine, galantamine and huperzine A have been shown to slow the progression of disease in AD. Tacrine which is the first drug in these compounds was approved by the FDA in the USA for the treatment of AD in 1993 [68]. Tacrine was withdrawn from the drug market very soon because of its hepatotoxicity and its subsequent use was restricted. However, the other three AChE inhibitors rivastigmine, donepezil and galantamine were approved as anti-AD drugs [69 - 71]. These drugs also have additional features that are not known at this time. For example, rivastigmine can inhibit BChE and galantamine has an impact on nicotinic acetylcholine receptors [72]. Donepezil is an inhibitor of β-secretase (BACE1) and also interacts with sigma-1 receptors [73].

Tacrine

Tacrine (Fig. **7**), which is both AChE and the BChE inhibitor, has beneficial effects on cognitive symptoms associated with Alzheimer's disease, such as memory, attention, reasoning, speech and simple tasks as a result of clinical trials and received FDA approval for Alzheimer's treatment. The higher substrate concentration results in the more competitive inhibitor for Tacrine, while the lower concentration causes a non-competitive one. Tacrine interact with the catalytic region of AChE, and consequently the ACh is prevented from binding to the active site. Although removed in 2013 because of its hepatotoxicity, tacrine has been used as synthetic starting compound to develop novel molecules which have several pharmacological activities [9, 67, 74, 75].

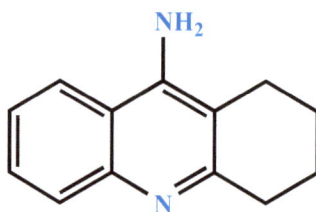

Fig. (7). Structure of Tacrine.

Physostigmine

Physostigmine (Fig. **8**) was the first cholinesterase inhibitor compound which was investigated for the treatment of AD. Physostigmine is isolated as a parasympathomimetic plant alkaloid from the seeds of the *Physostigma venenosum* plant. Although it passes through the blood-brain barrier (BBB), it is a short half-life compound with a narrow therapeutic index [76]. Physostigmine was used in Myastenia Gravis, glaucoma and delayed gastric emptying. However, the drug has not been approved, it is used as a standard in enzymatic analyzes instead of treatment of MG. In addition, due to the side effects and the disadvantages which mentioned above, the use of AD treatment was also discontinued. Rivastigmine, which is a carbamate derivative of this compound, has also been used in AD [77].

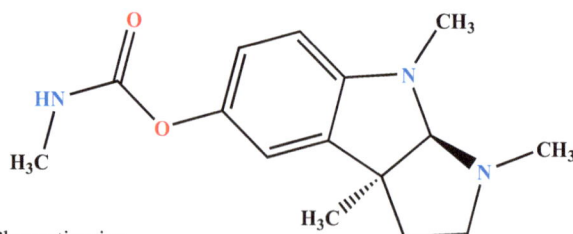

Fig. (8). Structure of Physostigmine.

Rivastigmine

Rivastigmine (Fig. **9**) is a ChEI with excellent features with respect to enzyme specific activity and lower risk of side effects. Rivastigmine has been approved for use in 60 countries, including the European Union and all member states of the United States. The rivastigmine, which inhibits both AChE and BChE, is a small molecule that can readily pass through BBB [70, 78].

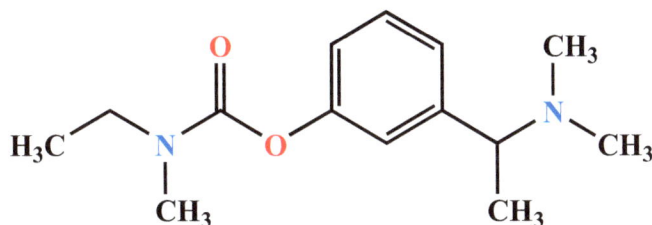

Fig. (9). Structure of rivastigmine.

Donezepil

Donepezil (Fig. **10**), approved for use in 1997, is now the most widely used AD drug. Donepezil is very effective inhibitor with high affinity to AChE. It interacts with the aromatic residue rings of the PAS and gorge of the enzyme by forming variety bonds [62]. Low selectivity for BChE is a consequence of the larger valley on BChE, and consequently, donepezil is simultaneously linked to peripheral anionic and catalytic sites [65, 69, 79].

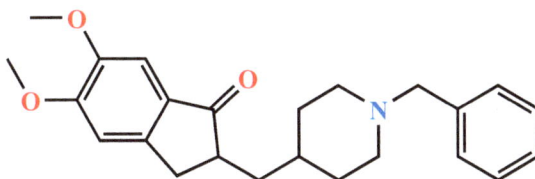

Fig. (10). Structure of Donezepil.

Galantamine

Several cholinesterase inhibitors are derived from natural products. Galantamine (Fig. **11**), one of these was isolated from the snowdrop plant (Galanthus spp.) [80, 81]. Galantamine is a competitive and reversible cholinesterase inhibitor. It is thought that galantamine enhances the cholinergic function by increasing acetylcholine concentration in the brain. The three-dimensional structure of the galantamine-(AChE) complex was determined by X-ray crystallography in 1999. There is no evidence that galantamine alters the course of the underlying dementia process. It was also shown that galantamine had effect on modulating nicotinic cholinergic receptors to increase the acetylcholine release [82].

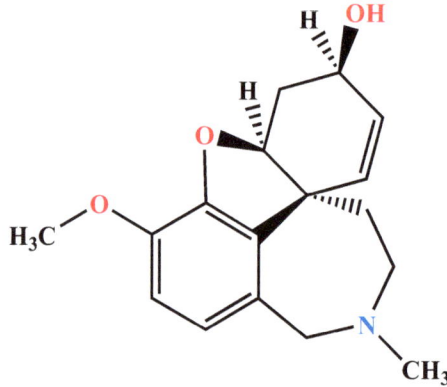

Fig. (11). Structure of Galantamine.

Huperzine A

Huperzine A Fig. (**12**) is an alkaloid which is isolated from Huperziaserrata and able to cross the BBB. Huperzine A, which inhibits AChE by binding to active gorge of enzyme, is a selective and reversible inhibitor. In a phase II study of huperzine A, although it was observed that patients did not have a significant benefit in cognitive activity with mild to moderate AD at 200 µg dose of drug as twice daily. However, it has been shown that further research is necessary to determine the effectiveness of current studies due to poor methodological quality. Huperzine A has also been reported to exhibit neuroprotective action against reducing oxidative stress and oxidative stress which is induced by β-amyloid [83 - 85].

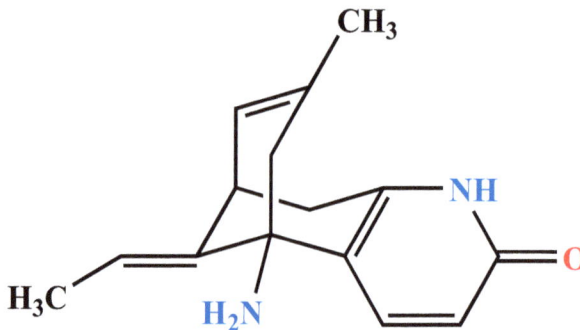

Fig. (12). Structure of Huperzine A.

Metriphonate

Metriphonate Fig. (**13**) is a long-acting and irreversible cholinesterase inhibitor which is originally used in the treatment of schistosomiasis. Short-term use has reduced risk of side effects, but long-term use has resulted in myasthenic crisis-like respiratory paralysis and neuromuscular transmission dysfunction. For this reason, the FDA application was discontinued and clinical trials were cut off during phase III. However, the compound has demonstrated a robust therapeutic effect on ADAS-cog (Alzheimer's Disease Assessment Scale-cognitive subscale) and other measures. Nevertheless, until the relationship between metriphonate and neuromuscular dysfunction is investigated further, metriphonate is no longer an option for AD therapy [86].

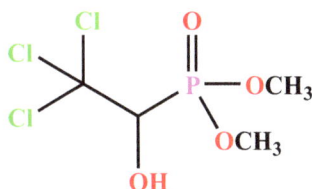

Fig. (13). Structure of metriphonate.

Recent Developed Cholinesterase Inhibitors

Currently, several FDA-approved medicines are available commercially for the treatment of Alzheimer's disease (AD). Among these, tacrine, donepezil, rivastigmine and galantamine are cholinesterase inhibitors. They can only offer palliative treatment for the disease, but the side effects of these drugs continue to be a source of concern. Obviously, the search for more effective drugs for the treatment of AD is one of the most dominant pharmacological targets and many other drugs are being tested in clinical trials, or both in the academic field and in the industry worldwide [9, 67, 75]. While the targeted pharmacological effects are achieved with these synthetic analogues which are in the developmental stage, at the same time hepatotoxicity and known gastrointestinal side effects are prevented. In synthetically developed analogues, there are risks such as inability to obtain derivatives with potent and BBB permeability properties which present in naturally derived cholinesterase inhibitors or obtain compounds with unexpected pharmacological properties [87, 88].

From these considerations, it has been reported that compounds containing the pyridazinone ring as the main structure exhibit AChE and BChE inhibitor activity. The activities of the compounds obtained in a study which 6- [(substituted-phenylpiperazinyl)-3 (2*H*)-pyridazinon-2-yl propionate and 6-[(substitute-

-phenylpiperazin-3(2*H*)-pyridazinon-2-yl propionohydrazide derivatives were synthesized as AChE / BChE inhibitor were compared to the galantamine, currently used in the treatment of Alzheimer's disease and ethyl 6-[4--3-trifluoromethylphenyl)-piperazin]-3(2*H*)-pyridazinon-2-yl propionate (Fig. **14**) was found to be the most active compound in inhibition of both the (AChE) and (BChE) enzymes [89].

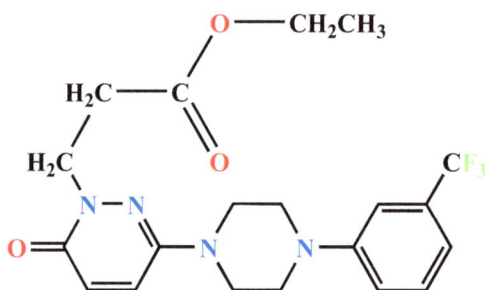

Fig. (14). Structure of ethyl 6-[4-(3-trifluoromethylphenyl)-piperazine]-3(2*H*)-pyridazinon-2-yl propionate.

The anticholinesterase activities of the compounds were evaluated by Ellman's method at concentrations of 0.05 mM, 0.1 mM, 0.2 mM, 0.25 mM and 0.5 mM in similar studies which 6-substituent-3(2*H*)-pyridazinon-2-acetyl-2- (substituted / nonsubstitued-benzene)hydrazone and N'-[(4-substituted-phenyl)sulfonyl-2-[4-(substituted-phenyl) 2*H*) -pyridazinon-2-yl acetohydrazide-derivatives were synthesized. It was found that 6-(3-chlorophenyl)-3-2-[4-(2- fluoro-phenyl--piperazin]-3(2*H*)-2-acetyl-2-(3-methoxy-4-hydroxybenzal)hydrazone and N'-[(--trifluoro-methyl-phenyl)sulphonyl]-2-[4-(2-fluoro-phenyl)-piperazine]-3 (2*H*)-pyridazinon-2-yl acetohydrazide (Fig. **15**) were the most active compounds and it was reported that N'-[(4-tri-fluoro-methyl-phenyl) sulfonyl]-2-[4-(2-fl-oro-phenyl)-piperazin]-3(2*H*)-pyridazinon-2-yl aceto-hydrazide has higher activity than galantamine [64, 90, 91].

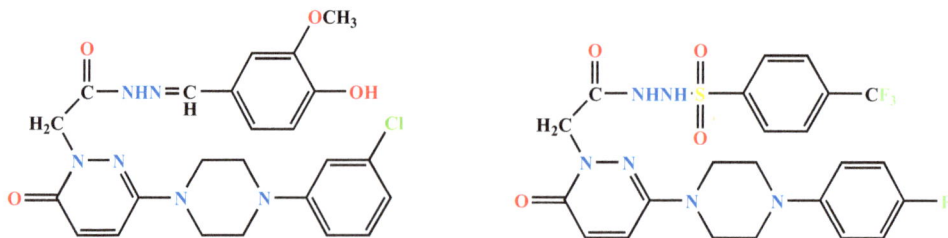

Fig. (15). Structures of 6- (3-chloro-phenyl)-3 (2*H*)-pyridazinon-2- acetyl-2- (3-meth-oxy-4- hydro- xy-benzal)hydra-zone and N'-[(4-tri-fluoro-methyl-phenyl) sulphonyl]-2- [4-(2-fluoro-phenyl)-piperazine]-3 (2*H*)-pyridazinon-2-yl aceto-hydrazide.

In another study made by *Xing et al.,* 2,6-disubstitued pyridazinone derivatives were synthesized and investigated the *in vitro* AChE / BChE inhibitor effects by using molecular modeling studies. As a result, the pyridazinone compound containing 6-ortho-tolylamino and N-ethyl-N-isopropylacetamide substituted piperidine groups Fig. (**16**) showed high AChE inhibitor activity and found to be AChE / BChE selective *in vitro*. Molecular modeling studies have also shown that the compound inhibited enzymes by binding to both the catalytic active site (CAS) and the peripheral anionic site (PAS) [92].

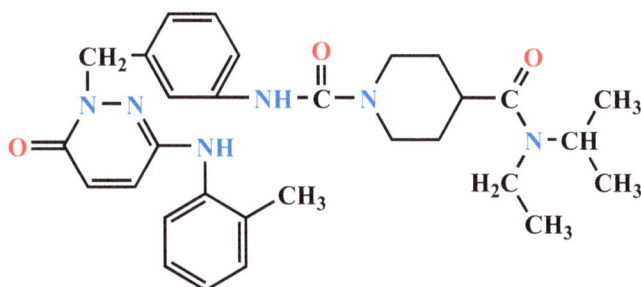

Fig. (16).   Structures of N^4-ethyl-N^4-iso-prop-yl-N^1-(3-((6-oxo-3-(o-tolyl-amino) pyridazin-1 (6*H*)-yl) methyl) phenyl) piperidine-1,4-di-carbox-amide.

In another study with pyridazinone derivatives, a series of 6-substituted-3(2*H*)-pyridazinone-2-acetyl-2-(p-substituted-benzal-hydrazone) derivatives were designed and synthesized as potent dual inhibitors of AChE and BChE. Most active AChE inhibitor compound showed 75.52% AChE inhibition and the most active BChE inhibitor compound showed 67.16% BChE inhibition at 100 μg/ml. 6- [4- (3,4-Di-choloro-phenyl) piperazine-1-yl]-3(2*H*)pyridazin-one-2-acetyl-2-(3-methyl-benzal)hydrazon was determined as a potent dual cholinesterase inhibitor. Through molecular docking studies it was theoretically evaluated the inhibition mechanism of 6-[4-(3,4-di-choloro-phenyl) piperazine-1-yl]-3 (2H)pyridazinone-2-acetyl-2-(3-methyl-benzal) hydrazon (Fig. **17**) for both enzymes in comparison with previously reported derivatives which differ in inhibition potency and tried to get insights into the factors that affect their receptor affinity in molecular level. It was found that the docking poses of the compounds were similar to donepezil and substitutions on phenyl-piperazine moiety might alter ligand efficacy [65].

Several strategies have been researched to develop disease modifying therapies (DMTs) since the receipt of tacrine as an initial AChEI for AD treatment in 1993. Nevertheless, in spite of the potentially huge commercial awards and numerous drug candidates entrance to clinical trials, effective treatment is not yet available [9]. As a result, it has also increased the importance of current and future academic studies in this area. Clinical failure is partly attributed to the complexity of this pathology which involves the multi-faceted interaction of various factors not exactly realized. This investigation has also established the principle for the current interest in Multitarget Directed Ligands (MTDLs), which a heterogeneous class of compounds designed to tackle multiple pathological events at the same time [93]. Therefore, the fact that AD is a multifactorial disease has revealed the single-molecule approach towards multiple targets. For this purpose, hybrid molecules with enhanced BBB permeability and targeting multiple receptors are also prepared.

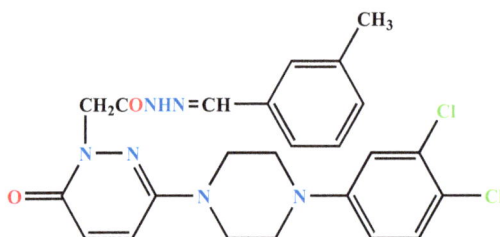

Fig. (17). Structures of 6- [4- (3,4-di-choloro-phenyl) piperazine-1-yl]-3(2*H*) pyridazinone-2- acetyl-2- (3-methyl-benzal) hydrazon.

Based on these strategies, a number of MTDLs have been developed as a result of the modification of commercial drugs and active scaffolds [94 - 96]. It has been shown that tacrine-based multi-target derivatives among these was able to shot several important targets in AD and to demonstrate versatile benefits both *in vitro* and *in vivo*. In fact, the tacrine skeletal structure is an optimal beginning to obtain highly active MTDLs. Considering all of these, a new MTDL class called tacripyrins, which was formed by the coexistence of tacrine and nimodipine as reference molecules with the profiles of (AChE) and calcium antagonism, respectively, was designed for the treatment of AD in recent years. The highest activity of the tacripyrin derivatives was (±) -p-methoxytacripyrin (Fig. **18**), which provides high selectivity for AchE. This compound also provided

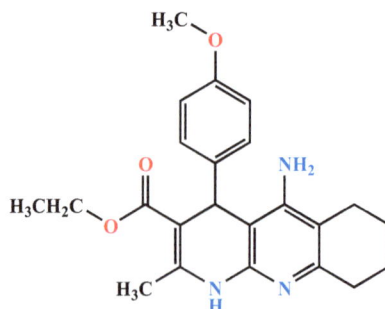

Fig. (18). Structure of (±)-p-methoxytacripyrin.

calcium uptake after potassium stimulation in SHSY5 cells, poor inhibition of the progregation effect of hAChE on the αβ peptide, and moderate inhibition of the self-aggregation of αβ [97].

Interest in calcium channel blockers is based on the fact that calcium levels regulate neuronal plasticity underlying learning, memory and neuronal survival. It is supposed that the intracellular calcium homeostasis disorder has an effect on both damages in neurons and loss of them [98].

The compound called memoquin (Fig. **19**) was also improved based on the polypharmacology model. Since it has multitargeted structure and marked antioxidant activity with *in vitro* neuroprotective effects [99], memoquin has arisen as a promising candidate. Some of other bifunctional ligands which have double gorge-spanning binding have been emerged as AChE and monoamine oxidase inhibitors, AChE inhibitors and antiinflammatory agents, or AChE inhibitors and reactive oxygen species (ROS) [100, 101]. Memoquin, a polyamine compound bearing a quinone structure, is the result of the MTDL design strategy. It is thought as a quite potent compound capable of inhibiting the enzyme AChE, β-amyloid aggregation, β-secretase, tauposporilation and free radical formation [100]. Memoquin is composed of the polyamineamidecaproctamine which is a familiar AChE inhibitor and the muscarinic M_2 autoreceptor antagonist which a synthetic coenzyme Q (CoQ) derivative. It is noticed that Coenzyme Q ameliorate cognitive and behavioral disorders in a clinical AD trial. The decline in cholinergic signaling has a crucial impact in mediating cognitive and behavioral disorders in the individuals suffer from AD. Administration of non-selective muscarinic antagonists may cause or intensify cognitive impairment in animals and people have AD. This suggests that the cognition may be modulated by muscarinic receptors directly [102]. Memoquin is the strongest inhibitor of human AChE. Compared with donepezil, it is ten times more potent inhibitor. It has been reported that Memoquin can modulate various AD-related mechanisms including preventing of AChE dual binding and β-amyloid aggregation inhibition, β-secretase inhibition, significant neuroprotective effects associated with antioxidant activity, and reduction of tau hyperphosphorylation [103].

Fig. (19). Structure of Memoquin.

Donepezil is an FDA-approved drug to palliate AD and a strong and selective inhibition of AChE [62]. The *in vivo* suitable pharmacokinetic profile and potent AChE inhibition of donepezil has contributed to making important research for improvement of donepezil analogues as potential therapeutics of AD [62, 69, 79]. AP2238 (Fig. **20**) is the firstly announced molecule which binds to both anionic sites of AChE. Although its potency against AChE is similar to donepezil, the beta-amyloid aggregation is higher [104].

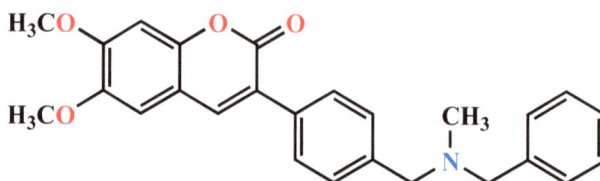

Fig. (20). Structure of AP2238.

Rizzo et al. reportedin their work on a series of hybrids developed from donepezil and AP2238 that the indanone nucleus from donepezil is attached to the phenyl-N-methylbenzylamino moiety from AP2238, and these hybrids show high anticholinesterase activity [105].

Özer et al. designed a new series of donepezil like compounds which are both predicted to inhibit cholinesterases and amyloid aggregation. This class of bifunctional compounds is a combination of a benzyl ring that is substituted with a different substituent attached to the 4-benzylpiperidine/piperazine and an N-acylhydrazone moiety. The benzene ring contains one or two methoxy/ethoxy groups (Fig. **21**). As a result of Ellman's test, these molecules were found to be moderate and non-selective inhibitors of both AChE and BuChE and that the IC_{50} values in micromolar range are 53.1-88.5 M for hAChE and 48.8-98.8 M for EqBChE [106].

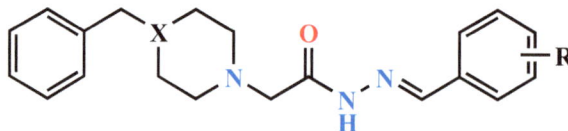

Fig. (21). Structure of 4-benzyl piperidin/piperazine derivatives.

A new set of benzo[e] [1, 2, 4]triazin-7(1H)-one and [1, 2, 4]-triazino [5, 6,1j,k]carbazol-6-onederivatives (Fig. **22**) have been identified as an AChE / BChE and amyloid aggregation inhibitor compound. These designed compounds carry the various alkylamine or arylalkylamine substituents on the 5th carbon of triazafluoranthenone and on the 6th carbon of benzotriazinone. The majority of the compounds exhibited an $A\beta_{1-40}$ anti-aggregation activity with IC_{50} values in the micro or submicromolar range ($A\beta_{1-40}$ IC_{50} = 0.37-65 M). The results of the Ellman analysis show that the molecules used in the experiment are low or medium levelinhibitors for AChE and BChE. The compound having the octamethylene chain from the triazafluoranthenone derivative compounds showed inhibition of $A\beta_{1-40}$ aggregations with a value of 1.4 M IC_{50}. This compound has been found to have a selective, very potent BChE inhibitor potency at IC_{50} = 25 nM and has been shown to be promising for further development as a potent anti-AD drug [107].

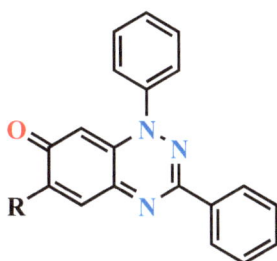

Fig. (22). Structure of benzo[e] [1, 2, 4]triazin-7(1H)-one and [1, 2, 4]-triazino [5, 6,1j,k]carbazol--onderivatives.

Ekiz et al. synthesized bromo-deneoquino-line derivative compounds and determined their potential of acetyl-cholinesterase (AChE), butyryl-cholinesterase (BChE) and human carbonic anhydrase cytosolic (hCA I and II) enzymes inhibition to investigate structure-activity relationship (SAR). Mono-pheny--indenoquino-lines are well-inhibited AChE and BChE enzymes at IC_{50} values of 37-57nM and 84-93 nM when compared to the starting materials and the reference compounds galanthamine and tacrine. However, these newly found compounds which are arylated indenoquino-line-based derivatives (Fig. **23**) were

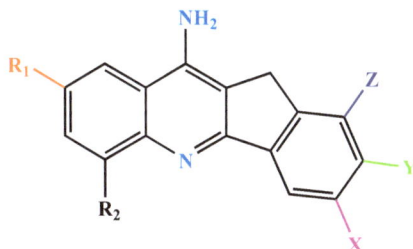

R_1: H, Br, Phenyl ; R_2: H, Br ; X: H, Br, Phenyl ; Y: H, Br, Phenyl ; Z: H, Br, Phenyl

Fig. (23). Arylated indenoquinoline amine derivatives.

found to be powerful inhibitors of enzymes with Ki values in the range of 37 ± 2.04 to 88640 ± 1990 nM for AChE, 267.58 ± 98.05 to 1568.16 ± 438.67 nM for hCA II at 120.94 ± 37.06 to 1150.95 for hCA, and 84 ± 3.86 to 144120±2910 nM. In conclusion, mono-phenyl-indenoquino-lines are potentially promising drugs used in the treatment of AD and 3,8-di-bromoenocinoline amine may be new hCA I and hCA II enzyme inhibitors [108].

Camps et al. have constructed a new donepezil-tacrine hybrid molecule sequence which interact at the same time as the peripheral, active, and midgorge binding regions of AChE. The synthesized compounds were investigated for whether they are capable to inhibit AChE-induced BChE, AChE and beta-amyloid aggregation. The compounds consist of a combination with 5,6-di methoxy-2- [(4-piperidinyl)methyl-1-idenone moiety of donepezil and the 6-chlorotacrine moiety (Fig. **24**). It has confirmed that all these new hybrids are really influential inhibitors of hAChE; also, these compounds showed crucial interference/ inhibition of beta amyloid aggregation and it was revealed that they were more influential comparing with the parent compounds which produced the hybrid molecules [62].

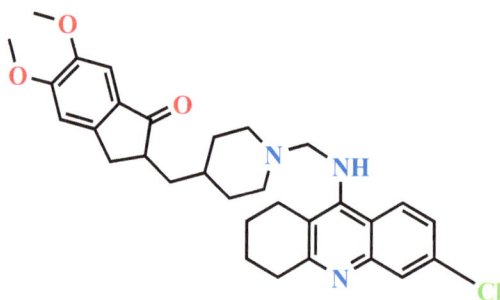

Fig. (24). Structure of donepezil-tacrine hybrid.

Multipotent approach has given tacrine-melatonin hybrids, AChE and MAO inhibitors or serotonin carriers, strong cholinesterase inhibitors with antioxidant and neuroprotective properties, galantamine-tacrine hybrids bound to cholinesterases and M_2 muscarinic receptors, NO donors, tacrine hybrids focusing on fluorescent tacrine and cumarine hybrids to AD therapy as hepatoprotective drugs [109, 110].

Both tacrine and PBT2 which is an 8-hydroxyquinoline derivative are known to inhibit cholinesterases and reduce beta-amyloid concentrations. On the basis of these compounds, *Fernandez-Bachiller et al.* designed and synthesized novel tacrine-8-hydroxyquinoline hybrids (Fig. **25**) in order to be used as possibly curative drugs in AD therapy and evaluated their activities. They found that these hybrids were more effective against AChE and BChE than tacrine. It was also reported that compounds exhibit toxicity at low cell levels and able to penetrate the CNS in experiments performed in the BBB model *in vitro*. The drug also exhibited antioxidant and copper complexing properties. Hybrid compounds with all these properties were thought to be potential therapeutic drugs *in vivo* [111].

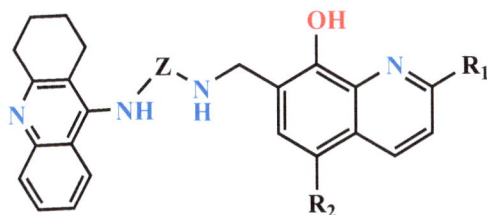

Fig. (25). Structure of tacrine-8-hydroxyquinoline hybrids.

Tacrine-cinnamic acid hybrids (Fig. **26**) also exhibited AChE and BChE inhibition with IC_{50} in the nanomolar range. SAR studies showed that a 6-carbon alkyl chain must be present as a linker for optimum activity in the molecule. Studies on these compounds showed that if a large/huge group like the extension of this linker or the benzyl group is added cholinesterases inhibition is reduced. Additionally, the results of the studies have shown that a group of these compounds inhibit spontaneously induced Aβ1-42 aggregation. Morris Water Labe test, an *in vivo* evaluation showed that cognitive impairment due to skopolamine was reduced and it was shown to be safe in assessing hepatotoxicity [112].

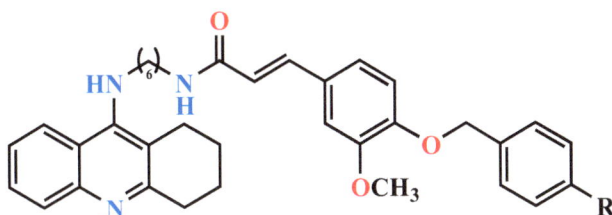

Fig. (26). Structure of tacrine-cinnamic acid hybrids.

Biochemical proofs argue that the activity of glutamatergic neurons is also dysfunctional in AD and that the cholinergic system and their co-dysfunction in AD pathology are important. For this reason, drug combinations concurrently targeting cholinergic and glutamatergic systems are suggested to be used in the treatment of AD as a recent standard [113]. The NMDA receptor has a crucial impact on the brain activities like memory and learning. The receptor is activated with two agonists, glycine and (S)–glutamate, which must be bound simultaneously. Clinically, over-activation of the NMDA receptor with glutamate in redundance has been shown to control uncontrolled Ca^{2+} ions in the neuron ending up excitotoxicity. Galantamine is known as a compound which both has a neuroprotective effect against glutamate toxicity through the activation of nicotinic ACh receptors, and is a significant inhibitor for AChE. Since the stimulation of this neuroprotective effect of the nicotinic acetylcholine receptor (nAChR), smokers are known to be less impressionable to Parkinson's disease [114]. Thus, galantamine and memantine work synergistically on the same excitotoxic cascade to ensure more synergical neuroprotective effect. From this hypothesis, *Lopes et al.* demonstrated that the neuroprotective impact of galantamine against NMDA-mediated neurotoxicity in primary cortical neurons contains principal nicotinic receptors [115]. Furthermore, they have indicated in their studies that subacute concentrations of galantamine and memantine may cause combined neuroprotection. As a result of these studies, a novel design method is presented to develop a MTDL with maximum efficiency and a novel group of hybrid molecules which combines the pharmacological activities of galantamine (AChE inhibitor) and memantine (NMDA receptor antagonist) which are two drugs used in the market is designed. New hybrid compounds are designed after double binding approach by linking two drug molecules from a polymethylene linker of variable length (Fig. **27**). Compounds bearing a hexamethylene spacer were found to be the most promising ones in the group. Consistent with molecular modeling studies, compounds were designed in AChE to provide ideal gap in order to interact with both PAS and CAS [116].

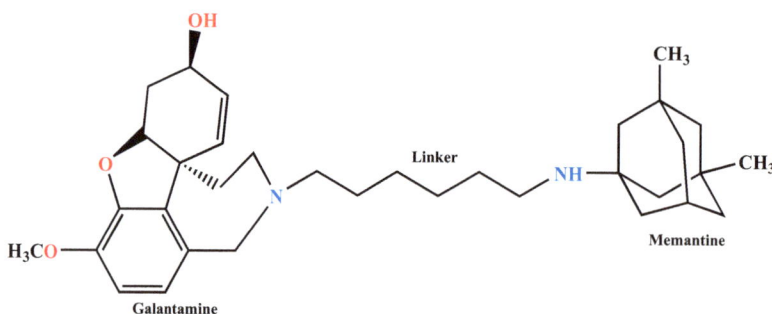

Fig. (27). Structure of galantamine-memantine dual-binding hybrid.

In a previously reported study, it was also found that chalcone and coumarin fragments were required for anticholinesterase activity and anti-aggregation properties of Aβ. For this purpose, chalcone and coumarin fragments were combined with various amine moieties to get double acting compounds to develop a new series of multifunctional compounds (Fig. **28**) [117].

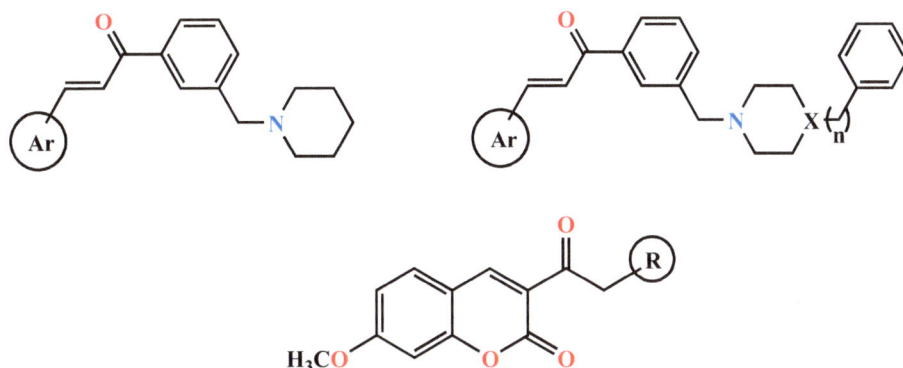

Fig. (28).   Structure of chalcone and coumarin hybrid.

All compounds tested were comparable or less potent to cholinesterases than galantamine which was the reference compound. The IC_{50} value of the most potent AChE inhibitor compound was 1.76 M. Chalcone derivatives with the strongest compound having an IC_{50} value of 8.27 M showed higher activity than inhibitor activity of coumarin against BChE. During the tests, all of the compounds were controlled whether they are capable to inhibit Aβ fibril formation in the thioflavin T fluorescence assay or not, and it is recognized that both chalcone and coumarin derivatives are moderately inhibitors of spontaneously induced Aβ aggregation [117, 118].

In addition to all these multi-target compounds and hybrids, natural antioxidants also help prevent AD and cell aging processes by protecting cells against oxidative damage. For this reason, in order to avoid AD and retain the cognitive activity, some antioxidants may be used [119]. Oxidative stress, free radicals, bio-metal toxicity and abnormal reactions promote the AD pathology.

Flavonoids are one of the the biggest molecules in plants and they have a broad range of biological activity [120]. These compounds which are polyphenolic are advantageous in the treatment of AD as a consequence of having strong antioxidant activity Oxidative stress is a consequence of the inequality in the ability of the biological system to produce and detoxify reactive oxygen species (ROS). The higher levels of ROS in the body may result in a higher rate of production or a lower performance in antioxidant enzyme activity. A decrease in antioxidant enzyme activity is detected in AD patients [121, 122]. As well as the formation of senile plaques, intense oxidative stress is also monitored in AD brains. Oxidative stress plays an important role in the pathogenesis of AD, in terms of the damage to vital cellular components, proteins, lipids and nucleic acids [123]. Protein carbonyl and 3-nitrotyrosine levels, which are indicators of oxidative damage to DNA and RNA, are high in AD brains [124]. Lipid peroxidation products are higher in most of the brains of patients with mild cognitive impairment or AD. The decrease in the activity of antioxidant enzymes such as superoxide dismutase (SOD) and catalase, as well as the occurrence of free radical damage, is also related to the progression of AD, and at the same time oxidative stress is associated with synaptic loss [125]. The oxidative stress hypothesis of AD suggests that ROS plays an important role in the onset and progression of the disease due to damage to neurons. However, the antioxidant activity of ROS is associated with its capability to clean free radicals by chelating redox-active transition metals, increasing reduced glutathione levels, and down-regulating inflammatory duration [126, 127]. In addition, some of the important features of these reactive oxygen species are that their absorption and distribution processes are quite easy, and they are converted to dihydrolipoic acid in a variety of tissues inclusive of the ones in the brain. Dihydrolipoic acid is also a powerful antioxidant. The disulfide cyclic moiety of lipoic acid potentially interact with the peripheral active site of AChE and inhibit Aβ aggregation started by AchE.

In a study based on these properties lipoic acid was coupled with tacrine to give the lipocrine molecule (Fig. **29**) as a new compound for MTDL [128].

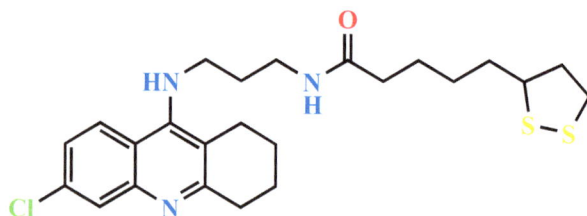

Fig. (29). Structure of lipocrine.

Lipocrine is a very potent inhibitor of nanomolar affinity to AChE. Kinetic studies also confirm the interaction of lipocrine with the peripheral region of AChE. The compound was found to be significantly higher than lipoic acid alone, in inhibiting Aβ aggregation induced by AChE and protecting cells against ROS formation. In another study, a series of lipoic acid compounds were prepared by combining lipoic acid, rivastigmin, and memoquin, and the synthesized compounds were compared for antioxidant, anticholinesterase and anti-aggregation activities. One of the compounds which was a hybrid of lipoic acid and memoquin was selected as the most active compound with strong AChE and moderate BChE inhibition and dose-dependent antioxidant activity. Nonetheless, AChE-induced Aβ anti-aggregation activity of the compound is found to be poor [99, 128].

CONCLUDING REMARKS

All these researches prove the ongoing value of cholinergic drugs as a crucial pharmacologic approach in Alzheimer's disease, especially when it is considered further combined therapies tackling the progression of the disease as well as the symptoms. Taking into account the multitude of therapeutic targets which was recommended for AD pharmacotherapy, the improvement of multiple target-directed ligands (MTDLs) has become a modern approach. The multifactorial nature of AD pathogenesis suggests that MTDLs, a broad spectrum of activity, must be designed. Double and very effective anti-AD drug candidates are achieved by binding structurally active parts that interact with secret targets. MTDLs, designed for this purpose as an important group of compounds in AD therapy, are cholinesterase inhibitors with additional features such as antioxidant activity, metal complexing properties, calcium channel antagonistic activity, excitotoxicity induced by anti-glutamate, as well as BACE1 inhibitor activity.

CONSENT FOR PUBLICATION

Not applicable.

CONFLICT OF INTEREST

The authors declare no conflict of interest, financial or otherwise.

ACKNOWLEDGEMENT

Declare None.

REFERENCES

[1]   Guyton AC, Hall JE. Medical Physiology. 10th ed., Philadelphia, Pennsylvania 2000.

[2]   Kumar V, Cotran RS, Robbins SL. Pathologic Basis of Disease. 7th ed., Philadelphia: Saunders 2004.

[3]   Brookmeyer R, Abdalla N, Kawas CH, Corrada MM. Forecasting the prevalence of preclinical and clinical Alzheimer's disease in the United States. Alzheimers Dement 2018; 14(2): 121-9.
      [http://dx.doi.org/10.1016/j.jalz.2017.10.009]

[4]   Rashid U, Ansari FL. Challenges in Designing Therapeutic Agents for Treating Alzheimer's Disease-from Serendipity to Rationality. Elsevier BV 2014; pp. 40-141.

[5]   Cummings JL, Morstorf T, Zhong K. Alzheimer's disease drugdevelopment pipeline: few candidates, frequent failures. Alzheimers Res Ther 2007; 6: 1-7.

[6]   Ballard C, Gauthier S, Corbett A, Brayne C, Aarsland D, Jones E. Alzheimer's disease. Lancett 2011; 377: 1019-31.
      [http://dx.doi.org/10.1016/S0140-6736(10)61349-9]

[7]   Ferri CP, Prince M, Brayne C, Brodaty H, Fratiglioni L, Ganguli M, *et al.* Global Prevalence of Dementia: A Delphi Consensus Study. Lancett 2005; 366: 2112-7.
      [http://dx.doi.org/10.1016/S0140-6736(05)67889-0]

[8]   Mendiola-Precoma J, Berumen LC, Padilla K, Garcia-Alcocer G. Therapies for Prevention and Treatment of Alzheimer's Disease. BioMed Res Int 2016.
        [http://dx.doi.org/10.1155/2016/2589276]

[9]   Anand P, Singh P. A review on cholinesterase inhibitors for Alzheimer's disease. Arch Pharm Res 2013; 36: 375-99.
        [http://dx.doi.org/10.1007/s12272-013-0036-3]

[10]   Tampi RR, Tampi DJ, Ghori AK. AChE inhibitors for delirium in older adults. Am J Alzheimer Dis 2016; 31(4): 305-10.
        [http://dx.doi.org/10.1177/1533317515619034]

[11]   Geldenhuys WJ, Darvesh AS. Pharmacotherapy of Alzheimer's disease: current and future trends. Expert Rev Neurother 2015; 15: 1-, 3-5.
        [http://dx.doi.org/10.1586/14737175.2015.990884]

[12]   Hampel H, Mesulam MM, Cuello AC, *et al.* The cholinergic system in the pathophysiology and treatment of Alzheimer's disease. Brain 2018; 141: 1917-33.
        [http://dx.doi.org/10.1093/brain/awy132]

[13]   Belluti F, Bartolini M, Bottegoni G, *et al.* Benzophenone-based derivatives: a novel series of potent and selective dual inhibitors of (AChE) and AChE-induced betaamyloid aggregation. Eur J Med Chem 2011; 46: 1682-93.
        [http://dx.doi.org/10.1016/j.ejmech.2011.02.019]

[14]   García-Font N, Hayour H, Belfaitah A, *et al.* Potent anticholinesterasic and neuroprotective pyranotacrines as inhibitors of beta-amyloid aggregation, oxidative stress and tau-phosphorylation for Alzheimer's disease. Eur J Med Chem 2016; 118: 178-92.
        [http://dx.doi.org/10.1016/j.ejmech.2016.04.023]

[15]   Terry AV. The Cholinergic Hypothesis of Age and Alzheimer's Disease-Related Cognitive Deficits: Recent Challenges and Their Implications for Novel Drug Development. J Pharmacol Exp Ther 2003.
        [http://dx.doi.org/10.1124/jpet.102.041616]

[16]   Maccioni RB, Farias GA, Morales I, Navarrete LP. The revitalized tau hypothesis on Alzheimer's disease. Arch Med Res 2010; 41(3): 226-31.
        [http://dx.doi.org/10.1016/j.arcmed.2010.03.007]

[17]   Mullane K, Williams M. Alzheimer's therapeutics: Continued clinical failures question the validity of the amyloid hypothesis—But what lies beyond? Biochem Pharmacol 2013; 85: 289-305.
        [http://dx.doi.org/10.1016/j.bcp.2012.11.014]

[18]   Hampel H, Prvulovic D, Teipel S, *et al.* The future of Alzheimer's disease: The next 10 years. Prog Neurobiol 2011; 95: 718-28.
        [http://dx.doi.org/10.1016/j.pneurobio.2011.11.008]

[19]   Perry EK. The cholinergic hypothesis-ten years on. Br Med Bull 1986; 42: 63-9.
        [http://dx.doi.org/10.1093/oxfordjournals.bmb.a072100]

[20]   Swerdlow RH. Alzheimer's disease pathologic cascades: Who comes first, what drives what. Neurotox Res 2012; 22: 182-94.
        [http://dx.doi.org/10.1007/s12640-011-9272-9]

[21]   Fotiou D, Kaltsatou A, Tsiptsios D, Nakou M. Evaluation of the cholinergic hypothesis in Alzheimer's disease with neuropsychological methods. Aging Clin Exp Res 2015; 27: 727-33.
        [http://dx.doi.org/10.1007/s40520-015-0321-8]

[22]   Farlow MR. Pharmacological treatment of cognition in Alzheimer's disease. Neurol 1998; 51: 36-44.
        [http://dx.doi.org/10.1212/WNL.51.1_Suppl_1.S36]

[23]   Bacalhau P, San Juan AA, Marques CS, *et al.* New cholinesterase inhibitors for Alzheimer's disease: Structure Activity Studies (SARs) and molecular docking of isoquinolone and azepanone derivatives.

Bioorg Chem 2016; 67: 1-8.
[http://dx.doi.org/10.1016/j.bioorg.2016.05.004]

[24]   Kasa P, Rakonczay Z, Gulya K. The cholinergic system in Alzheimer's disease. Prog Neurobiol 1997; 52: 511-53.
[http://dx.doi.org/10.1016/S0301-0082(97)00028-2]

[25]   Atri A, Frölich L, Ballard C, *et al.* Cummings. Effect of Idalopirdine as Adjunct to Cholinesterase Inhibitors on Change in Cognition in Patients With Alzheimer Disease Three Randomized Clinical Trials. JAMA 2018; 319(2): 130-42.
[http://dx.doi.org/10.1001/jama.2017.20373]

[26]   Blanco-Silvente L, Castells X, Saez M, *et al.* Discontinuation, Efficacy, and Safety of Cholinesterase Inhibitors for Alzheimer's Disease: a Meta-Analysis and Meta-Regression of 43 Randomized Clinical Trials Enrolling 16 106 Patients. Int J Neuropsychopharmacol 2017; 20(7): 519-28.
[http://dx.doi.org/10.1093/ijnp/pyx012]

[27]   Posadas I, López-Hernández B, Ceña V. Nicotinic receptors in neurodegeneration. Curr Neuropharmacol 2013; 11(3): 298-314.
[http://dx.doi.org/10.2174/1570159X11311030005]

[28]   Davis DG, Schmitt FA, Wekstein DR, Markesbery WR. Alzheimer neuropathologic alterations in aged cognitively normal subjects. J Neuropathol Exp Neurol 1999; 58: 376-88.
[http://dx.doi.org/10.1097/00005072-199904000-00008]

[29]   DeKosky ST, Ikonomovic MD, Styren S, *et al.* Upregulation of choline acetyltransferase activity in hippocampus and frontal cortex of elderly subjects with mild cognitive impairment. Ann Neurol 2002; 51: 145-55.
[http://dx.doi.org/10.1002/ana.10069]

[30]   Gilmor ML, Erickson JD, Varoqui H, *et al.* Preservation of nucleus basalis neurons containing choline acetyltransferase and the vesicular acetylcholine transporter in the elderly with mild cognitive impairment and early Alzheimer's disease. J Comp Neurol 1999; 411: 693-704.
[http://dx.doi.org/10.1002/(SICI)1096-9861(19990906)411:4<693::AID-CNE13>3.0.CO;2-D]

[31]   Koeppe RA, Frey KA, Snyder SE, Meyer P, Kilbourn MR, Kuhl DE. Kinetic Modeling of N-[11CJMethylpiperidin-4-yl Propionate: Alternatives for Analysis of an Irreversible Positron Emission Tomography Tracer for Measurement of (AChE) Activity in Human Brain. J Cereb Blood Flow Metab 1999; 19: 1150-1.
[http://dx.doi.org/10.1097/00004647-199910000-00012]

[32]   Kadir A, Darreh-Shori T, Almkvist O, *et al.* PET imaging of the *in vivo* brain (AChE) activity and nicotine binding in galantaminetreated patients with AD. Neurobiol Aging 2008; 29: 1204-17.
[http://dx.doi.org/10.1016/j.neurobiolaging.2007.02.020]

[33]   Pietrzak K, Czarnecka K, Mikiciuk-Olasik E, Szymski P. New Perspectives of Alzheimer Disease Diagnosis – the Most Popular and Future Methods. Med Chem 2018; 14: 34-43.
[http://dx.doi.org/10.2174/1573406413666171002120847]

[34]   Kuhl DE, Minoshima S, Fessler JA, *et al. in vivo* Mapping of Cholmergic Termina in Normal &ng, Alzheimer's Disease, and Parlunson's Disease. Ann Neurol 1996; 40: 3399-410.
[http://dx.doi.org/10.1002/ana.410400309]

[35]   Gibson GE, Peterson C, Jenden DJ. Brain acetylcholine synthesis declines with senescence. AAAS 1986; 213(4508): 674-6.

[36]   Schliebs R, Arendt T. The cholinergic system in aging and neuronal degeneration. Behav Brain Res 2011; 221(2): 555-63.
[http://dx.doi.org/10.1016/j.bbr.2010.11.058]

[37]   Mattson MP, Magnus T. Ageing and neuronal vulnerability. Nat Rev Neurosci 2006; 7: 278-94.
[http://dx.doi.org/10.1038/nrn1886]

[38] Sarter M, Turchi J. Age- and Dementia-Associated Impairments in Divided Attention: Psychological Constructs, Animal Models, and Underlying Neuronal Mechanisms. Dement Geriatr Cogn Disord 2002; 13: 46-58.
[http://dx.doi.org/10.1159/000048633]

[39] Hasselmo ME. The role of acetylcholine in learning and memory. Curr Opin Neurobiol 2006; 16(6): 710-5.
[http://dx.doi.org/10.1016/j.conb.2006.09.002]

[40] Lemke T, Williams DA, Roche VF, Zito SW. Foye's Principles of Medicinal Chemistry. Philadelphia: Lippincott Williams & Wilkins 2008.

[41] Toyohara J, Hashimoto K. α7 Nicotinic Receptor Agonists: Potential Therapeutic Drugs for Treatment of Cognitive Impairments in Schizophrenia and Alzheimer's Disease. Open Med Chem J 2010; 4: 37-56.

[42] Gundisch D. Gundisch D. Nicotinic Acetylcholine Receptors and Imaging. Curr Pharm Des. 2000;6:1143-1157(15)
[http://dx.doi.org/10.2174/1381612003399879]

[43] Zenko D, Hislop JN. Regulation and trafficking of muscarinic acetylcholine receptors. Neuropharmacology 2018; 136: 374-82.
[http://dx.doi.org/10.1016/j.neuropharm.2017.11.017]

[44] Scarr E. Muscarinic receptors: Their roles in disorders of the central nervous system and potential as therapeutic targets. CNS Neurosci Ther 2012; 18: 369-79.
[http://dx.doi.org/10.1111/j.1755-5949.2011.00249.x]

[45] Bentley P, Drive J, Dolan RJ. Cholinergic modulation of cognition: insights from human pharmacological functional neuroimaging. Prog Neurobiol 2011; 94: 360-88.
[http://dx.doi.org/10.1016/j.pneurobio.2011.06.002]

[46] Agatonovic-Kustrin S, Kettle C, Morton DW. A molecular approach in drug development for Alzheimer's disease. Biomed Pharmacother 2018; 106: 553-65.
[http://dx.doi.org/10.1016/j.biopha.2018.06.147]

[47] Colovic MB, Krstic DZ, Lazarevic-Pasti TD, Bondzic AM, Vasic VM. (AChE) inhibitors: pharmacology and toxicology. Curr Neuropharmacol 2013; 11(3): 315-35.
[http://dx.doi.org/10.2174/1570159X11311030006]

[48] Greig NH, Utsuki T, Yu Q, *et al.* A new therapeutic target in Alzheimer's disease treatment: Attention to (BChE). Curr Med Res Opin 2001; 17(3): 159-65.
[http://dx.doi.org/10.1185/03007990152673800]

[49] Begum S, Nizami SS, Mahmood U, Masood S, Iftikhar S, Saied S. In-vitro evaluation and in-silico studies applied on newly synthesized amide derivatives of N-phthaloylglycine as (BChE) inhibitors. Comput Biol Chem 2018; 74: 212-7.
[http://dx.doi.org/10.1016/j.compbiolchem.2018.04.003]

[50] Campanella L, Achilli M, Sammartino MP, Tomassetti M. Butyrylcholine enzyme sensor for determining organophosphorus inhibitors. J Electroanal Chem Interfacial Electrochem 1991; 3211(2): 237-49.
[http://dx.doi.org/10.1016/0022-0728(91)85599-K]

[51] Norel X, Angrisani M, Labat C, *et al.* Degradation of acetylcholine in human airways: Role of (BChE). Br J Pharmacol 1993; 108(4): 914-9.
[http://dx.doi.org/10.1111/j.1476-5381.1993.tb13486.x]

[52] Mesulam MM, Guillozet A, Shaw P. Widely spread (BChE) can hydrolyze acetylcholine in the normal and Alzheimer brain. Neurobiol Dis 2002; 9: 88-93.
[http://dx.doi.org/10.1006/nbdi.2001.0462]

[53]   Xie W, Wilder PJ, Stribley J, *et al.* Knockout of one AChE allele in the mouse. Chem Biol Interact 1999; 119-120(120): 289-99.
[http://dx.doi.org/10.1016/S0009-2797(99)00039-3]

[54]   Chatonnet A, Lockridge O. Comparison of (BChE) and (AChE). Biochem J 1989; 260(3): 625-34.
[http://dx.doi.org/10.1042/bj2600625]

[55]   Zhou Y, Wang S, Zhang Y. Catalytic reaction mechanism of (AChE) determined by born-oppenheimer AB initio QM/MM molecular dynamics simulations. J Phys Chem B 2010; 114: 8817-25.
[http://dx.doi.org/10.1021/jp104258d]

[56]   Dvir H, Silman I, Harel M, Rosenberry TL, Sussmana JL. (AChE): from 3D structure to function. Chem Biol Interact 2010; 187: 10-22.
[http://dx.doi.org/10.1016/j.cbi.2010.01.042]

[57]   Simoni E, Daniele S, Bottegoni G, *et al.* Combining Galantamine and Memantine in Multitargeted, New Chemical Entities Potentially Useful in Alzheimer's Disease. J Med Chem 2012; 55(22): 9708-21.
[http://dx.doi.org/10.1021/jm3009458]

[58]   Nachon F, Carletti E, Ronco C, *et al.* Crystal structures of human cholinesterases in complex with huprine W and tacrine: elements of specificity for anti-Alzheimer's drugs targeting acetyl- and (BChE). Biochem J 2013; 453: 393-9.
[http://dx.doi.org/10.1042/BJ20130013]

[59]   Leon J, Marco-Contelles J. A step further towards multitarget drugs for Alzheimer and neuronal vascular diseases: targeting the cholinergic system, amyloid-β aggregation and Ca2+ dyshomeostatis. J Curr Med Chem 2011; 18: 552.
[http://dx.doi.org/10.2174/092986711794480186]

[60]   Costanzo P, Cariati L, Desiderio D, *et al.* Design, synthesis, and evaluation of donepezil-like compounds as AChE and BACE-1 inhibitors. ACS Med Chem Lett 2016; 7: 470-5.
[http://dx.doi.org/10.1021/acsmedchemlett.5b00483]

[61]   Cheung J, Rudolph MJ, Burshteyn F, *et al.* Structures of human (AChE) in complex with pharmacologically important ligands. J Med Chem 2012; 55: 10282-6.
[http://dx.doi.org/10.1021/jm300871x]

[62]   Camps P, Formosa X, Galdeano C, *et al.* Novel donepezil-based inhibitors of acetyl- and (BChE) and (AChE) induced β-amyloid aggregation. J Med Chem 2009; 51(12): 3588-98.
[http://dx.doi.org/10.1021/jm8001313]

[63]   Bourne Y, Taylor P, Radić Z, Marchot P. Structural insights into ligand interactions at the (AChE) peripheral anionic site. EMBO J 2003; 22: 1-12.
[http://dx.doi.org/10.1093/emboj/cdg005]

[64]   Önkol T, Gökçe M, Orhan İ, Kaynak F. Design, synthesis and evaluation of some novel 3(2H)-pyridazinone-2-yl acetohydrazides as (AChE) and butyrylcholnesterase inhibitors. Org Commun 2013; 6: 55-67.

[65]   Özdemir Z, Yılmaz H, Sarı S, Karakurt A, Şenol FS, Uysal M. Design, synthesis, and molecular modeling of new 3(2H)-pyridazinone derivatives as (AChE)/ BChE inhibitors. Med Chem Res 2017; 26: 2293-308.
[http://dx.doi.org/10.1007/s00044-017-1930-x]

[66]   Brus B, Kosak U, Turk S, *et al.* Discovery, biological evaluation, and crystal structure of a novel nanomolar selective BChE inhibitor. J Med Chem 2014; 57: 8167-79.
[http://dx.doi.org/10.1021/jm501195e]

[67]   Mehta M, Adem A, Sabbagh M. New AChE inhibitors for Alzheimer's disease. Int J Alzheimers Dis
[http://dx.doi.org/10.1155/2012/728983]

[68]   Romero A, Cacabelos R, Oset-Gasque MJ, Samadi A. MarcoContelles J. Novel tacrine-related drugs as potential candidates for the treatment of Alzheimer's disease. Bioorg Med Chem Lett 2013; 23(7): 1916-22.
       [http://dx.doi.org/10.1016/j.bmcl.2013.02.017]

[69]   Chiu PY, Wei CY. Donepezil in the one-year treatment of dementia with Lewy bodies and Alzheimer's disease. J Neurol Sci 2017; 381: 322.

[70]   Birks JS, Chong LY, Grimley-Evans J. Rivastigmine for Alzheimer's disease. Cochrane Database Syst Rev 2015; 9: 1465-858.

[71]   Heinrich M, Lee-Teoh H. Galanthamine from snowdrop the development of a modern drug against Alzheimer's disease from local Caucasian knowledge. J Ethnopharmacol 2004; 92(2–3): 147-62.
       [http://dx.doi.org/10.1016/j.jep.2004.02.012]

[72]   Caplan B, Bogner J, Brenner L, *et al.* Brain Cholinergic Function and Response to Rivastigmine in Patients With Chronic Sequels of Traumatic Brain Injury: A PET Study. J Head Trauma Rehabil. 2018;33(1):25-32(8)

[73]   Machálková M, Schejbal J, Glatz Z, Preisler J. A label-free MALDI TOF MS-based method for studying the kinetics and inhibitor screening of the Alzheimer's disease drug target β-secretase. Anal Bioanal Chem 2018; 410(28): 7441-8.
       [http://dx.doi.org/10.1007/s00216-018-1354-6]

[74]   Fei M, Huang L, Luo Z, *et al.* O-Hydroxyl- or o-amino benzylamine-tacrine hybrids: Multifunctional biometals chelators, antioxidants, and inhibitors of cholinesterase activity and amyloid-Î2 aggregation. Bioorg Med Chem 2012; 20(19): 5884-92.
       [http://dx.doi.org/10.1016/j.bmc.2012.07.045]

[75]   Liu W, Wang H, Li X, *et al.* Design, synthesis and evaluation of vilazodone-tacrine hybrids as multitarget-directed ligands against depression with cognitive impairment. Bioorg Med Chem 2018; 26(12): 3117-25.
       [http://dx.doi.org/10.1016/j.bmc.2018.04.037]

[76]   Triggle DJ, Mitchell JM, Filler R. The Pharmacology of Physostigmine. CNS Drug Rev 1998; 4(2): 87-136.
       [http://dx.doi.org/10.1111/j.1527-3458.1998.tb00059.x]

[77]   Arens AM, Shah K, Al-Abri S, Olson KR, Kearney T. Safety and effectiveness of physostigmine: a 10-year retrospective review. Clin Toxicol (Phila) 2018; 56(2): 101-7.
       [http://dx.doi.org/10.1080/15563650.2017.1342828]

[78]   Işık AT, Bozoğlu E, Eker D. AChE and BuChE inhibition by rivastigmin have no effect on peripheral insulin resistance in elderly patients with Alzheimer disease. J Nutr 2012; 16(2): 139-41.

[79]   Gabr MT, Abdel-Raziqc MS. Structure-based design, synthesis, and evaluation of structurally rigid donepezil analogues as dual AChE and BACE-1 inhibitors. Bioorg Med Chem Lett 2018; 28(17): 2910-3.
       [http://dx.doi.org/10.1016/j.bmcl.2018.07.019]

[80]   Naharci MI, Ozturk A, Yasar H, *et al.* Galantamine improves sleep quality in patients with dementia. Acta Neurol Belg 2015; 115: 563-8.
       [http://dx.doi.org/10.1007/s13760-015-0453-9]

[81]   Hanazawa T, Kamijo Y, Yoshizawa T, Fujita Y, Usui K, Haga Y. Acute cholinergic syndrome in a patient with Alzheimer's disease taking the prescribed dose of galantamine. Psychogeriatrics 2018; 18(5): 434-5.
       [http://dx.doi.org/10.1111/psyg.12341]

[82]   Nakagawa R, Ohnishi T, Kobayashi H, *et al.* Long-term effect of galantamine on cognitive function in patients with Alzheimer's disease *versus* a simulated disease trajectory: an observational study in the clinical setting. Neuropsychiatr Dis Treat 2017; 13: 1115-24.

[http://dx.doi.org/10.2147/NDT.S133145]

[83]   Huang XT, Qian ZM, He X, *et al.* Reducing iron in the brain: a novel pharmacologic mechanism of huperzine A in the treatment of Alzheimer's disease. Neurobiol Aging 2013; 35(5): 1045-54.

[84]   Ding R, Fu JG, Xu GQ, Sun BF, Lin GQ. Divergent total synthesis of the Lycopodium alkaloids huperzine A, huperzine B, and huperzine U. J Org Chem 2014; 79(1): 240-50.
       [http://dx.doi.org/10.1021/jo402419h]

[85]   Wang R, Yan H, Tang XC. Progress in studies of huperzine A, a natural cholinesterase inhibitor from Chinese herbal medicine. Acta Pharmacol Sin 2006; 27(1): 1-26.
       [http://dx.doi.org/10.1111/j.1745-7254.2006.00255.x]

[86]   Mele T, Jurič DM. Metrifonate, like acetylcholine, up-regulates neurotrophic activity of cultured rat astrocytes. Pharmacol Rep 2014; 66(4): 618-23.
       [http://dx.doi.org/10.1016/j.pharep.2014.02.025]

[87]   Lao K, Ji N, Zhang X, Qiao W, Tang Z, Gou X. Drug development for Alzheimer's disease. review J Drug Target 2018; 164-73.
       [http://dx.doi.org/10.1080/1061186X.2018.1474361]

[88]   Anand A, Patience AA, Sharma N, Khurana N. The present and future of pharmacotherapy of Alzheimer's disease: A comprehensive review. Eur J Pharmacol 2017; 815: 364-75.
       [http://dx.doi.org/10.1016/j.ejphar.2017.09.043]

[89]   Özçelik AB, Gökçe M, Orhan İ, Kaynak F, Şahin MF. Synthesis and antimicrobial, (AChE) and BChE inhibitory avtivities of novel ester and hydrazide derivatives of 3(2H)-pyridazinone. Arzneimittelforschung 2010; 60(7): 452-8.

[90]   Utku S, Gökçe M, Aslan G, *et al.* Synthesis and *in vitro* antimycobacterial activities of novel 6-substituted-3(2H)-pyridazinone-2-acetyl-2-(substituted/ nonsubstituted acetophenone)hydrazone. Turk J Chem 2011; 35: 331-9.

[91]   Utku S, Gökçe M, Orhan İ, Şahin MF. Synthesis of novel 6-substituted-3(2H)-pyridazinone-2-a-etyl-2-(substituted/-nonsubstituted benzal)hydrazone derivatives and (AChE) and BChE inhibitory activities *in vitro*. Arzneimittelforschung 2011; 61: 1-7.
       [http://dx.doi.org/10.1055/s-0031-1296161]

[92]   Xing W, Fu Y, Shi Z, Lu D, Zhang H, Hu Y. Discovery of novel 2,6-disubstituted pyridazinone derivatives as AChE inhibitors. Eur J Med Chem 2013; 63: 95-103.
       [http://dx.doi.org/10.1016/j.ejmech.2013.01.056]

[93]   Fang J, Li Y, Liu R, Pang X, Li C, Yang R, *et al.* Discovery of multitarget-directed ligands against alzheimer's disease through systematic prediction of chemical–protein interactions. J Chem Inf Model. 2015;55:149–164

[94]   Wenzel TJ, Klegeris A. Novel multi-target directed ligand-based strategies for reducing neuroinflammation in Alzheimer's disease. Life Sci 2018; 207: 314-22.
       [http://dx.doi.org/10.1016/j.lfs.2018.06.025]

[95]   Guzior N, Wickowska A, Panek D, Malawska B. Recent development of multifunctional agents as potential drug candidates for the treatment of alzheimer's disease. Curr Med Chem 2015; 22: 373-404.
       [http://dx.doi.org/10.2174/0929867321666141106122628]

[96]   Sameem B, Saeedi M, Mahdavi M, Shafiee A. A review on tacrine-based scaffolds as multi-target drugs (MTDLs) for Alzheimer's disease. Eur J Med Chem 2018; 128(10): 332-45.

[97]   Chioua M, Buzzi E, Moraleda I, *et al.* Tacripyrimidines, the first tacrine-dihydropyrimidine hybrids, as multi-target-directed ligands for Alzheimer's disease. Eur J Med Chem 2018; 155: 839-46.
       [http://dx.doi.org/10.1016/j.ejmech.2018.06.044]

[98]   Zhang Y, Li P, Feng J, Wu M. Dysfunction of NMDA receptors in Alzheimer's disease. Neurol Sci 2016; 37(7): 1039-47.

[http://dx.doi.org/10.1007/s10072-016-2546-5]

[99]   Bolognesi ML, Cavalli A, Melchiorre C. Memoquin: A Multi-Target–Directed Ligand as an Innovative Therapeutic Opportunity for Alzheimer's Disease. ASENT 2009; 6: 152-62.

[100]   Sterling J, Herzig Y, Goren T, *et al.* Novel dual inhibitors of AChE and MAO derived from hydroxy aminoindan and phenethylamine as potential treatment for Alzheimer's disease. J Med Chem 2002; 45(24): 5260-79.
        [http://dx.doi.org/10.1021/jm020120c]

[101]   Gökhan-Kelekçi N, Yabanoğlu S, Küpeli E, *et al.* A new therapeutic approach in Alzheimer disease: Some novel pyrazole derivatives as dual MAO-B inhibitors and antiinflammatory analgesics. Bioorg Med Chem 2007; 15(17): 5775-86.
        [http://dx.doi.org/10.1016/j.bmc.2007.06.004]

[102]   Pan W, Hu K, Bai P, *et al.* Design, synthesis and evaluation of novel ferulic acid-memoquin hybrids as potential multifunctional agents for the treatment of Alzheimer's disease. Bioorg Med Chem Lett 2016; 26(10): 2539-43.
        [http://dx.doi.org/10.1016/j.bmcl.2016.03.086]

[103]   Capurro V, Busquet P, Lopes JP, *et al.* Pharmacological Characterization of Memoquin, a Multi-Target Compound for the Treatment of Alzheimer's Disease. PLoS One 2013; 8(2): e56870.
        [http://dx.doi.org/10.1371/journal.pone.0056870]

[104]   Rodrigues Simões MC, Dias Viegas FP, Moreira MS, Silva MF, Máximo Riquiel M, Mattos da Rosa P. Donepezil: An Important Prototype to the Design of New Drug Candidates for Alzheimer's Disease. Mini Rev Med Chem 2014; 14: 2-19.
        [http://dx.doi.org/10.2174/1389557513666131119201353]

[105]   Rizzo S, Bartolini M, Ceccarini L, *et al.* Targeting Alzheimer's disease: Novel indanone hybrids bearing a pharmacophoric fragment of AP2238. Bioorg Med Chem 2010; 18(5): 1749-60.
        [http://dx.doi.org/10.1016/j.bmc.2010.01.071]

[106]   Özturan-Özer E, Unsal-Tan O, Ozadali K, Küçükkılınç T, Balkan A, Uçar G. Synthesis, molecular modeling and evaluation of novel N'-2-(4-benzylpiperidin-/piperazin-1-yl)acylhydrazone derivatives as dual inhibitors for cholinesterases and Aβ aggregation. Bioorg Med Chem Lett 2013; 23(2): 440-3.
        [http://dx.doi.org/10.1016/j.bmcl.2012.11.064]

[107]   Catto M, Berezin AA, Lo Re D, *et al.* Design, synthesis and biological evaluation of benzo[e][1,2,4]triazin-7(1H)-one and [1,2,4]-triazino[5,6,1-jk]carbazol-6-one derivatives as dual inhibitors of beta-amyloid aggregation and acetyl/butyryl cholinesterase. Eur J Med Chem 2012; 58: 84-97.
        [http://dx.doi.org/10.1016/j.ejmech.2012.10.003]

[108]   Ekiz M, Tutar A, Ökten S, *et al.* Synthesis, characterization, and SAR of arylated indenoquinoline-based cholinesterase and carbonic anhydrase inhibitors. Arch Pharm Chem Life Sci. 2018;e1800167
        [http://dx.doi.org/10.1002/ardp.201800167]

[109]   Chen Y, Su J, Fang L, *et al.* Tacrineferulic acid-nitric oxide (NO) donor trihybrids as potent, multifunctional acetyl and BChE inhibitors. J Med Chem 2012; 55: 4309-432.
        [http://dx.doi.org/10.1021/jm300106z]

[110]   Messerer R, Dallanoce C, Matera C, *et al.* Novel bipharmacophoric inhibitors of the cholinesterases with affinity to the muscarinic receptors M1 and M2. MedChemComm 2017; 8: 1346-59.
        [http://dx.doi.org/10.1039/C7MD00149E]

[111]   Fernández-Bachiller MI, Pérez C, González-Muñoz GC, *et al.* Novel Tacrine–8-Hydroxyquinoline Hybrids as Multifunctional Agents for the Treatment of Alzheimer's Disease, with Neuroprotective, Cholinergic, Antioxidant, and Copper-Complexing Properties. J Med Chem 2010; 53(13): 4927-37.
        [http://dx.doi.org/10.1021/jm100329q]

[112]   Chen Y, Zhu J, Mo J, *et al.* Synthesis and bioevaluation of new tacrine-cinnamic acid hybrids as

cholinesterase inhibitors against Alzheimer's disease. J Enzyme Inhib Med Chem 2018; 33(1): 290-302.
[http://dx.doi.org/10.1080/14756366.2017.1412314]

[113] Parsons CG, Danysz W, Dekundy A, Pulte I. Memantine and cholinesterase inhibitors: complementary mechanisms in the treatment of Alzheimer's disease. Neurotox Res 2013; 24(3): 358-69.
[http://dx.doi.org/10.1007/s12640-013-9398-z]

[114] Lei zhao, Xiao'qin Cheng, and Chunjiu Zhong. Implications of successfully symptomatic treatment in Parkinson's disease for therapeutic strategies of Alzheimer's disease. ACS Chem Neurosci 2018.
[http://dx.doi.org/10.1021/acschemneuro.8b00450]

[115] Lopes JP, Tarozzo G, Reggiani A, Piomelli D, Cavalli A. Galantamine potentiates the neuroprotective effect of memantine against NMDA-induced excitotoxicity. Brain Behav 2013; 3(2): 67-74.
[http://dx.doi.org/10.1002/brb3.118]

[116] Peglow TJ, Schumacher RF, Cargnelutti R, *et al.* Preparation of bis(2-pyridyl) diselenide derivatives: Synthesis of selenazolo[5,4-b]pyridines and unsymmetrical diorganyl selenides, and evaluation of antioxidant and anticholinesterasic activities. Tetrahedron Lett 2017; 58(38): 3734-8.
[http://dx.doi.org/10.1016/j.tetlet.2017.08.030]

[117] Bag S, Ghosh S, Tulsan R, Sood A, Zhou W, Schifone C. Design, synthesis and biological activity of multifunctional α,β-unsaturated carbonyl scaffolds for Alzheimer's disease. Bioorg Med Chem Lett 2013; 23(9): 2614-8.
[http://dx.doi.org/10.1016/j.bmcl.2013.02.103]

[118] Marlyn C, Villamizar O, Carlos E, *et al.* Coumarin-Based Molecules as Suitable Models for Developing New Neuroprotective Agents Through Structural Modification. Elsevier BV 2018.

[119] Basha SJ, Mohan P, Yeggoni DP, Babu ZR, Kumar PB, Rao AR. New Flavone-Cyanoacetamide Hybrids with a Combination of Cholinergic, Antioxidant, Modulation of β-Amyloid Aggregation, and Neuroprotection Properties as Innovative Multifunctional Therapeutic Candidates for Alzheimer's Disease and Unraveling Their Mechanism of Action with (AChE). Mol Pharm 2018; 15(6): 2206-23.
[http://dx.doi.org/10.1021/acs.molpharmaceut.8b00041]

[120] Tu Y, Huang J, Li Y. Anticholinesterase, antioxidant, and beta-amyloid aggregation inhibitory constituents from Cremastra appendiculata. Med Chem Res 2018; 27: 857-63.
[http://dx.doi.org/10.1007/s00044-017-2108-2]

[121] Li Q, Tu Y, Zhu C, *et al.* Cholinesterase, β-amyloid aggregation inhibitory and antioxidant capacities of Chinese medicinal plants. Ind Crops Prod 2017; 108: 512-9.
[http://dx.doi.org/10.1016/j.indcrop.2017.07.001]

[122] Jalili-Baleh L, Babaei E, Abdpour S, *et al.* A review on flavonoid-based scaffolds as multi-targe--directed ligands (MTDLs) for Alzheimer's disease. Eur J Med Chem 2018; 152: 570-89.
[http://dx.doi.org/10.1016/j.ejmech.2018.05.004]

[123] Angelova PR, Abramov AY. Role of mitochondrial ROS in the brain: from physiology to neurodegeneration. FEBS Lett 2018; 692-702.
[http://dx.doi.org/10.1002/1873-3468.12964]

[124] Butterfield DA, Boyd-Kimball D. Oxidative Stress, Amyloid-Peptide, and Altered Key Molecular Pathways in the Pathogenesis and Progression of Alzheimer's Disease. J Alzheimers Dis 2018; 62: 1345-67.
[http://dx.doi.org/10.3233/JAD-170543]

[125] Mecocci P, Boccardi V, Cecchetti R, *et al.* A Long Journey into Aging, Brain Aging, and Alzheimer's Disease Following the Oxidative Stress Tracks. J Alzheimers Dis 2018; 62: 1319-35.
[http://dx.doi.org/10.3233/JAD-170732]

[126] Zhao Y, Zhao B. Oxidative Stress and the Pathogenesis of Alzheimer's Disease. Oxid Med Cell Longev 2013; 2013: 316523.

[http://dx.doi.org/10.1155/2013/316523]

[127] Markesberya WR. Oxidative Stress Hypothesis in Alzheimer's Disease. Free Radic Biol Med 1997; 23(1): 134-47.
       [http://dx.doi.org/10.1016/S0891-5849(96)00629-6]

[128] Rosini M, Simoni E, Milelli A, Minarini A, Melchiorre C. Oxidative Stress in Alzheimer's Disease: Are We Connecting the Dots? J Med Chem 204;57: 2821-31.

CHAPTER 6

Potential Biological Mechanisms with Prophylactic Action in Rapid Cognitive Impairment in Late-Onset Alzheimer's Disease

Ileana Marinescu[1], Puiu Olivian Stovicek[2,*], Dragoş Marinescu[3] and **Laurenţiu Mogoantă[4]**

[1] *Department of Psychiatry, University of Medicine and Pharmacy of Craiova, Romania*

[2] *Department of Pharmacology, "Titu Maiorescu" University of Bucharest, Faculty of Nursing Târgu Jiu, Romania*

[3] *Member of the Romanian Academy of Medical Sciences, University of Medicine and Pharmacy of Craiova, Doctoral School, Interim President of the Romanian Society for Biological Psychiatry and Psychopharmacology, Romania*

[4] *Department of Morphopathology, University of Medicine and Pharmacy of Craiova, Editor-i--chief of Romanian Journal of Morphology and Embryology, Romania*

Abstract: Given the alarming increase in the Alzheimer's Disease (AD) related costs and the number of patients, significant importance of the research of dementia in elderly people is represented by the early detection of dementia signs and identifying ways to slow down the cognitive decline that should be done by the entire medical community, not only by specialists in psychiatry, neurology and geriatrics. An integrated, multidisciplinary approach can only be achieved by recognizing some clinical-biological parameters that may represent an early signal for a potential onset of AD, or that may speed up the cognitive impairment. Dysfunctional neurobiochemical mechanisms in AD engaging in multimodal pathogenic processes of cognitive impairment, are correlated with the presence of clinical, imaging or biological markers. Identifying risk factors for the progression of cognitive impairment in AD will bring significant improvement in primary and secondary prophylaxis in late-onset AD with improved quality of patient life and a significant decrease in the cost of care associated with this pathology. The patient's assessment should consider multiple somatic comorbidities, associated with cognitive impairment: neurobiochemical vulnerabilities (acetylcholine, dopamine, serotonin and noradrenaline deficiency), traumatic brain injury, disruption of the blood brain barrier, insomnia, depression, cardiovascular diseases, diabetes, hepatic steatosis, infectious pathology. The etiology of late-onset AD and the rapid progression of cognitive decline are complex, multifactorial and incomprehensible, the genetic component is less involved, and is a real challenge for

* **Corresponding author Puiu Olivian Stovicek:** Department of Pharmacology, "Titu Maiorescu" University of Bucharest, Faculty of Nursing Târgu Jiu, Romania; Tel: 0040735204013; Fax: 0040351401336; E-mail: puiuolivian@yahoo.com

Atta-ur-Rahman (Ed.)

research on the pathology of cognitive impairment. These considerations make it difficult to diagnose early and develop effective therapeutic strategies.

Keywords: Alzheimer's disease, Acetylcholine, Amyloid-β, Biomarkers, Blood Brain Barrier, Cardiovascular Diseases, Cerebral Amyloid Angiopathy, Cognitive Impairment, Cognitive Reserve, Depression, Glymphatic System, Inflammation, Insomnia, Microbleeds, Mild Cognitive Impairment Syndrome, Neurobiochemical Vulnerabilities, Small Vessel Cerebral Disease, Traumatic Brain Injury, Vascular Risk Factors.

INTRODUCTIVE ARGUMENTS

This chapter aims at identifying risk factors for the rapid progression of cognitive decline in Alzheimer's Disease (AD), easy to use in the current practice of psychiatrists, neurologists or geriatricians, as well as physicians from other specialties (nutritionist, cardiologist, diabetologist, urologist, gastroenterologist, hematologist and others), who commonly diagnose and treat somatic comorbidities of elderly patients. An integrated, multidisciplinary approach can only be achieved by highlighting the importance in recognizing some clinical-biological parameters that may represent an early signal for a potential onset of AD, or that may speed up the progression of this condition.

Another category of medical specialists less involved in identifying risk factors in AD includes family medicine, ophthalmology and dentistry. In our opinion, these specialties have an important dimension in prophylactic activity due to the high addressability of the elderly who may have a potential risk to develop AD-like pathology with late onset. We are starting from the premise that, in the cabinet of family medicine or dentistry, pathological signs uncorrelated with other AD comorbidities can be identified. These risk factors can be confirmed relatively easily before the specialized psychiatrist, thus identifying more patients with potential risk for AD before clinical manifestations occur in the latent or prodromal stage. This may increase the hope in improving prevention plans in which we firmly believe and are trying to argue with a clinical and neurobiological model.

In our opinion, the prophylactic activity for preventing the rapid progression of cognitive impairment in AD, has been accepted in the medical world with a centripetal initial position in which the patient with suspicion or diagnosed with AD is at the center of the medical actions. Thus, the patient addresses specific comorbidities and the doctor has limited therapeutic goals, focusing on controlling, evaluating and treating such comorbidities without noticing the risk of progression of cognitive impairment. The sensitization of all medical specialties on the prevention possibilities of both late-onset AD and the rapid progress of cognitive deficits alters the strength ratio of the previous model in a centrifugal model. In this case, all medical specialties should give the elderly a special attention, in order to identify some potential clinical and biological markers that may precipitate the onset or accentuation of the deterioration-cognitive symptomatology.

We believe that the acceptance of the proposed model in current practice will bring a significant improvement in primary and secondary prophylaxis in late-onset AD with improved quality of patient life and a significant decrease in the cost of care associated with this type of pathology.

If in early-onset AD, cognitive impairment progresses through predominantly genetic mechanisms that are difficult to influence, in late-onset AD (sporadic form), rapid cognitive impairment is no longer a direct consequence of neurodegenerative pathogens, but is correlated with the degree of disconnectivity of cognitive neural circuits *via* multiple pathogenic mechanisms.

EPIDEMIOLOGICAL EVIDENCE

The late-onset AD prevalence reveals an alarming increase, significantly correlated with age. Thus, the data released by Evans since 1989 reveals a 3% prevalence for 65-74 age range, with an increase of over 6 times between 75-84 years (18.7%). After age 85, the prevalence rate increased to very high values, reaching 47.2% [1]. Data from Alzheimer's World Alzheimer's Report 2015 of Alzheimer's Disease International shows that, in 2015, 46.8 million patients had dementia, out of which 9.9 million were newly diagnosed cases (one new case every 3 seconds). Estimates show that the global number will increase to 74.7 million in 2030 and 131.5 million by 2050. It is alarming that this increase is 12-13% higher than the World Alzheimer's Report 2009 estimate. Early diagnosis in AD in the elderly and population aging, especially in countries with a less developed socioeconomic level, has altered the epidemiological situation, with the increase in these areas reaching over 200% in 2050 compared to 2015 [2].

Regarding the enormous costs involved in the alarming rise of the deterioration-cognitive pathology in AD, the same report shows that in 2015, worldwide dementia costs were at US $ 818 billion. It is estimated that in 2018, the costs would reach one trillion US $, and the amount would be double to $ 2 trillion in 2030. The annual incidence of AD incidence per 1000 inhabitants has a variability dependent on specific socio-economic and biological causes, and possibly genetic variations, depending on race. (10.5 for North America, 8.8 for West Europe, 9.2 for South America and 8.0 for China and Western-Pacific) [3].

In this context, a significant importance of the research of dementia in elderly people is represented by the early detection of dementia signs and the identification of ways to slow down the evolution of cognitive decline, as the major costs are related to: nursing home care and nursing care systems, following the cognitive deficit progression with the loss of self-care capacity, the decrease in the daily activities of the patients and the increase of the social stress to which the family members are subjected to. Starting from these epidemiological and economic premises, we consider that the identification of risk factors for the occurrence and evolution of AD should become a major and multidisciplinary concern, as neurodegenerative elements can have rapid progression, influenced by cardiovascular, metabolic, traumatic, toxic or infectious factors. Interpretation of AD pathology is going through a period of intense research on the biological and molecular mechanisms that can predict the risk of rapid progression of cognitive impairment.

Starting from this working hypothesis, we believe that identification and awareness of potential biological mechanisms should be done by the entire medical community, not only by specialists in psychiatry, neurology and geriatrics. This approach would allow for real prophylactic interventions to slow the progression of cognitive impairment. Fast cognitive impairment leads to loss of functional and social independence of patients and the impossibility of integrating into a therapeutic and rehabilitative program in the family environment. Losing family ties and fostering an institutionalized home nursing care program lead to an exponential increase in care costs, lower quality of life for patients and their families.

VULNERABILITIES CORRELATED WITH THE ONSET AND EVOLUTION OF AD

The etiology of late-onset AD and the rapid progression of cognitive decline are complex, multifactorial and incomprehensible, and is a real challenge for research on the pathology of cognitive impairment. These considerations make it difficult to diagnose early and develop effective therapeutic strategies (Fig. **1**).

Since most AD symptoms are confused with aging-specific manifestations, the physician should identify as early as possible risk factors and early signs correlated with cognitive decline. A mnemonic formula, easy to use in current practice, can be:

D - Drug use - illicit drugs or drugs that can influence cognition (anticholinergics, beta blockers)

E - Emotional disorders (depression)

M - Metabolic and endocrine disorders (dehydration, renal disorders, metabolic syndrome, diabetes)

E - Eye and ear disorders sensory deficiency of peripheral or central cause, Benson syndrome

N - Nutritional disorders (vitamin B12, D or folic acid deficiency, hyponatremia) and normal-pressure hydrocephalus

T - Tumor or trauma and other traumatic brain injury including family violence

I - Infection (urinary tract infection, acute bronchitis, HIV, herpes zoster)

A - Atherosclerosis (vascular component) and alcoholism (toxic component) [5, 6]

EVOLUTIONARY MODEL IN ALZHEIMER'S DISEASE

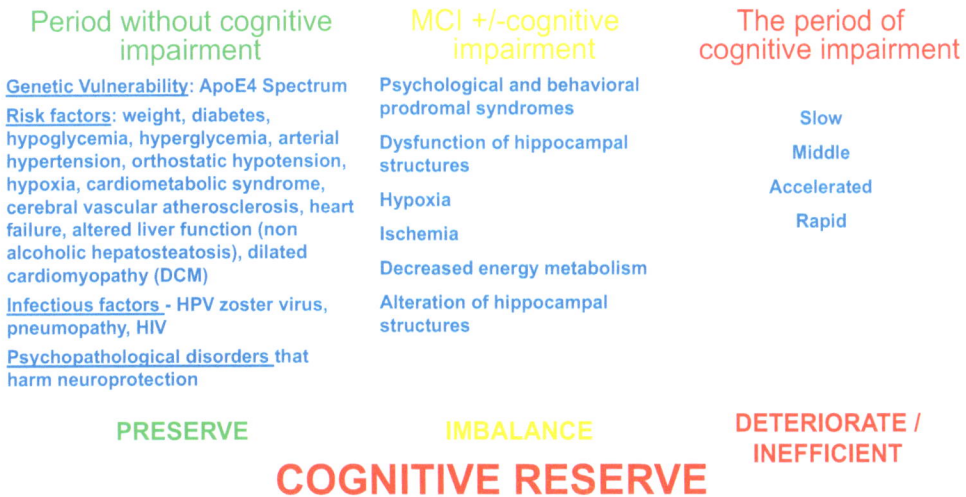

Period without cognitive impairment

MCI +/-cognitive impairment

The period of cognitive impairment

Genetic Vulnerability: ApoE4 Spectrum
Risk factors: weight, diabetes, hypoglycemia, hyperglycemia, arterial hypertension, orthostatic hypotension, hypoxia, cardiometabolic syndrome, cerebral vascular atherosclerosis, heart failure, altered liver function (non alcoholic hepatosteatosis), dilated cardiomyopathy (DCM)
Infectious factors - HPV zoster virus, pneumopathy, HIV
Psychopathological disorders that harm neuroprotection

Psychological and behavioral prodromal syndromes
Dysfunction of hippocampal structures
Hypoxia
Ischemia
Decreased energy metabolism
Alteration of hippocampal structures

Slow
Middle
Accelerated
Rapid

PRESERVE   **IMBALANCE**   **DETERIORATE / INEFFICIENT**

COGNITIVE RESERVE

Fig. (1). Evolutionary Model of Cognitive Decline in AD [4].

Genetic Vulnerability

This is significantly correlated with allelic variations of ApoE4 (apolipoprotein E4), which is associated with massive beta amyloid (Aβ) deposits in cognitive structures, resulting in an increase in the cognitive impairment rate. The variation of ApoE2 is correlated with vascular vulnerability for cerebral development of amyloid angiopathy. This condition is associated with the phenomenon of global vascular hypoperfusion at the cerebral level, through deposits of Aβ at the structures of large, medium and small cerebral vessels.

ApoE4-associated amyloid angiopathy increases the risk of ischemic stroke [7], whereas ApoE2 is associated with hemorrhagic vascular accidents, including subarachnoid hemorrhage [8]. The genetic changes of ApoE4 and ApoE2 are more common in elderly people, which is why these ApoEs represent the genetic background of vulnerability of the sporadic form of AD.

Familial Early-Onset AD, which usually occurs before the age of 65, accounts for only 1% of all AD cases. Genetic vulnerability is related to mutations in the genes encoding APP, presenilin 1 or presenilin 2. These changes affect the activity of the γ-secretase complexes involved in Aβ formation and result in aberrant increase in Aβ level [9].

Other data from the literature show that, of all cases of AD diagnosed, 5-10% are of a familial type. The genetic link of this rare form of AD is demonstrated by the identification of the three autosomal dominant genes at the level of chromosomes 1, 14 and 21 [5].

One can also mention the Down syndrome, because it is known as the relationship between the existence of a Down syndrome at the familial level and the risk of AD development [10].

The common presence of Aβ in these two disorders, has led to the formulation of the amyloid cascade hypothesis: patients with Down syndrome develop AD neuropathology by the age of 35 years; the amyloid precursor protein (APP) gene, which encodes Aβ, is located on chromosome 21; AD-causing mutations have been identified in the APP gene and in presenilin 1 and 2 genes, which encode components of the γ-secretase enzyme responsible for one of the Aβ generating cleavages of APP [11].

The γ-secretase complex is located in the cell membrane, where it cleaves different transmembrane proteins involved in the chemical signaling pathways that link the cell to the outside of the cell. Of these, the Notch signaling pathway [12] is essential for the normal maturation and division of hair follicle cells and other skin cell types, and is also involved in normal immune function. Several mechanisms are known to affect immune system's involvement in AD pathogenesis. The γ-secretase complex also regulates the processing of APP at the cerebral or systemic tissue level by dissolving them into smaller peptides, such as the soluble amyloid protein precursor protein (sAPP) and several versions of the amyloid peptide β (versions of β amyloid peptide). Although the functions of these compounds are not fully understood, sAPP is believed to have cell growth stimulation properties and may play a role in neuronal cell formation both before and after birth [13].

However, it is not clear whether these genetic changes trigger AD or represent only a risk factor. Since approximately 50% of cases of AD have not revealed alleles of ApoE4 or mutations in the genes responsible for encoding APP, presenilin 1 or presenilin 2 proteins, it is accepted that AD's etiology is also involved with other risk factors, some of which are already accepted, others are in confirmation [14].

The Vulnerability of the Brain Vascular System

The Most Specific Vascular Vulnerability is considered cerebral amyloid angiopathy (CAA), which is characterized by massive deposition of Aβ in the cortical and leptomeningeal arteries, deposition that favors the extraction of the red blood cells due to microhemorrhagia (Fig. **2**).

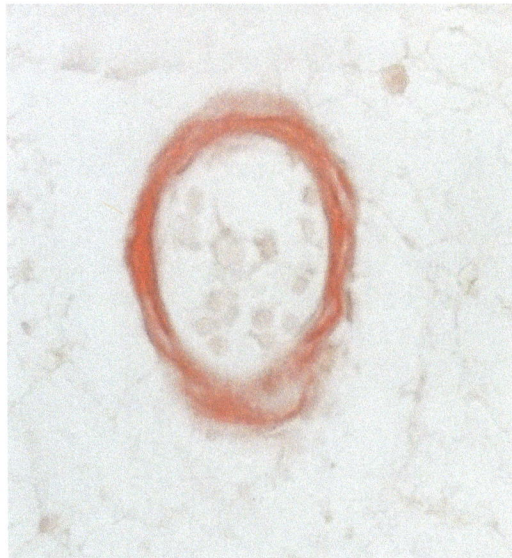

Fig. (2). Postmortem microscopic image of a patient with AD and cognitive impairment. Micro arteriole in the left front, Congo red staining, which identifies massive amyloid deposition at endothelial level and extravasation of red blood cells. Magnification 40x. Image used with permission from Daniel Pirici.

Interestingly, CAA has a form of early onset and familial spectrum and a late-onset form (sporadic form) associated with ApoE4 and largely determines cognitive impairment in AD. Potential clinical and neuroimaging markers of CAAs that may occur in the asymptomatic or prodromal phase of the disease may have a predominantly neurological aspect. Because of this, they are often associated with cerebral vascular disease without linking to the risk of further development of AD. The most common manifestations with this clinical profile are lobar or lobar intracerebral hemorrhage, microbleeds or small cortical infarction. The imaging marker of this pathology is the accumulation of hemosiderin.

The prevalence of cerebral microbleeds (CMB) in the general population increases from 6.5% in people aged 45 to 50 years, 17.8% in those aged 60 to 69, and 38.3% in subjects aged 80 and over [15, 16].

In addition to healthy subjects, CMBs have also been confirmed in neuroimaging exams in patients with ischemic stroke, intracerebral hemorrhage, traumatic brain injury and diffuse axonal injury, all of these pathologies being considered risk factors for AD [17].

Microbleeds were recognized in 29% of patients with AD and were multiple (>1) in 48% of cases. 92% of the patients had a lobar localization, while 57% of microhemorrhages were seen in the occipital lobes [18].

Cortical neural mass can also be identified by recurrence (cortical or repeated intracerebral infarction or hemorrhage), the old infarcts being identified as cortical superficial siderosis (cSS), which is a neuroimaging risk marker for the role of CAA in triggering intracerebral hemorrhage and acceleration of the cognitive deterioration rate [19, 20].

Small vessel brain damage can be caused by two mechanisms, one Aβ dependent (CAA) that works in the cerebral lobes and the other non-amyloid (hypertensive vasculopathy), which causes lesions in some deep brain areas (basal ganglia, thalamus, brainstem) [15].

Nonlobar microbleeds are significantly and independently associated with cardiovascular mortality whereas CAA-type micro-lesions can be considered markers for stroke-related mortality, independent from cardiovascular risk factors. Also, the number of CMB significantly increases the death risk [17].

From a clinical point of view, depending on the number and location of microhemorrhages, the cognitive deficit is accentuated and memory, attention, motor speed is affected. In the context of a mild cognitive impairment syndrome (MCI), amyloid-type micro-bleeds frequently result in decreased cognitive performance, while cardiovascular micro-lesions have a major effect on cognitive decline [21].

Microhemorrhages may also cause dysfunctional motor changes that amplify the difficulties of caring for these patients. Neurological recovery significantly increases the cost of care and, in particular, increases cognitive impairment and the risk of complications of somatic complications of the infectious type (pneumopathies, cutaneous infections, bedsores). The cerebral vascular component plays an important role in precipitating / accelerating the cognitive decline of AD in elderly people [22].

Neurovascular unit dysfunction is, in turn, an important factor in the progression of neurodegenerative elements in all forms of dementia including that in AD [23].

This dysfunctionality precipitates the deposition of Aβ in the leptomeningeal arterioles, compressing the perivascular space through the development of CAA, and at blood–brain barrier (BBB) blocking Aβ drainage, potentially causing excessive accumulation in the brain. This creates a veritable CAA-controlled vicious circle through perivascular compression mechanisms that maintain high levels of Aβ in the brain space that will accentuate CAA process [24].

Hyperhomocysteinemia is an important biological marker that signals the presence of stroke in accelerating cognitive impairment and hemorrhagic risks for CAA [25].

Excessive levels of Aβ at the cerebral level inhibit the normal functioning of neurovascular unit by mechanisms of blocking control processes exerted on molecular adhesion, a process controlled by the interaction normally existing between leukocytes and endothelial treatment mechanisms. Thus, the excess of Aβ amplifies also hyperphosphorylation of tau proteins, which accentuates neurovascular unit dysfunction, compromising the efficiency of BBB [26].

Maintaining the functional stability of BBB is an important target for strategies to prevent rapid cognitive impairment, strategies that need to be launched early, especially in the elderly and risk factors for AD. These strategies aim at controlling and limiting inflammatory processes, metabolic disorders in the brain as well as avoiding TBI or toxic and infectious processes [27].

NonSpecific Vascular Vulnerabilities are secondary to atherosclerotic processes. In the prodromal phase and the MCI syndrome in AD there are several elements of easy identification of atherosclerotic processes through diagnostic means accessible to both the specialist doctor and the family doctor. They are represented by:

Ophthalmologic Eye Exam (Ophthalmoscopy)

The ophthalmologic examination can highlight and confirm the pathogenesis of visual sensory deficits that can be corrected, or pathogenesis that depends on the cortical projection area. Visual acuity deficits, oculomotor nerve paralysis, altered color perception, or amblyopia condition can often be caused by non-ophthalmic causes. In these situations, the peripheral ophthalmologic examination may be normal. These deficiencies are due to the alteration of the visual transmission pathways (optic chiasm pathology or components of this structure anterior or posterior crossing), as well as the damage to optical areas at the level of cortical projection areas (central blindness, Benson syndrome).

The most common pathogenic mechanism causing changes in visual perception is the cerebral vascular mechanism, ischemic or hemorrhagic. The evaluation of retinal vascular structure through eye exam gives data almost identical to vascular status of small cerebral vessels. CAA can be anticipated in the elderly and MCI syndrome by identifying angiopathy and angiosclerosis by this specialty exam. Frequent pathogenic conditions such as diabetic angiopathy, systemic atherosclerosis, high blood pressure can be expected.

It is considered that the retina is a true "Window to AD" due to common features with the brain (embryological origin, response to injury, immunological aspects) [28]. Due to the identification of Aβ deposits in eyes from both human AD and transgenic mouse models, the retina may be included in investigations for screening of AD. The advantages for this examination include its low cost, easy accessibility and the non-invasive nature of the tests [29].

Raising the awareness of general practitioner and ophthalmologists to direct these patients to a psychiatric specialist after a minimal MMSE screening can lead to early diagnosis and specific therapeutic interventions that may delay cognitive deficits. When multiple risk factors are summed up, early identification of retinal angiopathy (stage 1) may be a significant prediction indicator. Retinal angiosclerosis (stage 2) requires multidisciplinary assessment and multisystemic therapy, being associated with somatic multisystem pathology in most cases (Fig. **3**).

We did not insist on the severe progressive phases because the prophylactic measures decrease significantly in importance at this stage. A particular feature in confirmation of the senile systemic amyloidosis is the presence of vitreous amyloidosis and retinal amyloidosis.

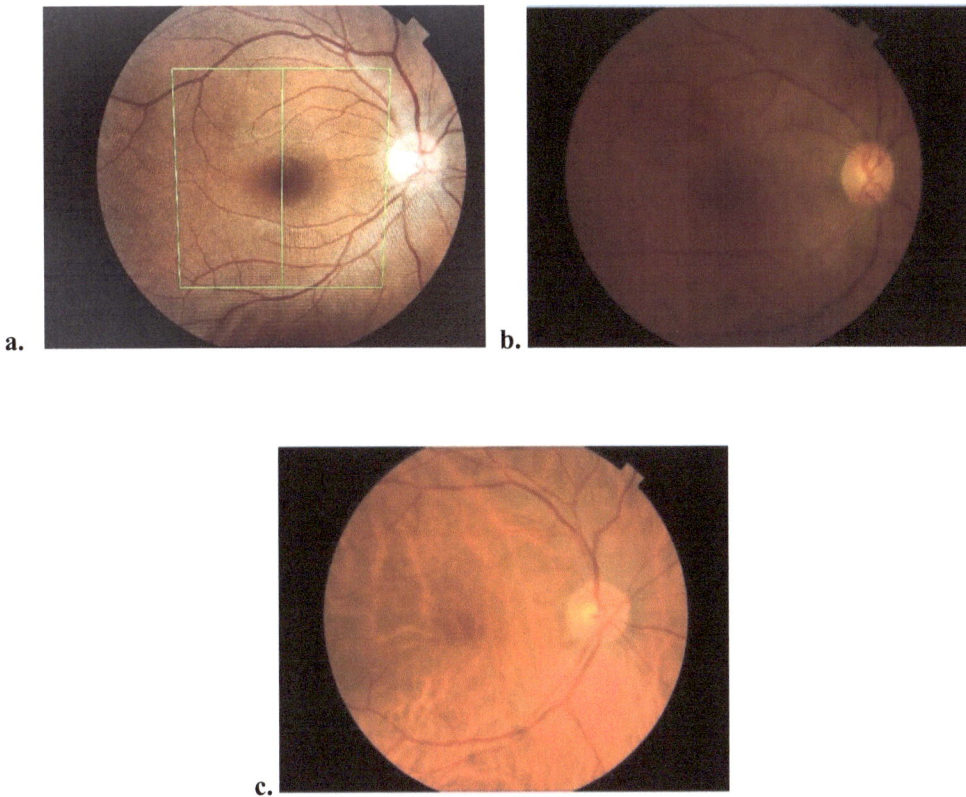

Fig. (3). Ophthalmologic eye exam. **a**: normal right eye. **b**: retinal angiopathy, right eye (stage 1). **c**: retinal angiosclerosis, right eye (stage 2). Image used with permission from Andreea Tanasie.

Transcranial Doppler Examination

Another valuable indicator that can be used in predicting cognitive impairment in AD is related to the velocity of cerebral flow in the Willis polygon and the pulsatile index. A risk for AD development is considered the lower the rate and the higher the index, regardless of the time of the evaluation, while the MCI stage may indicate a major risk of having the disease itself with a the rapid progression of cognitive decline. The mechanism underlying these indicators is related to clearance-Aβ. Perivascular and glymphatic pathways for the Aβ elimination are influenced by the systolic pulse. Also, cerebrovascular motility alterations negatively influence the elimination of Aβ on these two pathways and indirectly cause cerebral hypoperfusion [30].

The Doppler evaluation can identify early endothelial dysfunction and lesion manifestations at this level, either at the level of the carotid vessels or in the anterior large cerebral arteries. Endothelial dysfunction changes have been always associated with a proinflammatory components, accompanied by activation of the glial system [31]. The presence and progression in the dynamic evaluation of these elements suggest the amplification of cerebral hypoperfusion and destructive mechanisms at the neuronal level in the cerebral, frontal and parietal-temporal cortex (Figs. **4-8**).

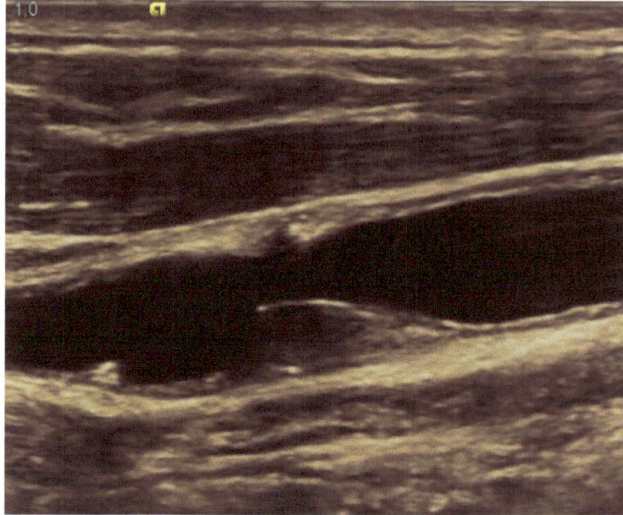

Fig. (4). Internal carotid artery stenosis in a patient with cognitive impairment and atherosclerosis. Image used with permission from Dragoș Camen.

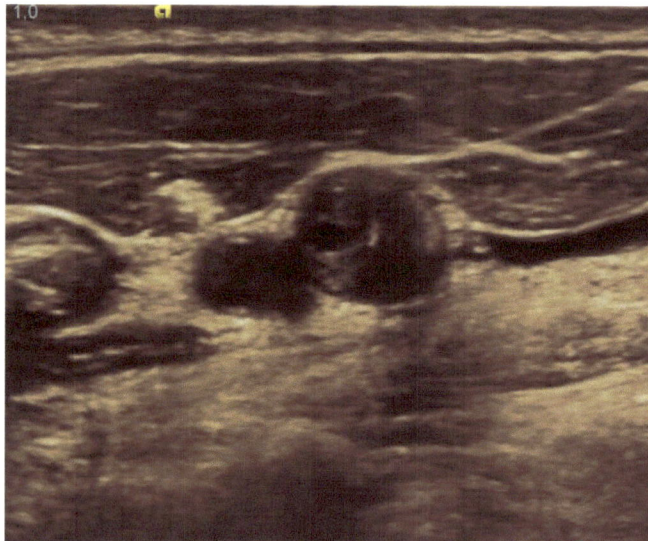

Fig. (5). Narrowing of the arterial lumen with modifications of the endothelial basal membrane - upper crack and lower rupture. Image used with permission from Dragoș Camen.

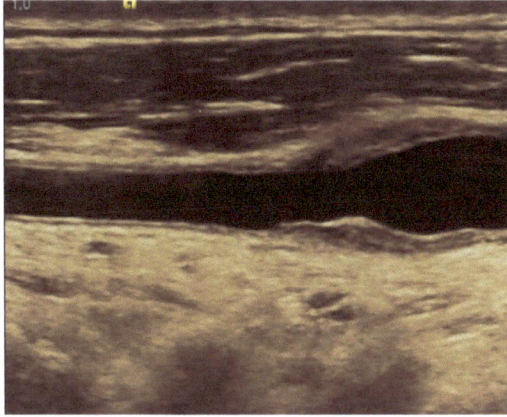

Fig. (6). Severe atherosclerotic modifications without subendothelial blood infiltrates. Image used with permission from Dragoş Camen.

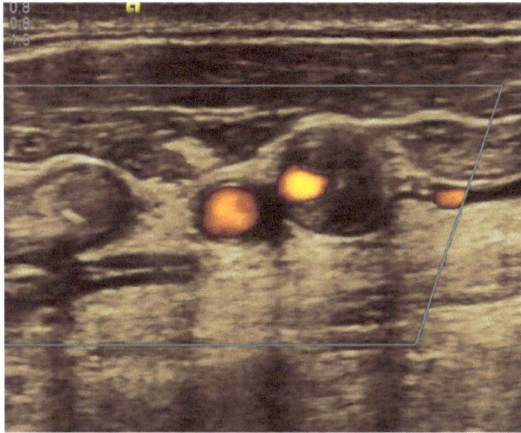

Fig. (7). Severe atherosclerotic modifications with subendothelial blood infiltrates. Image used with permission from Dragoş Camen.

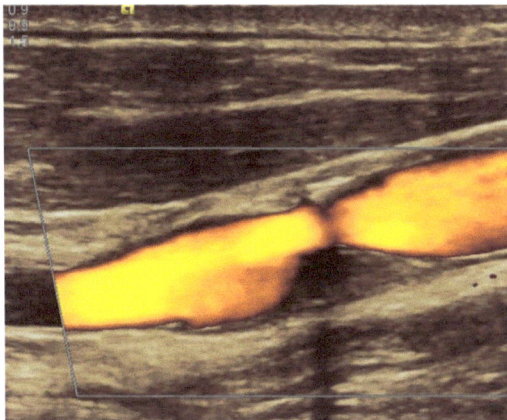

Fig. (8). Internal carotid artery (ACI) thrombosis. Image used with permission from Dragoş Camen.

The same type of alterations can also be highlighted in the Doppler evaluation of posterior large arteries, basal artery and vertebral arteries, correlated with posterior cortical hypoperfusion of the occipital lobe and arteries providing the irrigation of the hippocampus, thalamus and limbic system. The practical execution of such a paraclinical investigation is more difficult, being dependent on the accessibility offered by the particular anatomical conditions of the skull, allowing visualization of the monitored arteries.

Cervical spondylosis-type manifestations with vertigo-like disorders and tinnitus are common in at least 2/3 of elderly patients and are attributed exclusively to vertebral osteoarticular manifestations of degenerative type, thus ignoring the possibility of an early diagnosis of the vascular component that may precede MCI syndrome or AD.

Information related to the obliteration of the internal carotid artery can be obtained surprisingly, in panoramic dental radiography (Fig. **9**). The involvement of the dentist in prophylaxis of cognitive changes in AD with vascular component becomes an important link in this strategy [32].

On the other hand, the presence of periodontitis elements in elderly people highlights the lack of oral hygiene that can become an important marker of cognitive status, suggesting the appearance of a cognitive impairment if, prior to the assessment, the family claims that the patient had oral hygiene skills. This suspicion can be confirmed by the progression in time, from one dental examination to another, of hygiene deficiency and also of cognitive abilities, as the dentist may observe amnestic phenomena or orientation difficulties. A second pathogenic condition identified in the periodontitis-AD relationship is the chronic inflammatory process, which can sometimes evolve to severe manifestations in the oral cavity by gram-negative germ infections, which in turn suggests a deficient immune status (Fig. **10**).

Periodontitis is a chronic inflammatory disease that significantly decreases the cognitive process [33], through a systemic reactive type amyloidosis mechanism, and by constantly maintaining the chronic inflammatory process characterized by high levels of reactive C protein, interleukin IL6, α TNF [34, 35].

The Component of Genetically Controlled Vascular Vulnerability is Represented by the presence of Cerebral Autosomal Dominant Arteriopathy with Subcortical Infarcts and Leukoencephalopathy (CADASIL), the most common genetic form for small vessel disease (SVD) [36].

The presence of the genetic spectrum of CADASIL syndrome associated with rapid cognitive impairment is related to elevated levels of homocysteine (HCY)

[37]. This biological marker signifies the significant increase in oxidative stress and inability to regulate at this level, amplifying oxidative harmful action of reactive oxygen species (ROS). In this context, oxidative damage and excitotoxicity increase, especially through mitochondrial dysfunction. CADASIL markers of endothelial dysfunction, which may be correlated with clinical outcome, are represented by C-reactive protein (CRP), circulating progenitors cells (CPC), circulating endothelial cells (CEC) [36].

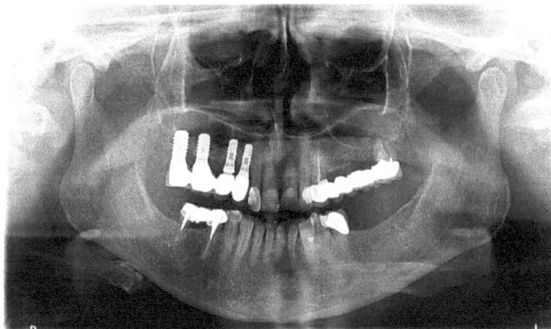

Fig. (9).   Panoramic radiograph which highlights atheroma in right carotid artery. Image used with permission from Adrian Camen.

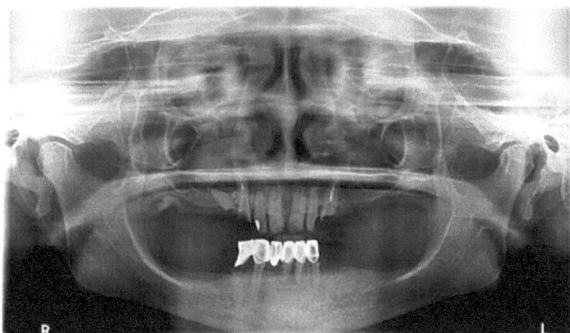

Fig. (10).   Aspects of periodontitis in an elderly patient with cognitive impairment. Image used with permission from Adrian Camen.

CRP signals the maintenance of the inflammatory process and is associated with cognitive decline [38, 39] and CPC and CEC are associated with endothelial dysfunction and the level that compensates for neuronal deficits following repeated cerebral ischemia or micro-bleeds through the development of vascular compensatory circuits [36].

Interestingly, this disease with the expressed genetic component (the familial form) occurs in young age being theoretically the form of early onset of cerebral leukoaraiosis. As in AD, there is a late-onset leukoaraiosis or the sporadic form in which genetic control is little expressed (the CADASIL gene spectrum involved in elderly patients).

We suggested the increase of research activity through studies to confirm or not, the correlation of the CADASIL pathogenic form, the CADASIL gene spectrum and the two clinical forms of AD. The similarity of this type can be also identified for Parkinson's disease or systemic amyloidosis, which is why there is a suspicion that with age genetic abnormalities become aggressive due to the genetic spectrum, activated by multifactorial, biological and epigenetic mechanisms, triggering pathogenic mechanisms of the neurodegenerative type.

Neurobiochemical Vulnerability

Loss of homeostasis in cerebral neurotransmission subsystems that can no longer provide the functioning of compensatory mechanisms under conditions of alteration of the baseline pattern of specific circuits. This type of vulnerability favors the progression of neurodegenerative elements. In our opinion, it is dependent on both vascular and systemic (cardio-cerebrovascular) changes and the modification of the functional relationship between neural structure, astrocytes, neurovascular unity.

Progression of vascular vulnerability and, implicitly, neurodegenerative factors is significantly influenced by cerebral neurobiochemical alterations associated with cognitive impairment (acetylcholine, dopamine and noradrenaline deficiency).

The hypothesis of AD's cognitive deficit as well as the progression of β-amyloid elements is supported by the acetylcholine deficiency theory as a baseline element of the progression of cognitive dysfunction in AD [40, 41].

Cognitive cholinergic circuits have a specific structure in the brain, there is a Meynert basal nucleus and many interneurons that act in a manner diffused to all cortical and subcortical neural structures. Interneuronal cholinergic-type mechanisms have a marked vulnerability, which can be modified by intervention of drug-type blocking factors (anticholinergic medication) [42] or vascular factors destroying heterologous neuronal structures (serotonergic, dopaminergic and noradrenergic) constituting relay stations for cholinergic interneurons. Also, metabolic, toxic or traumatic factors can cause axonal dysfunctions that disrupt the acetylcholinergic connectivities of the corticosteroids at the level of the cortical projections [43, 44].

Cholinergic interneurons control the levels of major neuromediators that are involved in the pathogenic mechanisms of major psychiatric disorders, given the relationships between acetylcholine - serotonin in depressive disorder, acetylcholine - dopamine in schizophrenia with positive or negative symptoms or relationships between excess cholinergic transmission and depressive episode of disease bipolar treatment in contrast to acetylcholine deficiency in the manic episode [45].

A greater imbalance between levels of acetylcholine controlled by neural structures and those controlled by astrocytes causes the activation of the glutamatergic system and the involvement of excitotoxic mechanisms (Fig. **11**).

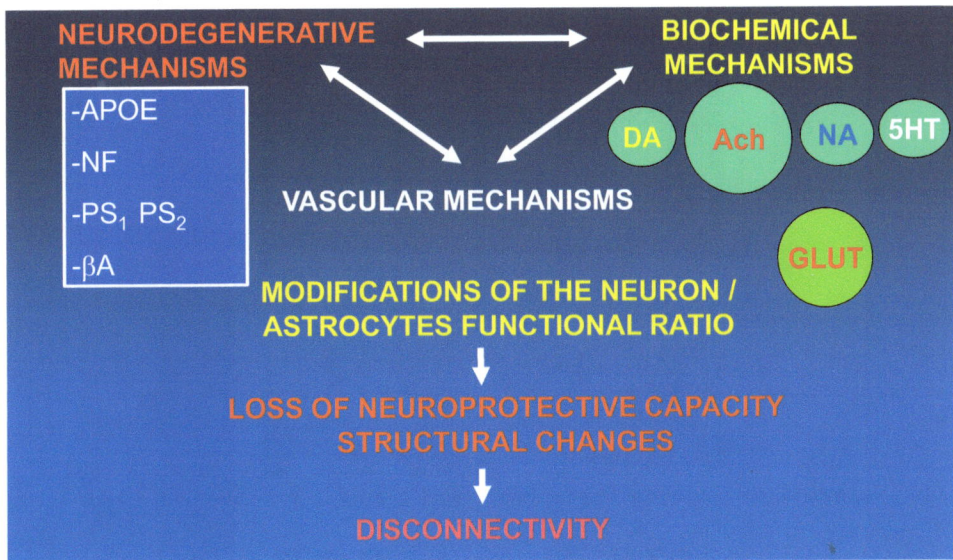

Fig. (11). Dysfunctional neurobiochemical mechanisms in AD that engage in multimodal pathogenic processes. APOE = Apolipoprotein E; NF = neurofibrils; PS_1 = presenilin 1; PS_2 = presenilin 2; DA = Dopamine; Ach = Acetylcholine; NA = Noradrenaline; 5HT = Serotonin; GLUT = Glutamate.

A bimodal role can be attributed to biochemical mechanisms, acetylcholine, noradrenaline, dopamine, or serotonin deficiency, causing vasoconstriction of small cerebral vessels with the appearance of small vessel disease.

Cerebrovascular disease is commonly associated with CAA and with massive Aβ deposition. We notice that small cerebrovascular disease can be announced by a valuable neuroimaging marker of enlargement of the intergiral sulcus. This indicator is frequently ignored by both clinicians and neuroradiologists, especially when evaluating a patient sent by a psychiatrist is evaluating. The CT scan images may highlight this marker correlated with small vessel disease, the CAA process, and the massive Aβ deposition (Figs. **12** and **13**).

The presence of Benson syndrome characterized by alteration of acuity and visual functions by central mechanisms often brings this type of patients to the ophthalmology cabinet. Despite the integrity of the visual function in the ophthalmologic examination, the patient and his / her family urgently require ophthalmologic reconsultation or even surgical interventions, with a strong belief of diagnosis error from the ophthalmologist, a situation in which a malpractice conflict may be triggered [46]. Recognition of this type of patient and guidance to a neuroimaging assessment and interdisciplinary neuropsychiatric consultation can identify the true pathogenic support of visual deficits, which is Benson's syndrome or posterior cortical atrophy [47].

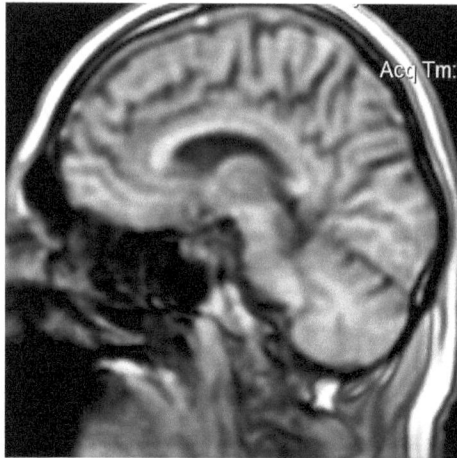

Fig. (12). Nonspecific atrophy of the left occipital pole suggesting a particular form of AD - Benson syndrome.

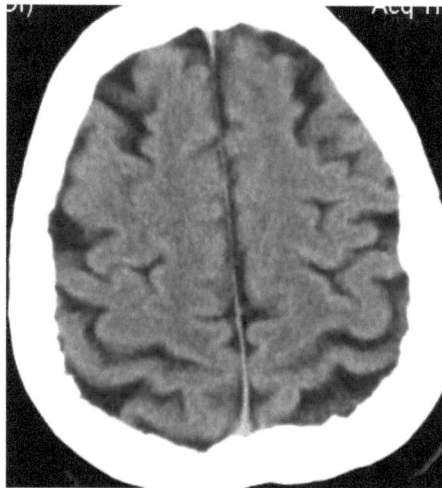

Fig. (13). Enlargement on the entire brain surface of the intergiral sulcus.

Small cerebral vessel dysfunction affects pyramidal neural structures and may present significant neuroimaging indicators long before the minimal cognitive deficits (MCI) or neurological deficits are detected. Small vessel disease can be considered as a permissive factor in stimulating mechanisms controlled by ApoE4, with the rapid progression of Aβ deposits.

The deficiency of the above-mentioned brain neurotransmitters may determine the psychiatric facet of the prodromal states of AD:

- Dopamine deficiency induces a form of dopamine-dependent depression characterized by adynamia, apathy, anhedonia, to which the extrapyramidal neurological phenomenon, including the restless legs syndrome, is associated in a variable proportion.

- Noradrenaline deficiency has a clinically appearance dominated by depression, psychomotor inhibition, loss of interest to the outside world, stuporous state, mimicking a serious somatic disease, which alerts the family. Cognitive deficiency is important, with the appearance of a "pseudo-dementia" with noradrenaline deficiency.

- Serotonin deficiency has a clinical appearance of an anxiety depression, rebel insomnia, and hostilities, with a strong behavioral component and increased suicide risk.

All of these neurotransmitter deficits generate compensatory hyperactivity of glutamate stimulating amino acid, with hyperglutamatergia being favored by the primary deficiency of gamma-aminobutyric acid (GABA) (Fig. **14**).

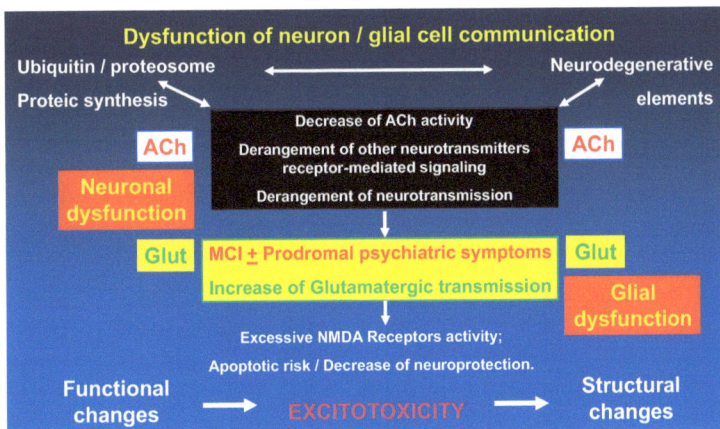

Fig. (14). Cellular, biochemical and neuropathological multisystemic mechanisms that cause neuronal excitotoxicity.

These imbalance mechanisms of GABA-glutamate ratio increase in a first phase neuronal excitability, especially in the area of the cingulate gyrus hippocampus, the cerebral tonsil and the temporal medial lobe. Hyperexcitability and neurobiochemical multisystemic dysfunction can be signaled by changes in electroencephalogram (EEG) with the appearance of irritant waves or very fast frequencies (gamma rhythm). Gamma rhythm is considered by neuro-electrophysiologist as a potential indicator of neurodegenerative phenomena. The specificity of encephalographic markers increases in behavioral sleep disorders with changes in REM latency. Exceeding this pathogenic stage causes excitotoxic mechanisms generated by glutamate. Hyperactivity of the NMDA type of glutamate triggers apoptotic mechanisms with significant neuronal destruction, causing disconnective processes.

Cognitive Vulnerability, Cognitive Reserve and Cognitive Circuits - the Correlation between Structural Brain Changes and Progression of Cognitive Deterioration in Alzheimer's Disease

The major progress of early diagnosis in AD was achieved when MCI mechanism was recognized as a potential clinical entity announcing the subsequent installation of AD. Petersen's initial data were based on descriptions of the neurocognitive disorders depicted in 1988 by Reisberg et al, who first introduced the term MCI [48].

In the MCI syndrome, amnesia was situated as a central element, which may affect a single or multiple cognitive domains. Since the notion was launched, the interest has been relative, the medical world being concerned with diagnosing genetic risk and identifying potential neurobiological or neuroimaging markers to turn the potential risk into certainty. In the first decade, 1990-1999, the number of articles published on this topic was very small. The first diagnostic criteria were established in 1999: memory complaint, preferably corroborated by an informant; memory impairment documented according to appropriate reference values; essentially normal performance in non-memory cognitive domains; generally preserved activities of daily living; not demented [49].

After establishing the criteria, the interest in researching the correlations between the earliness of MCI syndrome and the risk of switching to AD dementia cognitive impairment increased. The number of publications devoted to this research theme has increased exponentially, from about 50 in 1999 to 900 in 2007 [50].

Longitudinal studies highlight the fact that there is a long period of years when the accumulation of Aβ, confirmed by neuroimaging images and its presence in the CSF, is positive without delineating the clinical signs of MCI syndrome. This clinically silent period has "neurodegenerative" support, represented by the progressive accumulation of Aβ. It is also called the "prodromal period of AD" or latency of cognitive deterioration manifestations, this pathophysiological stage being also considered "AD without dementia".

Thus, the premises are for the clinical interpretation of three clinical forms:

- AD without cognitive impairment, characterized by identifying Aβ deposits and positive markers at the CSF for amyloid and your proteins

- AD with incipient cognitive impairment - "MCI syndrome", which once overcome, causes the dementia to be installed. Thus, a very little-studied link can be made with other neurodegenerative diseases that can be accompanied or not by important cognitive impairment: dementia from Parkinson's disease, multiple sclerosis, amyotrophic lateral sclerosis.

- Proper AD with early, medium or severe cognitive impairment [50].

We notice that the speed of transition from one clinical phase to another, in the late form of AD, is predominantly dependent on multisystemic mechanisms and less on the genetic factor. The final phase of AD occurs after 10-15 years of evolution.

Differently, AD with early onset has a mainly genetic pathogenic influence, of neurodegenerative type, with rapid progression, the terminal phase being in 3-5 years.

The longitudinal studies of AD with late onset (sporadic form) revealed the tendency of progressive accentuation in the aggressiveness of the neurodegenerative elements, this trend being considered as a real objective marker of the evolutionary risk, of the rapid transition to the phase with severe cognitive deficits.

The issue of AD's cognitive decline is correlated with a multifactorial pathogenic model, covering the three distinct stages in evolutionary terms.

1. The First Stage is represented by the period when the pathogenic risk mechanisms are present without any cognitive disturbances that can be fit within the criteria of the MCI syndrome. This period can be defined as a period of cognitive deficiency latency, a period that can be assumed in terms of identifying potential pathogenic mechanisms and where the major role is played by risk factors.

2. The Second Stage in which the cognitive deficit is identified (MCI stage). The risk factors at this stage are those that influence the progression of cognitive decline, depending on individual cognitive resilience. Also, maintaining the functionality of cognitive circuits through connectivity loops may in turn be dependent on preserving the structural integrity of emerging zones for stem cells (the subventricular granular area and the granular area of the dentate gyrus). In this context, we note the importance of maintaining an effective relationship between neuroprotection capacity, neurogenesis and neurodestructive mechanisms, in correlation with risk factors. At an older age, there is a decline in neurogenic capacities, which is why the use of stem cell transplant therapies should be made at this stage of latency, knowing that vascular, metabolic and toxic risk factors compromise neurogenesis.

This component of the multifactorial pathogenic model brings to mind the importance of conserving and even developing the cognitive resilience capacity represented by the cognitive reserve. This hypothesis can be correlated with the low intellectual level, considered as a risk factor for AD in older age. On the other hand, the cognitive reserve in individuals with a high intellectual level leads to a significant delay of diagnosis and limitation of the possibilities of therapeutic intervention and improvement of the evolution of this disease.

The neurobiological model of the cognitive reserve highlights the importance of prophylactic activity in order to maintain functional cognitive brain circuits by intensifying the molecular, cellular and systemic modulation activities. The analysis of risk factors supports the involvement of several pathogenic mechanisms and requires a differentiated therapeutic approach [51]. The synchronization of the cognitive deficit with the pathogenic pattern of this dysfunction can be done in several stages:

a. Using cognitive training techniques that require continuous stimulation and cognitive training.

b. Correction of sensory deficits to allow a correct collection of information for cognitive processing [52]. Untreated sensory deficits, visual (cataracts senile) [53] or auditory (hearing loss) [54] can generate aberrant hyperstimulation of cortical projection areas by the intervention of hyperglutamate mechanisms that eventually cause excitotoxic or apoptotic neuronal damage.

At the same time, physical activity (physio-kinesiotherapy) can maintain energy balance in the mitochondria by involving neuroprotective factors represented by brain-derived neurotrophic factor (BDNF), and by reducing the production of ROS [55], but also in the neuron by involving ubiquitin proteasome system (UPS). The role of the UPS in synaptic plasticity, could be linked to the early phase of AD, which is marked by synaptic dysfunction, as well as to the late stages of the disease, characterized by neurodegeneration [56].

c. Combining physical activity with cognitive training and correcting sensory disabilities that can rebalance existing biochemical deficits between the pre- and post-synaptic poles of neurons involved in cognitive circuits and may increase neurogenesis. These compensatory mechanisms can be accomplished by inverse stimulation, from motor to brain level, as well as by reducing oxidative stress.

d. Rehabilitation of neurogenesis should be considered in the inefficient conditions of the methods applied in the previous 3 stages. This rehabilitation is represented by the use of exogenous neurotrophic factors and avoiding decompensation moments related to comorbid somatic disorders, after which it is possible to decide the use of stem cell transplantation techniques. The neurogenesis rehabilitation brings homeostatic type benefits to the presynaptic pole by rendering the neuron-astrocytes unit functional and by increasing the secondary signaling of cyclic adenosine monophosphate (cAMP).

This functionality determines the decrease of glutamate aggressiveness at the postsynaptic multisystem interconnections through a potential increase in GABA level. In this way, by reducing the excitotoxicity and aggressiveness of the glutamate system, the synaptogenesis is improved, the transport capacity of the neurotrophic factors at the axonal level increases and the synaptic re-plasticizing at the dendritic level is stimulated.

Targets pursued by neurogenesis rehabilitation strategies, respectively, synaptogenesis and neurogenesis with preservation of neuron-astrocytes ratio, can be effective in the following circumstances [57]:

a. The existence of an acceptable stem cell potential in emergence areas. It should be taken into account that hippocampal atrophy or sclerosis (Fig. **15**) and ventriculomegaly (Fig. **16**), are indicators that predict the ineffectiveness of such an approach.

Sensory deficits are not irreversible due to deterioration of the analyzer, transmission pathways or brain area because of pathogenic tumor or lesion conditions.

b. The impossibility of practicing physical activity, which is dependent on the individual reserve of the functioning of the peripheral skeletal neuromuscular system, of the myocardium, as well as of the preservation of the liver, renal and pancreatic functions.

All of these functions are negatively disturbed by the presence of systemic amyloidosis, which has several forms: light chain amyloidosis (AL), reactive systemic amyloidosis, or senile amyloidosis [58].

The positive effects of sensory, cognitive and motor stimulation, have been demonstrated through cellular and molecular research. Physical activity, including the use of running wheels by patients, has shown significant efficiency in mental activity and behavior [57].

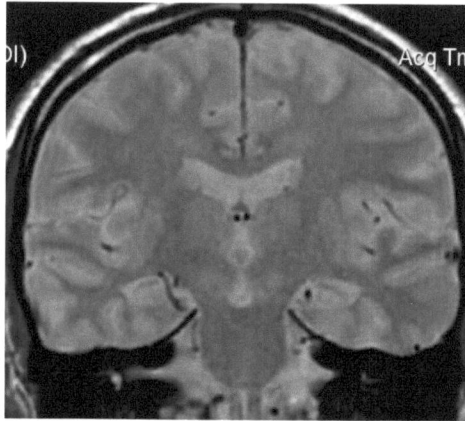

Fig. (15). MRI image with bilateral hippocampal atrophy in a patient with early AD.

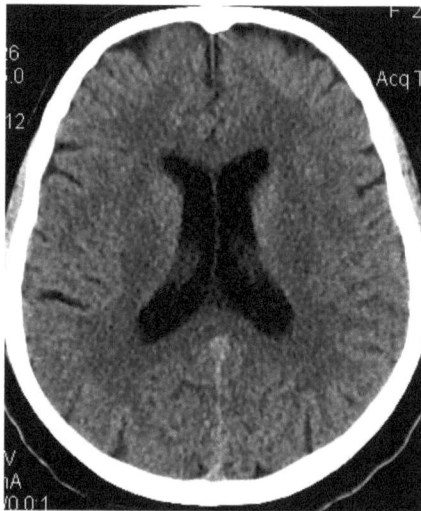

Fig. (16). Image of ventriculomegaly in a patient with moderate AD.

All of these multisystem dysfunctionalities may be considered premorbid AD comorbidities by pathogenic mechanisms of the amyloidosis type, often ignored in everyday practice. We mention this particularity as there are still insufficiently sustained but pathogenic hypotheses. Schroder in 1999 [59], identified post-mortem in 20 patients with amyloid light-chain (AL) or primary amyloidosis and amyloid A amyloidosis (AA), amyloid vascular changes in the choroid plexus in 17 out of 20 cases. This confirms the suspicion of our working group that structural alteration or BBB dysfunction can be associated with calcification of the choroid plexus becoming a major neuroimaging marker that can signal the imminence of installing the transformation of amyloid pathology into Aβ aggression (Fig. **17**).

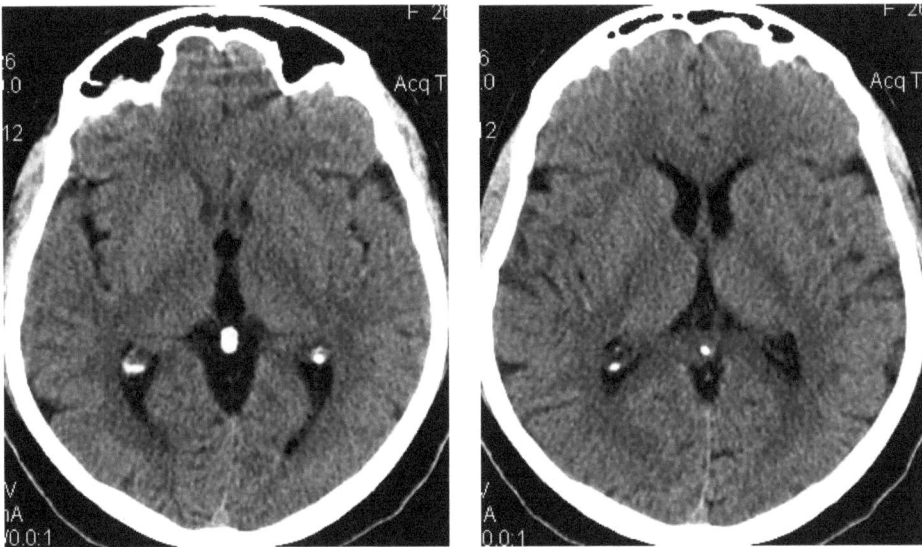

Fig. (17). (a and b) Patient with MCI suspected of AD with calcifications of choroid plexus with discrete ventriculomegaly and right temporoparietal atrophy.

One of the causes of systemic amyloidosis involvement in the pathogenic AD model is due to the amyloid fibrillogenesis, process in which the cross-β structural motifs of the amyloid precursor protein, can increase the aggregation of cerebral Aβ plaques and amplify the progression of cognitive decline [60].

β-amyloid protein fibrillogenesis is considered as a complex, central molecular mechanism that transforms β-amyloid precursor proteins into protofibrils through a polymerization process, and protofibrils component peptides ($A\beta_{1-40}$, $A\beta_{1-42}$) can be identified as markers of risk for AD at cerebrospinal fluid (CSF) level. This mechanism has become a target in pharmacological research to identify molecules with inhibitory effect for the fibrillogenesis process. We mention that β-amyloid protein is a normal component in cerebral structures, in the CSF and plasma, where soluble, residual variants are identified. Their level changes with the risk of developing AD. Alteration of the polymerization process causes the transformation of these residues into insoluble variants that make up a real fibrillary "net" in which Aβ will be deposited, strengthening the senile plaque [61].

Fibrillogenesis inhibitors can block the deposition of amyloid plaques at the extracellular level while retaining the functionality of cellular structures at the central (neurons) and peripheral (hepatocytes, nephrons, and myocytes) cellular structures. This creates a real vicious circle in which the cerebral neurodegenerative process contributes to peripheral dysfunction and suffering, which in turn alter cerebral homeostasis by accelerating the neurodegenerative process. Two amyloidosis processes are identified:

- Central, at the cerebral level

- Peripheral, following the alteration of cellular mechanisms of transformation of β-amyloid protein structures. Peripheral amyloidosis mechanisms may also be the consequence of the genetic vulnerability induced by the systemic amyloidosis spectrum.

By accepting these new perspectives in AD pathogenesis, functional disorders of cerebral homeostasis support systems can be disrupted by two types of comorbidities:

- Somatic amyloidosis that may be the consequence of central mechanisms (the excess of central Aβ reaches the periphery and causes cellular damage to several organs).

- Actual somatic ones or non-amyloidosis.

The association of somatic disorders can disturb cerebral homeostasis, which favors the progression of neurodegenerative mechanisms and the rapid cognitive decline in AD. The importance of this pathogenic differentiation results from the fact that the systemic amyloidosis process, especially senile amyloidosis, determines an unfavorable prognosis through therapeutic non-responsiveness to the usual therapies for non-amyloidosis comorbidities [62].

An example is cardiac amyloidosis with a poor prognosis, regardless of the mechanism of immunoglobulin amyloidosis, familial amyloidosis, senile systemic amyloidosis, secondary amyloidosis [63].

3. The Third Stage of AD is the one with medium or severe cognitive deficit, where control of risk factors is difficult to achieve. For this reason, interventions for limiting evolution and switching to the terminal phase of the disease are minimal or impossible. Based on the theoretical premises we support, the prophylactic strategies of rapid progression of cognitive deficits must be instituted from the prodromal stage of the disease through specific pharmacological or non-pharmacological measures.

Progression in the prodromal phase of neurodegenerative elements, but also in the MCI syndrome stage, can be influenced by many factors that determine an accelerated or slow rate of evolution and which, in our opinion, can be significantly influenced to delay the deteriorating process. Depending on the preservation of a balance between neurodegenerative cerebral aggression and neuroprotective and neurogenic mechanisms, the rate of cognitive impairment may be slow or rapid (Fig. **1**).

From our point of view, we consider that there is an associated somatic pathology, independent of AD, and a somatic pathology that shapes the vulnerability for a generalized amyloidosis process (dilated cardiomyopathy, nonalcoholic steatosis, renal amyloidosis with heavily controllable high blood pressure, systemic amyloidosis with altered diencephalic neuroendocrine function and cardiometabolic syndrome - obesity, diabetes).

Somatic pathology prior to AD and independence from the amyloidosis process can precipitate cognitive impairment phenotypes by pathogenic factors linked to functional cerebral perfusion ineffectiveness, similar with other pathogenic processes (heart failure following a valve lesion, dilated cardiomyopathy present in systemic amyloidosis).

In conclusion, independent somatic diseases or those caused by systemic amyloidosis have an identical dysfunctional mechanism at cerebral level, causing ineffective blood perfusion (chronic cerebral hypoperfusion syndrome), chronic metabolic deficit (hypoglycemia) or oxygen deficiency (hypoxia).

In the case of associated somatic diseases independent from the amyloidosis process, cognitive dysfunction and progression of neurodegenerative elements may also be influenced by the inappropriate use of therapeutic drug strategies:

- Beta blockers that alter the noradrenergic system and cause lesions of the locus coeruleus. Frequently, patients with this type of therapy due to noradrenaline deficiency may experience attention disorders, apathy, adynamia, depression and anhedonia associated with sleep disturbances and aberrant behaviors during sleep. There is sufficient evidence of progression of cellular neurodegenerative elements, Aβ progressing during insomnia. Insomnia is a risk factor for accentuating Aβ deposits [64].

- Anticholinergic medication or the use of drugs that pharmacologically block M1-M5 muscarinic receptors. These drugs produce the classic pattern of cognitive impairment caused by the blockade of cholinergic mechanisms [41, 65].

These mechanisms gain major importance when there is a depressive episode in the patient's history requiring psychopharmacological therapy because both antidepressants and antipsychotics have the ability to block muscarinic receptors.

Translational studies on possible cholinergic mechanisms involved in the progression of cognitive dysfunction in AD on the animal model have shown the following aspects: the progression of Aβ deposition is favored by the destruction of cholinergic interneurons terminations, while functional maintenance of cholinergic interneurons diminishes the aggressiveness of Aβ [66].

There is medication with an indirect "detectable atropine like activity" that can associate cognitive impairment in the long-term use but also side effects that can consequently result in drug therapies: dizziness with visual disturbances and diffused swallowing and consecutive speech, the ocular and buccal mucosal dry; dysuria and urinary blockage requiring catheterization; tachycardia and various cardiac arrhythmias; marked anxiety accompanied by psychomotor agitation; confusion, delirium [67]. The most commonly used drugs of this type are: cimetidine, prednisolone, digoxin, furosemide, warfarin, ranitidine, codeine, captopril [68].

Other classes of drugs with anticholinergic effects, commonly used in the pathology of elderly patients are:

- Glucocorticoid cortisone medication.

- Commonly used antihistamine medication for the elderly with chronic dermatological and respiratory diseases.

- Medication leading to a reduction in cerebral blood flow: excessively used antihypertensive medications or inappropriate associations of several classes of antihypertensive substances that favor orthostatic hypotension. The minor symptoms associated with hypotension (dizziness and steady-state equilibrium disturbances in orthostatism) may be clinical markers that predict the progressive deterioration of executive cognitive processes and the speed of information processing through cognitive circuits. The predictive value may correlate with the history of these somatic disorders requiring antihypertension medication and cerebral structural status prior to hypotension.

- Medication disrupting hydro-electrolytic balance and pH - thiazide diuretics. The ionic changes following this type of imbalance and which are associated with dysfunctional cognitive potential are represented by hypokalemia, hypomagnesemia and hyponatremia.

- Anticonvulsant medication represented by phenobarbital and first generation of antiepileptic drugs.

- Cephalosporin class antibiotics.

- Anti-inflammatory and antialgic medication (ibuprofen, indomethacin, tramadol).

Vulnerabilities Generated by Neuropsychiatric Pathological Conditions
Traumatic Brain Injury (TBI)

Cerebral traumas with alteration of brain function have a high frequency in the elderly, most of them as a result of car accidents or domestic accidents. Few epidemiological studies point to the fact that at the old age there is a predisposition towards low intensity, unique or repeated TBI, which in conditions of structural cerebral and cerebrovascular vulnerability may have severe consequences in the medium and long term.

The link between TBI and cognitive deficit is based on the following mechanisms:

- TBI increases deposition of Aβ deposits and amplifies CAA, especially in the presence of ApoE4 spectrum [69].

- Functional disturbance of neurons and connections with the astroglial system, favoring extracellular amyloid deposition with blocking functional communication capabilities between neuron and astrocytes, making a true pathological triangle between TBI, microglial activation and Aβ deposition (Fig. **18**) [70].

TBI acts through the following mechanisms:

- The specific mechanism is represented by *diffuse axonal distress* with significant dysfunction in the anterograde and retrograde transport of the axons, a mechanism associated with the amplification of axodendritic or axo-axonic disconnectivities between cerebral neurons belonging to a single neurotransmission system (*unimodal disconnectivity*) or between neurons multiple neurotransmitter systems that act compensatory (*multimodal disconnectivity*).

- In the dementia resulting from medium or severe cranial injuries, single or repeated, post mortem revealed the important contribution of multiple protein changes that cause cognitive impairment, changes involving Aβ, tau proteins, Lewis body inclusions, TDP-43, leading to the hypothesis of multiple neuropathogenic mechanisms [71].

- The occurrence of post-traumatic encephalopathy that can be evidenced by EEG changes or by clinical manifestations of epileptic type (post-traumatic epilepsy) [72].

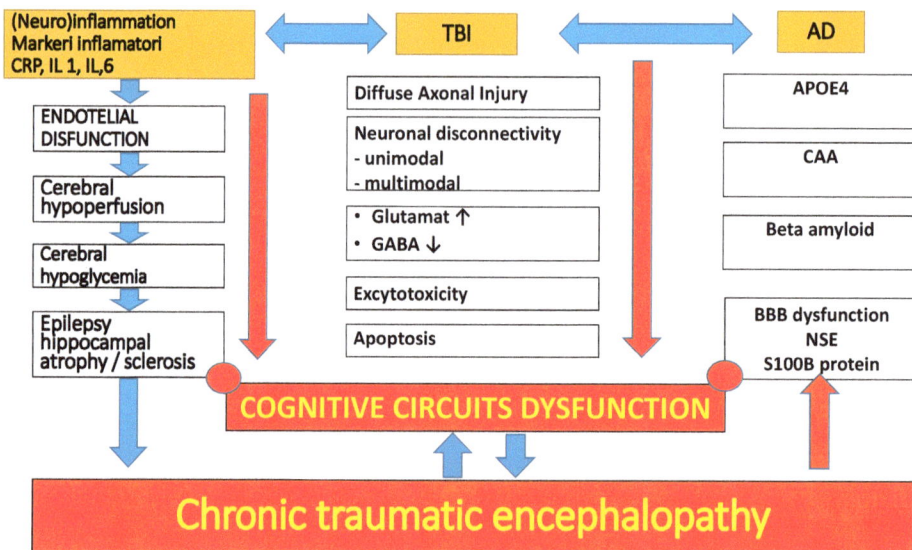

Fig. (18). Multifactorial model of cognitive impairment in AD, after TBI.

- "Chronic traumatic encephalopathy", characterized by common elements with AD, which suggests potential common neuropathogenic mechanisms: accumulations of Aβ, diffuse senile plaques, neurofibrillary degeneration, accumulation of tau-immunoreactive astrocytes [73].

- Excitotoxic mechanisms which in turn amplify Aβ deposition. This peptide has engrams the neuroprotective compensatory action but acting aberrantly in the TBI situation, effectively stifling the neuronal and astrocytes communication connections, favoring triggering of neuronal apoptosis.

- The intervention of hyperglutamatergic mechanisms is associated with the amplification of cognitive dysfunction, and progression may become even faster in the conditions of identification of hippocampal sclerosis.

In the conditions of older people, the brain microtrauma generated by family violence, which is frequently associated with changes in behavior at the incipient / dormant stages of AD, is of major importance. Social, educational, economic components and family violence amplifies interpersonal stress, becoming a factor that precipitates AD evolution and significantly reduces home care. It can be discussed in this context of specialized prophylactic measures to increase cognitive resilience by diminishing the social, economic and educational stress factors and addressing people specialized in social psychological assistance.

Assessments of brain structure changes after craniocerebral trauma were evaluated in our working group post-mortem, in elderly patients by histopathological examination. The following anomalies were highlighted:

1. Hemorrhagic type abnormalities:

- Meningeal hemorrhage favored by the atherosclerotic process or amyloid angiopathy (Fig. **19**).

- Intraparenchymal diffuse hemorrhage with axonal degeneration, favoring disconnective mechanisms and dysfunctional cognitive circuits (Figs. **20** and **21**).

- Cortical microhemorrhage, similar to micro-lesions, with important consequences on cognitive rapid deterioration processes by altering the speed of information processing and synthesis at the cortical level (Fig. **22**).

- Perivascular hemorrhage in Virchow-Robin spaces similar to CAA-induced and small cerebrovascular disease-induced changes (Fig. **23**).

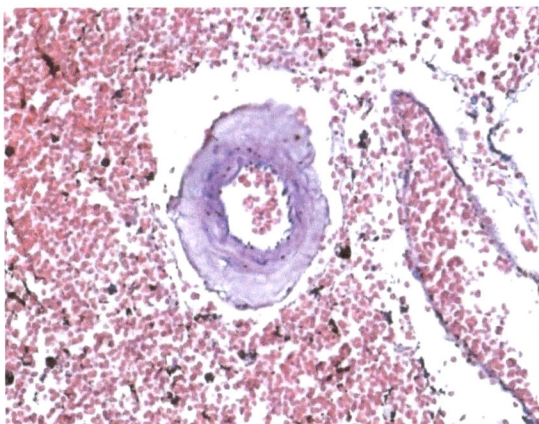

Fig. (19). Meningeal hemorrhage occurred in a patient with arteriolosclerosis (see the central arteriole).

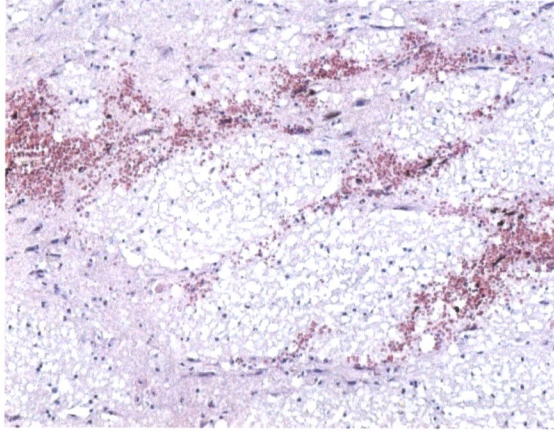

Fig. (20). Diffuse intraparenchymal hemorrhage associated with axonal degeneration.

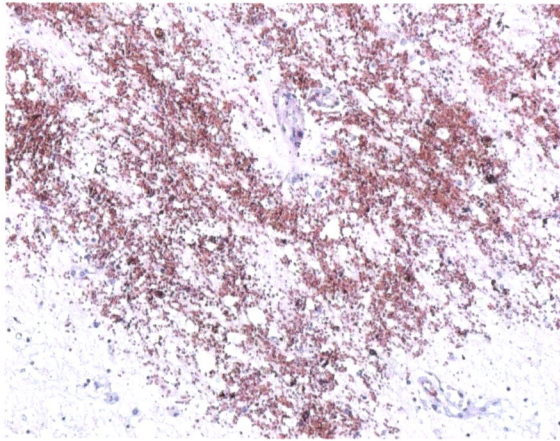

Fig. (21). Diffuse intraparenchymal hemorrhage.

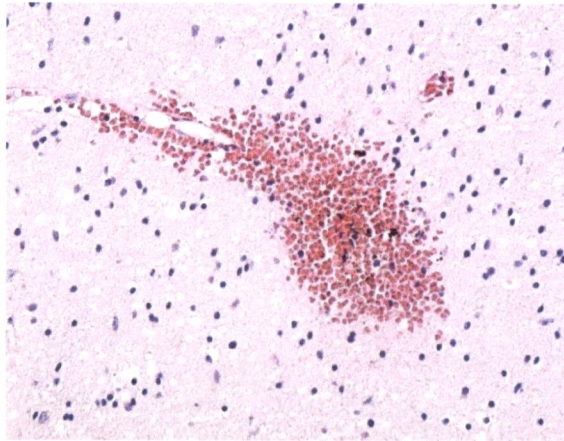

Fig. (22). Areas with microhemorrhagic point in cerebral cortex.

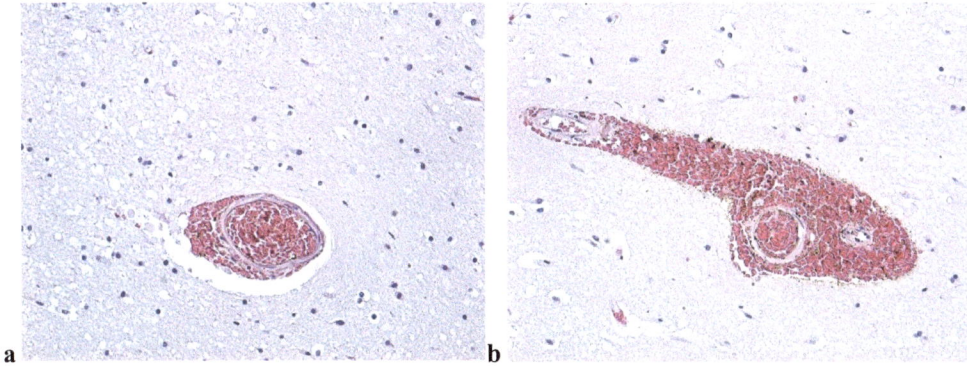

Fig. (23). (a and b) Perivascular hemorrhage through the Virchow-Robin space (basically the blood diffuses through the perivascular sheath which is made of lax tissue at relatively large distances from the hemorrhagic focal site).

2. ischemic abnormalities with neuronal injury with edema and pyknosis, similar to lesions induced by acute ischemia processes following a transient ischemic attack or chronic ischemia through various mechanisms of blood flow and blood flow lowering mechanisms (Fig. **24**).

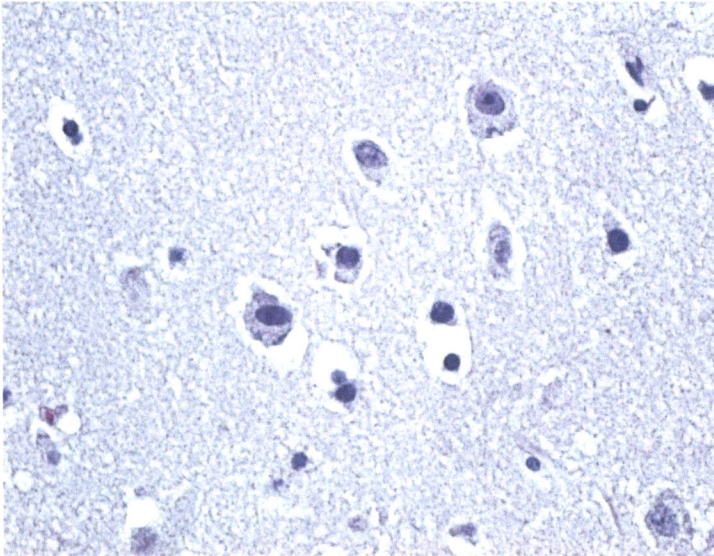

Fig. (24). Neural distress characterized by ischemic neurons (red neurons), perineuronal edema, tachicrom nucleus, nuclear pyknosis, condensation of neuroplasma.

3. Neuronal degeneration with alterations of white matter, of the vacuolar, spongiform type (Fig. **25**).

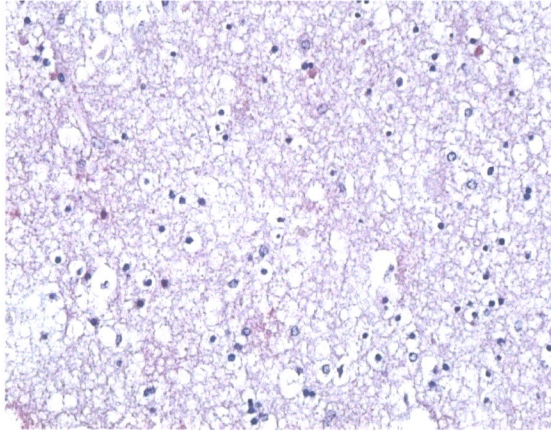

Fig. (25). White matter with signs of neuronal degeneration.

The traumatic history of elderly with small or medium intensity may produce severe alterations over medium and long periods, reason why they are considered at impact, without anticipating the consequences at that time. The cerebral vulnerability in the elderly is amplified by the process of cerebral atherosclerosis, metabolic dysfunction, use of anticoagulant medication, a history of orthostatic hypotension.

The severe traumatic conditions that occur in elderly people by car or household accidents require hospitalization, intensive care, polymedication. The biological vulnerability in the elderly determines the acute psychiatric symptomatology, confusion state, delirium, after the critical phase is overcome. These symptoms are a consequence of severe disturbance of brain homeostasis and therapeutic intervention with antipsychotic or sedative substances (benzodiazepines) increases the risk of triggering the neurodegenerative process and rapid cognitive decline.

Insomnia, Sleep Disturbances and Behavioral Disorders in Sleep (REM Sleep Behavior Disorder - RBD)

Insomnia in the elderly is one of the most common reasons why this category of patients is reaching to the family doctor. It is estimated that about 5% of the elderly population has insomnia, an incidence that increases significantly over 85 years [74].

The chronic nature of insomnia among the elderly is also amplified by comorbid conditions in which somatic disorders do not meet the criteria of chronic organ or multiple organ failure with significant dysfunctions. The important part for the family doctor is algal pathology and residual symptoms after an acute somatic episode including a transient ischemic attack. Another cause may be pharmacological, elderly people with a tendency towards self-medication, and due to the accumulation of residual somatic small dysfunctionalities that at one point had an acute episode during their lifetime, the pharmacological intervention becomes polymedication. The most common drugs involved in pharmacologically induced insomnia are: caffeine, nicotine, antidepressants, antiparkinsonian drugs (levodopa), beta-blockers, bronchodilators, anticholinergics, corticosteroids, H2 inhibitors, stimulant laxative [75].

Insomnia is a risk factor for cognitive decline by favoring the progression of neurodegenerative elements and the amplification of stress factors and, implicitly, hyperactivity of the hypothalamic-pituitary-cortical adrenal axis (HPA). An increased release of glucocorticoids causes an apoptotic aggression in the hippocampus and the frontal cortex by excessive glutamatergic activation at the extracellular level [76].

This mechanism was demonstrated on animal models (Wistar rats at which hypercortisolemia was induced by administering dexamethasone at the dose of 0.20 mg / kg / day). The destructive effects on neuronal structures were identified after 14 days of administration and confirmed the aggressiveness of excess glucocorticoids on the frontal cortex (Fig. **26**) and the hippocampus (Fig. **27**) [77].

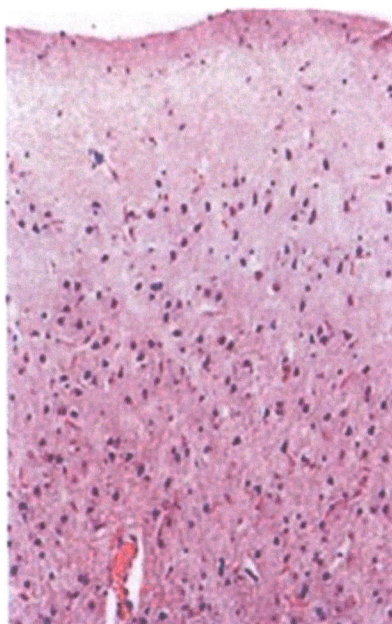

Fig. (26). neuronal destruction with pyknosis in frontal cortex after administration of Dexamethasone (HE staining, magnification 200X) [77]. Image used with permission from the main author and the editors of the RMJE.

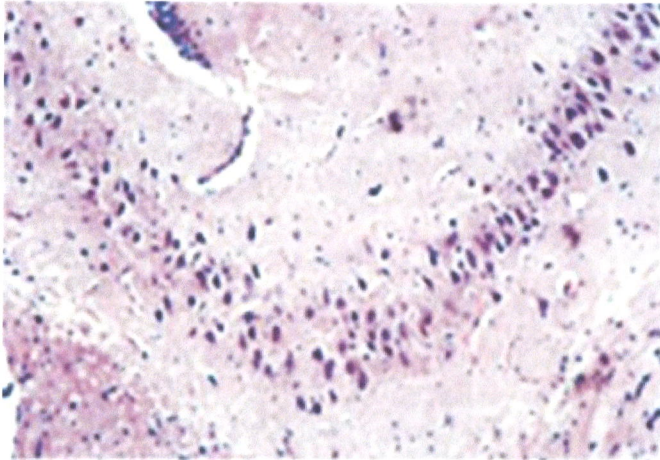

Fig. (27). Hippocampus with multiple vacuolization and neural loss (HE staining, magnification 200X) [77]. Image used with permission from the main author and the editors of the RMJE.

Hippocampal atrophy may be correlated with the decline of declarative memory, while frontal atrophy may be clinically correlated with decreasing of working memory and anhedonia apathy phenomena, thus confirming the indirect potential of cortisol activations to generate the disconnective process of brain structures involved in cognition ecosystems. The perception of insomnia by the elderly has a bimodality of interpretation: Sleep disorder has a somatic or pharmacological cause and the second is the fear from the elderly of death during sleep (thanatophobia).

The presence of insomnia, anxiety and agitation create additional stress in the familial microgroup or care staff, which leads to the establishment of a pharmacological therapy, often of the benzodiazepine type. Approximately 25% of the elderly use benzodiazepines. Insomnia and anxiety are often interpreted as part of the picture of a depressive disorder, especially when associated with frontal lobe symptoms: apathy, anhedonia, adynamia. Interestingly, in elderly patients with depressive disorder benzodiazepines are used in 37% of patients, while antidepressants in 27% of cases [78].

There is evidence to confirm the excessive use of benzodiazepines in the elderly without taking into account the risks of this type of medication: addiction and the occurrence of discontinuation syndrome, bronchoplegia, relaxant effect on the skeletal muscles that increase the risk of home accidents with severe consequences (head trauma or basal fractures or femoral neck). Decrease of attention may be in turn, especially in the early stage of MCI for the in people who are still on the job and using benzodiazepine treatment, the main cause of car accidents or cognitive errors that can cause work accidents or computational errors (economic crimes). In these situations, a rigorous forensic analysis is required, because long-term benzodiazepines are considered a significant risk factor for cognitive deficits, dementia and AD [79 - 81].

Benzodiazepine discontinuation syndrome, imposed by an acute pathologic condition, can greatly complicate the clinical picture and may become a cause of malpraxis. The main symptoms of the discontinuation syndrome are: paradoxical symptoms (sleep disorders, anxiety, irritability, panic attack) and somatic cardiovascular symptoms (palpitations, tachycardia), neurological (tremors of the extremities, headache, osteoarticular pain and muscle stiffness) metabolic (sweating that may suggest hypoglycemia in a patient with diabetes) [82].

Interestingly, excessive use of benzodiazepines in advanced age for sleep induction and control and also anxiety can cause and accentuate sleep apnea and bronchoplegia, complicating the progression of somatic diseases.

The appearance of sleeping discomfort has been identified in the elderly, some with severe consequences. Forensic expertise highlighted and was subsequently confirmed as REM sleep behavioral disorder. This entity is considered as a clinical, electrophysiological and neurobiological marker for predicting the development of neurodegenerative pathology. We report these issues, since RBD is not a rare entity, with an incidence of 1.15% for idiopathic form in the elderly population and 2% for all forms of RBD [83]. The idiopathic form of RBD is associated with α-synuclein and within the AD may be an important clinical predictor of the variant with the extrapyramidal component (Dementia with Lewy bodies). The precipitation in the evolution to the neurodegenerative form can be confirmed by some indicators: marked EEG slowing on spectral analysis; decreased striatal [123I] FP-CIT, a radioligand for the imaging of dopamine transporters; impaired color vision (marked EEG slowing) [84, 85]. Polysomnographic analysis highlights the decrease in sleep time with muscular atony in the RBD, being considered as an important marker of elevated muscle activity in REM [86].

The development of electrophysiological investigation techniques, including polysomnography, makes this type of investigation a common one. Data becomes more interesting, looking at the link between RBD mechanisms and benzodiazepine, as these drugs caused diminished muscle activity during REM sleep. Progression of cognitive deficits may be associated with EEG-type changes that correlate with diminished cholinergic activity. RBD is significantly associated with the decrease in cholinergic transmission from the early stages of neurodegenerative affections, and there is a confirmation of the overall decrease in cholinergic efficacy at the subcortical and cortical level. EEG analysis will reveal in patients who develop neurodegenerative diseases an absolute increase in all cortical areas of delta and theta rhythms compared to previous examinations [87 - 89].

If the first mechanism described above can be interpreted as a multifactorial pathogenic mechanism but non-specific to AD, the second mechanism is specific, as the deficiency of sleep increases the risk of excessive deposition of Aβ.

The influence of insomnia or sleep disorders on Aβ clearance is another mechanism that can anticipate the onset or accentuation of a cognitive impairment through EEG evaluation. PET imaging evaluations used to measure brain Aβ burden (ABB) in healthy, performed after a sleepless night compared to a resting night sleep (baseline), showed a significant increase in Aβ deposits in the hippocampus and right thalamus *versus* baseline. These changes were clinically correlated with mood worsening but were not related to genetic risk (ApoE genotype) for AD. Correlation of ApoE genotyping with subcortical ABB demonstrates that different AD risk factors can independently affect cerebral Aβ aggregation [90, 91].

Aβ deposits can induce circadian sleep-wake disturbances, and on the other hand, poor sleep may increase the risk of Aβ aggregation, the link between sleep and Aβ being therefore bidirectional. For example, obstructive sleep apnea can increase Aβ deposition by effects on hypoxic stress and inflammation or may increase the amount of Aβ by increasing vigilance. Levels of cognitive and physical activity have a bidirectional relationship with both sleep-wake patterns and AD and can therefore intensify the feedback loop between poor sleep and AD [92, 93].

The correlations between the values of Aβ and sleep modifications (evaluated through EEG) should allow us to begin tearing apart the details of sleep abnormalities appearing with the onset of AD pathology, as well as the direction in the relationship between sleep and Aβ deposits. Since sleep interruption increases the risk of further AD onset, then this supports the importance in identifying and treating the patients with sleep disorders, especially obstructive sleep apnea [92].

Depression

Depressive disorder in the elderly may have a major importance in triggering neurodegenerative mechanisms and cognitive impairment. From a clinical point of view, a depressive episode before the age of 65 years or recurrent episodes of depression are important, since older age onset of depression is seldom [94].

Depressive disorder has a complex, multifactorial pathogenicity (Fig. **28**), and it is recognized that the pharmacological therapy approach may have an efficiency of about 45-50% for complete remission. Incomplete remissions are usually the rule in the development of a depressive disorder over the age of 65 years. Through complete remission we understand both the decrease in the intensity of depression symptomatology with values in the Hamilton Rating Scale for Depression (HRSD) below 7, as well as the biological markers for depression: proinflammatory cytokine and endothelial dysfunction.

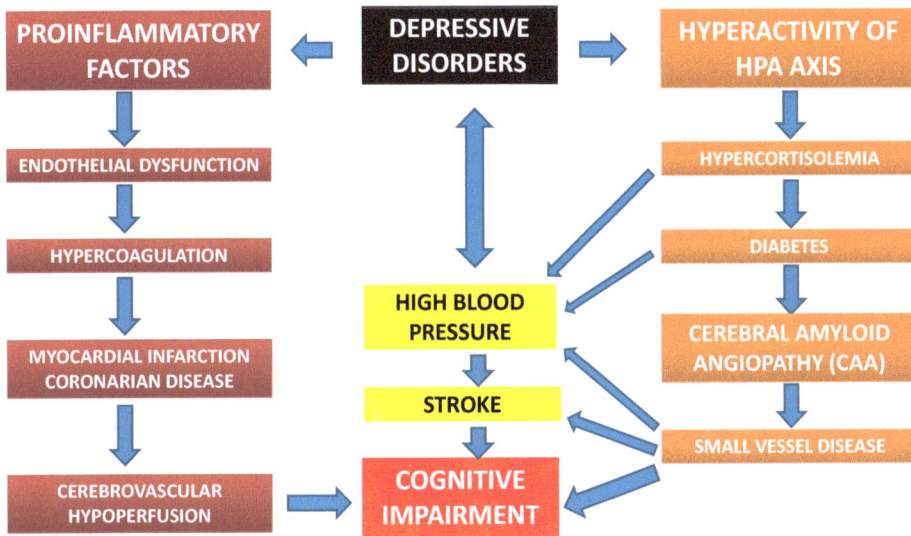

Fig. (28). Multifactorial pathogenesis of depression and risk factors involved in progression of cognitive impairment following incomplete remissions of previous depressive disorder.

The common occurrence in current practice is that of a symptomatic remission with the maintenance of multisystemic imbalances and factors that potentiate predominantly vascular mechanisms.

The biological model of incomplete remissions is supported by the presence of biological markers (Fig. **29**), anticipating the potential risks of development of stroke and cardiac stroke, resulting in a decrease in cerebral blood perfusion that drives cascade mechanisms of the neurodegenerative type, favoring cerebral vascular amyloidosis, disconnectivity of cognitive circuits and rapid progression of cognitive impairment.

Depression prior to AD diagnosis is an important independent risk factor for the rapid progression of cognitive decline and aggravation of somatic comorbidities

[95]. Frequently, in elderly depression, residual symptoms are dominated by insomnia, a phenomenon that aggravates the risk of rapid deterioration. In the conditions of depression during the latency of AD, it is necessary to avoid pharmacological therapies with antidepressant or antipsychotic drugs that have anticholinergic effects, as well as excess benzodiazepines.

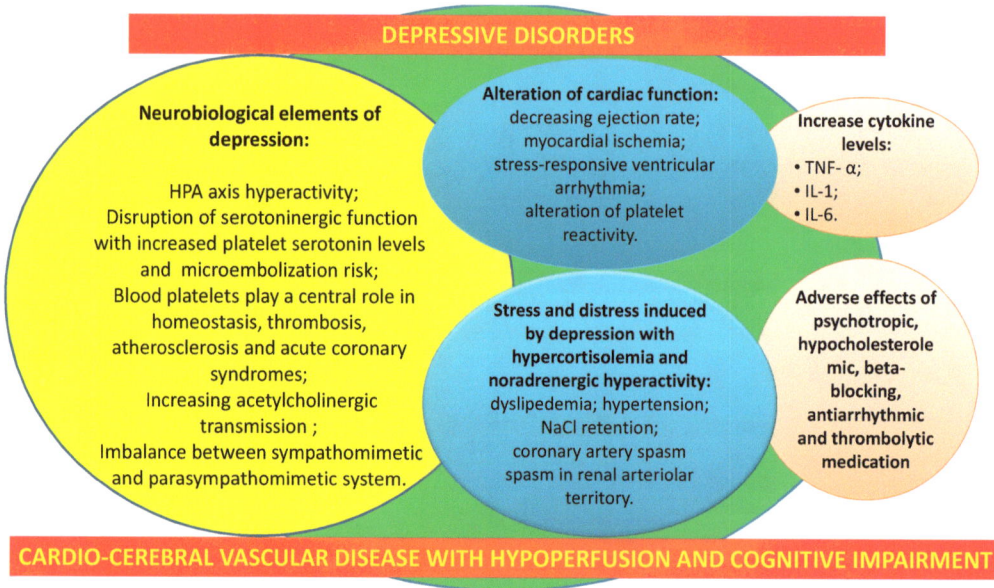

Fig. (29). Markers that can predict cognitive impairment in a depressive disorder.

Vulnerabilities Consecutive to Comorbid or Independent Somatic Pathological Conditions

Cardiovascular (Cardiopathy, Hypertension and Hypotension)

Clinically identified cardiovascular disease associated with depressive symptoms, with the presence of specific biologic markers can represent another alarm signal for the clinician, signaling rapid cognitive deterioration, as well as an important cardiac risk [96].

Potential biological risk indicators for ischemic cardiopathy associated with depression are: platelet factor 4 (PF4) - a small cytokine known as chemokine (C-X-C motif) ligand 4 (CXCL4); platelet/endothelial cell adhesion molecule 1 (PECAM-1); P-Selectin; E-Selectin; thromboxane B2; 6-ketoprostaglandin-f1α; β-thromboglobulin [97 - 99].

Cardiovascular disorders due to the frequency and consequences of cerebral circulation, cardiac pump failure, result in cerebral vascular hypoperfusion phenomena associated with accentuating cognitive deficits due to lack of metabolic and oxygenation (cerebral ischemic syndrome) or cerebral hyperperfusion phenomena exaggerated blood pressure or increased pressure on the arterial wall, causing the occurrence of microbleeds or cerebral hemorrhagic vascular disease.

The frequency of this type of pathogenic condition in the advanced larynx is greatly increased, which is why specific cardiologic therapy is frequently prescribed, which in turn can cause cognitive dysfunction secondary to medication.

The main risk factors for cognitive impairment identified in cardiovascular diseases inducing hypoperfusion are: reduced cardiac output (hypotension, heart failure, hypoxia and increased vascular resistance), mitral valve thickening, aortic valve thickening, atrial fibrillation, hypertension with stroke risk, aortic stiffening, left ventricular hypertrophy, coronary artery disease [100].

Dilated cardiomyopathy (DCM) is given a particular importance are, a frequent but underdiagnosed affection. This may be related to the spectrum of systemic amyloidosis, becoming an independent risk factor associated with rapid cognitive decline, with poor comorbid somatic progression and even sudden premature death (Figs. **30-33**).

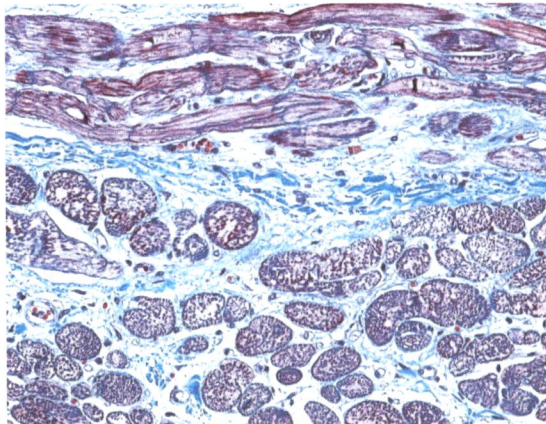

Fig. (30). Dilated myocardial picture showing longitudinal and transverse sections of myocardial fibers. We can observe the rarefaction, number reduction and thickening of the remaining myofibrils, their non-homogeneous arrangement in myocardiocytosis and vacuolization of the sarcoplasm. Interstitial connective tissue contains increased amounts of fibril collagen and a small number of blood vessels. Trichrome staining Goldner - Szeckeli, magnification 200X.

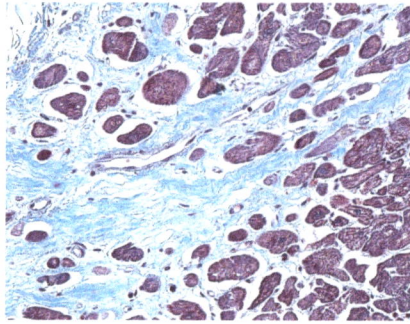

Fig. (31). Myocardium with high content of collagen-rich interstitial connective tissue in a patient with dilated cardiomyopathy and moderate ischemic cardiopathy. Trichrome staining Goldner - Szeckeli, magnification 200X.

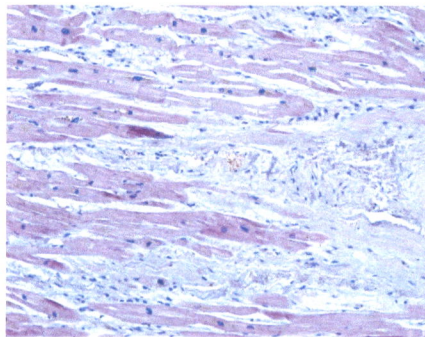

Fig. (32). Myocardial image of muscle fiber reduction and increase in the amount of conjunctive matrix, especially collagen fibers, in a case of severe ischemic cardiopathy. Staining Hematoxylin eosin, magnification 200X.

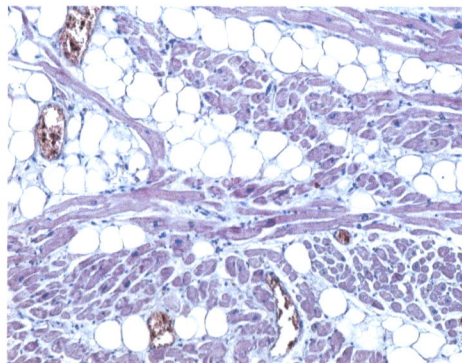

Fig. (33). Myocardial wall showing the increase in the amount of conjunctive matrix and disorganization of myocardial fibers by adipocytes in the myocardial stroma in a person with dilated cardiomyopathy and obesity grade II. Trichrome staining Goldner - Szeckeli, magnification 100X.

Cardiac amyloid is an example of the infiltrative cardiomyopathies, characterized by the deposition of Aβ. Cardiac involvement is common and represents a major source of associated morbidity and mortality with amyloidosis. Electro cardiography and echo cardiography are very important in preventing heart failure, main manifestation of amyloid cardiomyopathy, which can be concomitant, or not, with other symptoms of systemic amyloidosis [101].

Dilated cardiomyopathy has various pathogenic conditions but in old age it is associated with the systemic atherosclerotic process, contributing to alteration of cerebral vascular structure, but also by accumulations of Aβ which may favor the phenomena of destruction of myocytes and external compression of the coronary vessels. This favors the possible acceleration of the risk for acute coronary artery disease (myocardial infarction) and the risk of impairment of the independent control system in cardiac rhythm, with the occurrence of severe arrhythmias. Evolution of this cardiac disease is precipitated in the conditions of patient's dependence on alcohol consumption and vitamin B1 deficiency.

Type II Diabetes

Type II diabetes increases the risk of AD by several mechanisms:

- Insulin resistance, which can increase Aβ aggression.

- Diabetes mellitus associates diabetic angiopathy which causes endothelial dysfunction, a phenomenon that favors the development of amyloid angiopathy and BBB disruption.

- The alternation between episodes of hypoglycemia and hyperglycemia following minimal cognitive dysfunction and disturbing the social and family relationships of the elderly, which does not allow for self-care measures and the neglect of hypoglycemic therapy.

- Hippocampal lesions and mitochondrial dysfunctions in neurons, hypoglycemia potentiating the rapid cognitive impairment rate.

- The uncontrolled use of hypoglycemic medication in case of cognitive impairment from AD can lead to self-induced hypoglycemic accidents, and in the presence of depressive symptomatology and early diagnosis of AD in a person with a high cognitive level, hypoglycemic medication can be used in autolysis [102].

For these reasons, for AD patients, hypoglycemic medication should be done under strict supervision to avoid this type of serious injury.

Non-alcoholic Hepatic Steatosis

One of the hepatic disorders that can precipitate neurodegenerative phenomena is non-alcoholic hepatic steatosis (Fig. **34**), which can induce or accelerate cognitive impairment through the peripheral clearance of amyloid-type elements or by dysfunction of liver enzymes involved in the activation of pharmacological compounds. Consequently, the effects are of the neurotoxic type or cause the ineffectiveness of specific anti-dementia drugs.

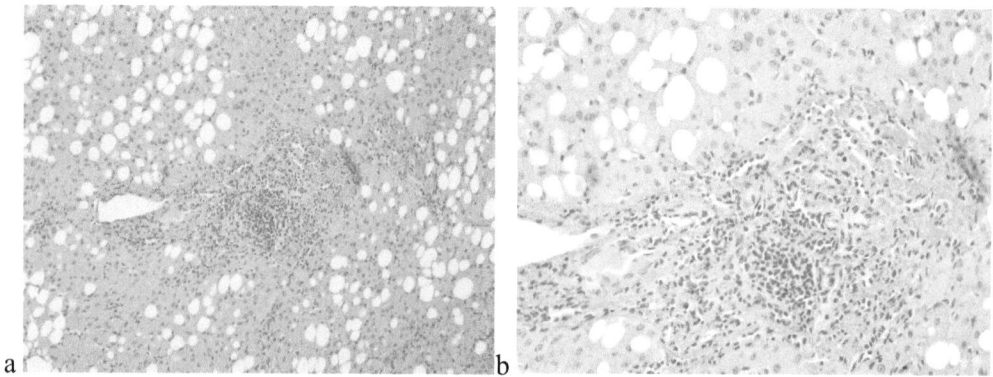

Fig. (34). Macronodular and micronodular pancreatic steatosis with abundant chronic inflammatory infiltration in the spatial port that diffuses in the lobules. Hematoxylin-eosin staining; a: magnification 10X; b: magnification 20X. Image used with permission from Daniel Pirici.

Hepatic steatosis following viral infection plays also an important role in favoring cognitive impairment in patients with AD and this type of hepatic comorbidity. The action of neurodegenerative factors may be potentiated by depressive phenomena secondary to interferon-specific medication or excessive cortisone therapy, a therapy that frequently associates neuronal destruction at the hippocampus level, destroying the emergence of cognitive circuits.

On the animal model, the aggressive effect of high doses of dexamethasone on the rat hippocampus was demonstrated by the amplification of stress factors and hyperactivity of the hypothalamic-pituitary-cortical adrenal axis (HPA).

A particular yet insufficiently clarified aspect is the existing relationship between hepatic transplantation elderly and the progression of cognitive impairment elements associated with neurodegenerative elements of the AD spectrum. Correlations have been made with the ApoE4 genetic spectrum, the difficult issue being to find some evidence to explain the role of the donor as compared to the recipient or the risk of genetic modification simultaneous in the recipient and donor.

A second option is correlated with the long duration of anesthesia that favors cerebral hypoxia and mitochondrial alterations at the neuronal level, or with specific treatments that include immune-modulators or methotrexate immunosuppressive substances, a substance recognized for hippocampal neurotoxic aggression [103].

This mechanism is accentuated by associating high doses of exogenous glucocorticoids. Endogenous glucocorticoid mechanisms following HPA hyperactivity are the consequence of stress and depression following transplant.

Infectious Pathology

Infectious pathology may represent a risk of rapid progression of cognitive decline in the elderly leading to triggering neurodegenerative mechanisms that frequently takes AD aspect. The most common pathological conditions of infectious etiology are viral infections (herpes zoster, cytomegalovirus, viral pneumopathy, HIV infection). In patients at risk of AD because immune mechanisms are deficient, infections amplify chronic neuroinflammation and neurodegenerative mechanisms that cause progressive loss of neurons and cognitive impairment [104].

HIV infection is widely discussed in the deteriorating pathology since the last generation antiviral medication has led to a high survival rate, constituting a significant group of patients presenting with a debilitating cerebral vulnerability specific to HIV, "AIDS dementia complex" - frequent feature of HIV disease before antiretroviral therapy, whose progression was stopped by antiviral medication [105]. Because comorbidities increase the risk of cognitive impairment, the patients must remain on the antiviral medication, but they must also receive specific, individualized treatment for comorbidities [106].

On the other hand, the sexual disinhibition of AD elders with AD in the prodromal phase leads to aberrant sexual behavior and contamination with sexually transmitted diseases, including HIV. Specific antiviral medication also has a recognized neurotoxic effect that can accentuate neuronal destruction and, implicitly, cognitive impairment. Increased risk of infections or tumors, and a side effects of antiretroviral drugs, contributes to more severe neurologic and cognitive symptoms: HIV-associated neurocognitive disorders (HAND) [105].

Poor immunity of the elderly frequently causes skin infections, of the zoster type, and persistent hyperalgesia after this condition may suggest the posterior cerebral vascular system vulnerability with thalamic dysfunction and disconnectivity of the thalamo-cortical circuits. This type of persistent algal hypersensitivity can be an important marker of predicting rapid cognitive impairment (Fig. **35**).

Other infectious diseases encountered in the elderly requiring prolonged care in nursing homes are [107]:

- Chronic or pulmonary infections of the nosocomial type that have an important place in the pathology of the elderly, triggered by a traumatic condition (femoral neck fracture) or a neurological disabling condition (stroke). They have a long-lasting evolution and are predominantly encountered at the nursing home level, with the most frequent evolution leading to significant cognitive and somatic deterioration and even death.

- Urinary tract infections (Escherichia coli, Klebsiella, proteus) most frequently favored by the use of prolonged catheterization.

- Chronic bacterial pneumopathy favored by prolonged decubitus.

- Tuberculosis of the elderly with an increased incidence worldwide in the last decade.

- Multiple chronic skin infections following herpes zoster infection.

- Gastrointestinal infections with prolonged diarrheal syndrome causing hydro-electrolytic imbalances, the main bacterial cause being Clostridium Difficile infection.

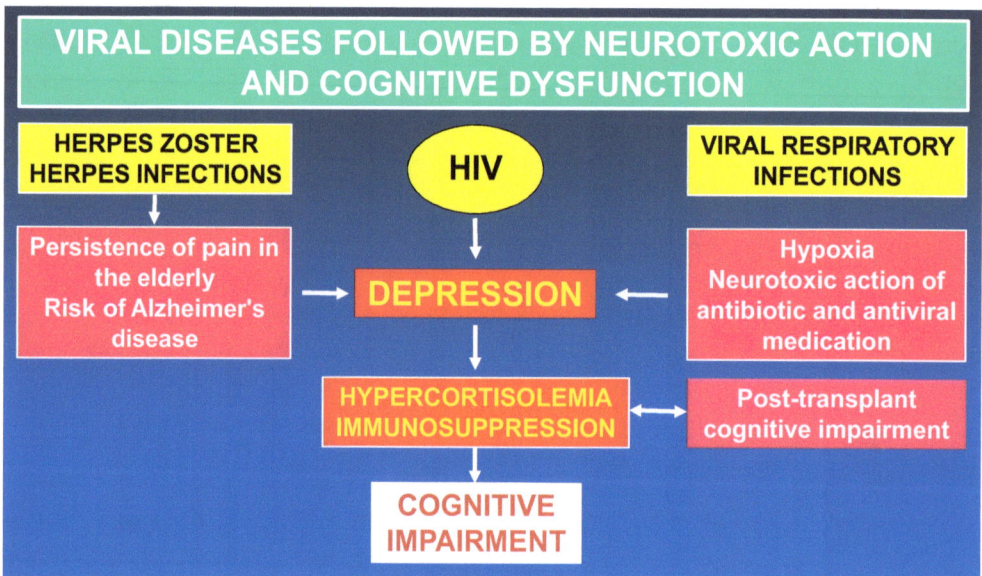

Fig. (35). Correlations between infectious pathology and cognitive impairment.

All of these infectious pathogens with severe evolution risk require special prophylaxis measures adapted to the somatic status and cognitive reserve of the patient and permanent monitoring by specialized personnel. In addition, a balanced emotional climate must be maintained that does not increase the stress in these patients with immune deficiency, as well as an ambient atmosphere as close as possible to the one previous to nursing home care.

A recent research directive is fungal infections in the progression of cognitive deficits in AD, being highlighted in the central nervous system, fungal cells at intracellular and extracellular levels. Using antibodies was observed immunoreactivity for Candida glabrata, Candida famata and Candida albicans [108]. Fungal cell aggression may be correlated with the overall decrease in immunocompetent mechanisms, increased BBB permeability and the general infectious context mentioned above. Excessive use of antibiotic therapies, H2 receptor antagonists and the combination of immunosuppressants (corticosteroids or other immunosuppressants used especially in transplantation) may increase the risk of aggressive fungal infections [109].

Increased risk for cognitive impairment in the elderly, induced by infectious pathology, including fungal infections, may be favored by hospital-acquired infection or nosocomial infections. Their prophylaxis may be an important target for several clinical specialties involved in the treatment of comorbid conditions of various etiologies. Avoiding hospitalizations and finding alternative treatment methods to maintain the patient within the family, can help slow the cognitive deterioration [110].

Somatic Pathological Vulnerabilities in Old Age Correlated with Amyloidosis

1. Reactive systemic amyloidosis is correlated with increased production of the serum amyloid-A protein (SAA), precursor of amyloid-A (AA) protein, a mechanism that can be triggered by chronic pathogenic condition, with a prolonged evolution of infectious or inflammatory type, conditions favoring activation of the microglial system and neuronal destruction but also myelin type dysfunction, generating axonal dysfunction that accentuates disconnectivities in neuronal systems including the cognitive system [111, 112].

Triggering reactive systemic amyloidosis mechanisms can be favored by immune deficiencies of the elderly and some comorbid medical conditions (diabetes, obesity, cachexia, malnutrition, vitamin deficiency, heart failure and chronic cerebral circulatory insufficiency with multiple exacerbation episodes.

The most important chronic inflammatory conditions with autoimmune or infectious component involved in the degradation of reactive systemic amyloidosis are the following diseases:

- Rheumatoid arthritis with increased incidence in the elderly and is frequently associated with depression and with biological indicators that support the chronic inflammatory process: C-reactive protein, TNF alpha, interleukin-1, cytotoxic T-lymphocyte antigen 4 immunoglobulin [113]. Cognitive impairment is associated with rheumatoid arthritis due to the constant presence of proinflammatory indicators that predict the rate of rapid cognitive impairment. The rhythm can be further augmented by the use of medication specific to this condition (methotrexate, cyclosporine A, glucocorticosteroids, non-steroidal anti-inflammatory drugs) [114]. This correlation is important because the diagnosis of rheumatoid arthritis prior to AD diagnosis may become a risk indicator for the development of cognitive impairment.

- Gastrointestinal disorders associated with reactive amyloidosis and rapid cognitive decline are Crohn's disease, inflammatory bowel disease, celiac disease.

- Chronic nephropathy correlated with senescent nephropathy [115], primary glomerular disease [116] and nephritic syndrome [117].

All of these conditions can produce secondary stroke mechanisms that accentuate cognitive dysfunction in patients with AD.

PARTICULAR PATHOGENIC ASPECTS THAT MAY ALLOW PROPHYLACTIC STRATEGIES FOR RAPID COGNITIVE IMPAIRMENT AND SLOW NEURODEGENERATIVE TYPE PROCESSES

Relationship between Late-Onset AD and Systemic Amyloidosis

Early onset AD form associates changes in presenilin 1 and presenilin 2. For the elderly there is data involving only presenilin 1, in forms with rapid cognitive deficits [118]. Both the ApoE4 spectrum and presenilin 1 and 2, favor the development of neurodegenerative elements represented by Aβ deposits, neurofibrillary degeneration and senile plaques.

Another genetic component that raises problems in interpretation of pathogenic mechanisms in AD is related to the protein component - the amyloid precursor protein. It is extremely interesting that this protein is involved in systemic amyloidosis and is correlated with ApoA1 [119], which is associated with HDL decrease (hypoalphalipoproteinemia), hypertriglyceridemia, and lowering the cholesterol transport enzyme (Lecithin-cholesterol acyltransferase - LCAT), the enzyme that plays a key role in reverse transport cholesterol [120].

The high protein levels of amyloid A in the blood, serum amyloid A protein (SAA) as well as the serum amyloid P protein (SAP) are components of all types of amyloid fibrils and are therefore present in all amyloid deposits and plaques, including those in AD [58, 111, 121].

In our opinion, the reassessment of the protein alterations in systemic amyloidosis but also in AD raises suspicion of pathogenic mechanisms such as proteinopathies. This hypothesis can be supported by several arguments:

a. Protein deficiencies and neurodegenerative mechanisms are different in the AD situation, the major protein deficits being related to Aβ, tau proteins, α-synuclein and TAR DNA-binding protein 43 (TDP-43) [122]. Protein changes may be neurobiological markers in CSF assessments or neuroimaging markers. The link between the significance of protein biologic markers in the CSF and the functional integrity of BBB may call into question their use as indirect indicators of BBB disruption.

b. The amyloid-β precursor protein has a physiological and pathogenic mechanism still insufficiently deciphered, recognizing the neuroprotective capacity and role of this protein in neurodevelopment or neurogenesis, following the destructive neuronal events (but also the neurotoxic effect linked to a high probability of intermediate metabolites of this protein structure).

TDP-43 determines the transformation of Aβ into amyloid oligomers whose aggregation with TDP-43 aggregation accentuating neurotoxicity. The presence of the TDP-43 protein is an indirect indicator of frontal and temporal brain structures involvement, setting the premises for an intermediate neurodegenerative pathology between dementia in AD where the structure of the parietal lobe and frontotemporal dementia is involved, the so-called lobular TDP dementia [123].

The neurotoxic effect of TDP-43-coupled oligomers favors the alteration of temporal lobe structures and the development of hippocampal astrogliosis and epileptic manifestations. In fact, this type of manifestation is more common in fronto-temporal dementia than in AD, and psychotic schizophrenia such as psychosis is more commonly cited in this type of dementia, although as presented later, temporal lobe epileptic manifestations and schizophrenia like psychosis are also present in AD but underdiagnosed. This underdiagnoses may result in unfavorable pharmacological therapies that can precipitate the evolution of the neurodegenerative process from AD and the rapid passage into a severe cognitive impairment. Identification of increases in TD-43 and interleukin 1A can avoid delaying a correct diagnosis [124, 125].

The amyloid-β precursor protein is involved in the following physiological mechanisms:

- Maintaining neuronal structures homeostasis with rebalancing signaling between pre and post synaptic neurons, while ensuring neuroprotection of structures of recurrent protein structures (G protein).

- Controls the evolution and movement of stem cells both in neurodevelopment processes and in repair neurogenesis mechanisms or in cell stem transplant [126 - 128].

Reparative neurogenesis in AD pathology is correlated with two cerebral areas: subcutaneous and ventricular granulation and CA1 dentate gyrus (hippocampal Ammon's horn 1). Recognizing this type of pathogenic mechanism, identification of ventriculomegaly and hippocampal atrophy (the neuroimaging marker present since the MCI phase) cancels out the physiological role of the β-amyloid precursor protein. The removal of this role suggests excessive processing of this protein through Aβ production mechanisms, significantly increasing the aggressiveness of Aβ deposits, probably predominantly in the mentioned areas, involved in neurogenesis. A prediction indicator of rapid cognitive decline in AD is the progression of ventriculomegaly and hippocampal atrophy. We mention that hippocampal atrophy and alteration of glucose metabolism at this level are indicative of high significance and specificity of the imminent transition from the MCI stage to the AD itself. Excess β-amyloid precursor protein can rapidly pass through dysfunctional BBB, and increase of this protein in the cerebrospinal fluid (CSF) becomes another important biological indicator of rapid cognitive decline [129].

c. Mechanisms for activating electrical synapses with epileptic manifestations.

The process of fibrillogenesis following the mechanism of β-protein polymerization leads to an increase in protein membrane imbalance and to the amplification of the electrically-active membrane potentials with glutamatergic hyperactivity, a mechanism leading to epileptic manifestations. They are present in about 1/3 of the AD cases, being correlated with the disconnective mechanisms between the limbic system and the temporal cortex. Hippocampal lesions and accumulation of β-amyloid precursor protein are significantly associated with temporal lobe epilepsy and the presence of proinflammatory mechanisms identified by the presence of interleukin 1A [130]. This epileptic-type pathology may embody a pseudo-psychiatric, as often temporal lobe epilepsy has the clinical expression of "schizophrenia-like psychosis". Intervention with antipsychotic medication deters the progression of neurodegenerative phenomena and irreversible neuronal destruction. The appearance of "schizophrenia-like psychosis" phenomena in an elderly person requires a complex evaluation: neuroimaging, electroencephalography, neurobiochemical. Increase in the level of proinflammatory factors and the presence of ventriculomegaly and hippocampus atrophy may lead to the diagnosis of neurodegenerative process [131].

Epileptic manifestations may be associated with hippocampal sclerosis, representing a massive CA1 neuronal destruction and a progression of astrogliosis-like replacement processes. Hippocampal sclerosis is estimated at 18% in patients over 90 years and young patients in 9.2% of cases. This is associated in 86% of cases with the increase in TDP-43, and is expressed by the rapid cognitive impairment on several cognitive domains as a precipitating factor in the onset of severe dementia [132]. Besides hippocampal sclerosis, TDP-43 is associated with decreased episodic memory efficiency. Precipitation of epileptic manifestations and cognitive decline may be associated with decreased cerebral metabolism for glucose, cerebral vascular hypoperfusion with hypoxia, and intensification of inflammatory or immune processes in the conditions of association with an infectious type pathology (herpes zoster manifestations, HIV). Consequently, these three mechanisms confirm the multifactorial etiopathogenesis of systemic amyloidosis and the implication of genetic vulnerability, which apparently seems to be different from the late-onset AD gene condition.

The paradox that is recognized by most scientific evidence is the recognition of the increase in the frequency of systemic amyloidosis with age, but no real connections with the pathogenic mechanisms of Alzheimer's disease.

If we start from the assumption of the genetic vulnerability of sporadic AD and the existence of demonstrated associations of cognitive decline with the genetic spectrum of ApoE4, the same connection can be made with the vascular vulnerability of systemic amyloidosis under conditions of renal, liver or cardiac focus, phenomenon that could be the bridge between the two entities.

Systemic amyloidosis in young age is included in the rare disease category, which may be considered similar in terms of rare disease on rare AD (rarely onset AD). Existence in AD with early onset of at least two models of genetic cohort (early onset AD in the Irish family and Dutch family) raises the problem of different genetic spectrum pathology even within the clinical entity defined as AD. In systemic amyloidosis, a hereditary form with specific genetic mutations is recognized [133]:

- *Hereditary form can occur at the age of childhood* (early childhood associated with other genetic patterns [134].

The existing link between systemic amyloidosis and the genetic pathology of immunoglobulins made it possible to identify primary forms of familial amyloidosis with predominantly hematological clinical manifestations - monoclonal gammopathies. Under this vision, since 1986, Morie Gertz has made a link between the genetic support of the systemic amyloidosis spectrum and several hematological diseases (multiple myeloma, Waldenstrom's macroglobulinemia). Primary amyloidosis AL in families [135].

- *Variations of ApoA1* cause manifestations with asymptomatic cholestatic liver disease, renal failure, and testicular involvement with hypogonadism [136]. The spectrum of the ApoA1 mutation becomes extremely important in major psychiatric pathology (schizophrenia, bipolar disorder) due to the fact that antipsychotic medication produces cholestasis and risk of acute liver failure under non-genetic risk identification. Organic brain manifestations of systemic or AD amyloidosis may be associated with psychotic schizophreniform episodes in which this type of medication should be avoided.

A stage finding, resulting from the literature, outlines a genetic vulnerability of AD to the apolipoprotein E spectrum, while systemic amyloidosis is associated with the ApoA1 spectrum.

- *The senile form of systemic amyloidosis* is associated with an increase in the multiorgan pathology and the multisystemic component of the disease. Systemic amyloidosis is a proteinopathy that forms the fibrillary "net" necessary for Aβ deposition in AD, it can be considered as a real risk factor for sporadic AD development. The multisystem pathogenic hypothesis of AD and the systematization of somatic disorders and their mechanisms of production, following systemic amyloidosis, are arguments that can confirm our assumption that classifies the somatic risk factors in dependent or independent factors of systemic amyloidosis [137].

The systemic amyloidosis process may also be secondary to chronic inflammatory processes, a pathogenic mechanism that can be considered in the context of non-systemic amyloidosis as a risk factor [121].

Biomarkers of diagnosis and evolutionary risk in systemic amyloidosis:

- Growth differentiation factor-15 (GDF-15) is an important prognostic marker in systemic light chain (AL) amyloidosis with multiorgan pathology, especially the heart and kidneys. An increase level in this indicator shows an early mortality risk, poor outcome in heart failure from other etiologies, increased risk of renal failure in diabetes or higher risk of progressing to dialysis [138]. GDF-15 is a potential biomarker for mitochondrial dysfunction, and has been associated with cognitive impairment, 20% of cases having an evolution from normal cognition to MCI or dementia [139].

- In terms of a multiorgan pathology, N-terminal pro-B-type natriuretic peptides (NT-proBNP) and troponin T are recommended to be used along with GDF-15, as a biological marker of cardiac and renal response, risk of dialysis or treatment response. High level of NT-proBNP could be an important marker for early diagnosis of cardiac amyloidosis, which cannot be identified with the echocardiographic exam [138, 140].

- The high plasmatic levels of the mid-regional pro-atrial natriuretic peptide (MR-proANP) and midregional proadrenomedullin (MR-proADM) are strong correlated with the progression from the prodromal MCI to the clinical AD, they may be used as markers for screening in patients at risk for AD [141].

Systemic amyloidosis, and especially the spectrum of this pathology, can be considered as a risk factor for rapid progression of cognitive deficits in elderly, not through classical neurodegenerative mechanisms controlled by ApoE4 spectrum, Aβ, and vascular components of the CAA angiopathy spectrum, but secondary to multifactor changes produced by systemic amyloidosis. This hypothesis may be a solution in case of resistant therapies for comorbid disorders that normally influence the functioning of the cerebral system and maintain energetic homeostasis of the brain. In elderly patients, there are cumulative cardiovascular somatic comorbidities, diabetes, renal failure and secondary hypertension, and other chronic inflammatory conditions including arteriopathies that do not respond to medication and standardized pharmacological strategies. We are questioning whether these diseases are not part of the non-cerebral systemic amyloidosis, becoming the primary pathogens of the cerebral energy imbalance. This dysfunctionality essentially resets the neurodegenerative mechanisms and their aggressiveness favors the disconnective mechanisms at the level of the interneural systems, altering the cognitive circuits, on the predominantly subcortical routes. Once triggered, the neurodegenerative phenomenon acts specifically at the vascular structures causing lesions in the cortical neural structures.

Knowing this possible pathogenic mechanism and especially the possibility of using biological markers can be a valid prophylactic strategy for slowing down the cognitive deterioration process in AD.

The Glymphatic System

Involvement of the glymphatic system and the aquaporin-4 (AQP4), the most commonly water channels in the brain, is important as support for perivascular lesions at the BBB and Virchow-Robin spaces, which provide at the CSF level recirculation and Aβ clearance. Glymphatic system is a way to remove molecular waste or excess extracellular fluid from the brain, by facilitating exchanges of solutes between cerebrospinal fluid (CSF) and brain interstitial fluid (ISF).

In addition, the role of BBB, and in particular its functional disruption associated with alteration of the glymphatic system, is a risk factor in the evolution of AD with late onset. Disruption of glymphatic pump in the nucleus basalis of Meynert and the hippocampus is correlated with cognitive impairment, by accumulation of molecular debris or protein aggregation do to failure of local clearance (Figs. **36** and **37**) [142].

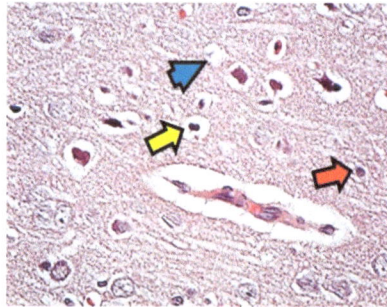

Fig. (36). In the central area a collapsed capillary is seen surrounded by pericapillary edema and fragmentation of astrocytic end feet. Accumulation of molecular debris from disintegration of astrocytic end-feet containing AQP 4 water receptors. Surrounding the capillary there are neurons in different stages of apoptosis. Apoptotic neuron (blue arrow), partially apoptotic neuron (yellow arrow), healthy neuron (red arrow). Hematoxylin and eosin staining, magnification 200x [142]. Image used with permission from the main author and editorial staff of the Romanian Journal of Psychopharmacology.

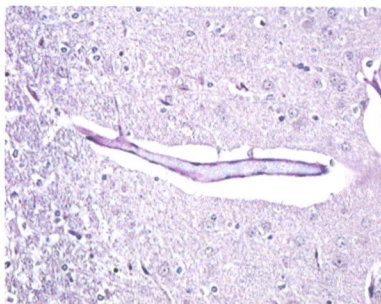

Fig. (37). Perivascular edema with neuronal and dendritic fragmentation causing molecular debris that cannot be cleared by the glymphatic system. PAS hematoxylin staining, magnification 200x [142]. Image used with permission from the main author and editorial staff of the Romanian Journal of Psychopharmacology.

Involvement of the glymphatic system in the alteration of neurovascularity is correlated with the acetylcholine deficiency hypothesis that is the basis in the anti-dementia pharmacological strategies increasing cerebral acetylcholinergic levels by blocking destructive enzymes of neuronal (acetylcholinesterase) or astrocytic (butyrylcholinesterase) enzymes. It highlights also the Meynert basal nucleus damage (Figs. **38** and **39**).

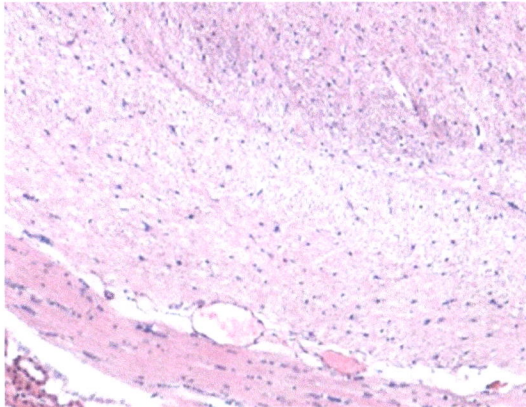

Fig. (38). Nucleus basalis of Meynert: changes in capillary structures with pericapillary edema and various stages of apoptosis of the cholinergic interneurons.Glymphatic system failure leads to accumulation of molecular waste resulting from neuronal disintegration leading to a vicious circle of further apoptosis. Hematoxylin and eosin staining, magnification 100x [142]. Image used with permission from the main author and editorial staff of the Romanian Journal of Psychopharmacology.

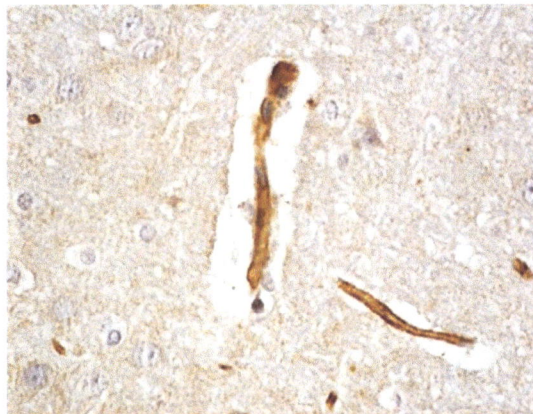

Fig. (39). Profound anatomic changes of the neurovascular unit with pericyte detachment from endothelial cells and astrocyte damage with accumulation of protein aggregates such as glial fibrillary acidic protein that cannot be cleared by the glymphatic system due to AQP 4 receptors loss. Endothelial specific CD 3 staining anti-glial fibrillary acidic protein antibody (GFAP), magnification 200x [142]. Image used with permission from the main author and editorial staff of the Romanian Journal of Psychopharmacology.

Disruption of the Blood Brain Barrier (BBB)

Disruption of the BBB is an important indicator for functional impairment of brain neurovascular unit in Alzheimer's disease and who are associated with cognitive impairment. Functional alteration of neurovascular unit was correlated with the following biological markers: increased neuron-specific enolase (NSE), myelin basic protein (MBP) and S100B protein in peripheral blood associated with neuronal lesions; increases in homocysteine values have been associated with cerebrovascular disease. We note that the functional alteration of BBB significantly influences the efficacy, but also the adverse effects of anti-dementi--specific drugs. The use of biological markers for BBB disorder in the clinic and drug therapy of Alzheimer's disease can improve the prognosis of this condition [143 - 145].

The interaction between BBB and the glymphatic system becomes a decisive factor for the penetration capacity of Aβ at the level of brain struggles and implies the massive deposition to the detriment of Aβ transition in CSF. There are 2 mechanisms that can be involved:

- BBB transport mechanisms, which in turn are dependent on cerebral blood flow and the capacity of intervention of the hepatic homeostatic circuits. Pathology of non-alcoholic hepato-steatosis or hepatic amyloidosis may alter the hepatic filter and favor the progression of amyloid deposits at the cerebral level.

- Mechanisms of the glymphatic system involving disturbance of interstitial fluid and CSF exchanges due to functional deterioration of AQP4 receptors, water release control channels, controlled at this level by the astrocytes. The glymphatic system can manage the amount of interstitial paravenous fluid in the Virchow-Robin interstitial spaces and may aggravate the return circulation flow through the lymphatic system both at the hepatic and cerebral level, by affecting BBB [146].

We question whether this mechanism and BBB disruption may be a factor that intensifies Aβ deposition and by involving mechanisms of systemic amyloidosis can penetrate the brain.

THEORETICAL PREMISES BASED ON THE NEUROBIOLOGICAL MODELS THAT MAY BE BASIS FOR ACTIONS TO PREVENT THE COGNITIVE DECLINE

At the center of the biological models that will be the basis of the preventive strategies for slowing the rate of cognitive deficit progression in AD, there must be the lack of pathogenic homogeneity of AD cognitive impairment. This pathogenic diversity is also suggested by the fact that massive Aβ deposition may occur in the elderly without cognitive alterations or ischemic vascular lesions in non-strategic territories that may be silent in terms of motor, sensory or cognitive consequences. From this perspective, the detection of small cerebrovascular syndrome is a predictive value that can be influenced by prophylactic strategies [77].

This disease can be recognized by the following neuroimaging markers: enlargement of Virchow-Robin spaces, decrease in cerebral gyrus volume and impressive growth of inter gyrus fissures or delimitations fissures between the brain lobes (Sylvian sulcus).

Vulnerability of the cerebral vascular system in conjunction with cerebral amyloid angiopathy with small cerebrovascular disease leads to the appearance of micro-bleeding, whose location on strategic areas may have similar consequences to ischemic manifestations. Cerebral micro hemorrhages may be favored by the excessive use of antiplatelet medication, aspirin, or anticoagulant drugs and may be prompted by the presence of hypertension or depressive disorders that increase the proinflammatory mechanisms and endothelial dysfunction.

Asymptomatic micro hemorrhages can be detected late, their appearance on neuroimaging evaluations being signaled by cortical siderosis Fig. (**40**). These vascular changes caused by multiple microbleeds are associated with rapid cognitive decline, but can be anticipated by potential neuroimaging markers (Fig. **41**).

A background of genetic vulnerability can be discussed, where main role is held by the spectrum of E2 and E4 apolipoproteins. Conservation of cognitive neuronal circuits in the elderly is dependent on the functional and structural integrity of the hippocampus, cingulate cortex, cerebral tonsil, temporoparietal and frontal cortex, each of which may be involved in various neurocognitive disorders.

The main pathogenic mechanisms that can be influenced by specific prophylactic therapeutic strategies in order to preserve the structural and functional integrity of the mentioned brain structures and their interconnectivity are represented by:

- Depressive disorder neurobiological mechanisms, a disease with high incidence in the elderly (9-12%), associating psychotrauma events and social stress that cause hyperactivity of the hypothalamic-pituitary-adrenal axis. Consequently, it increases the level of endogenous cortisol that will amplify hippocampal atrophy. Depression through its multifactorial pathogenic mechanisms can be considered as a risk factor independent of neurodegenerative mechanisms that influence the rapid progression of cognitive impairment by: endless immune and immune imbalance; increased proinflammatory processes and microbial hyperactivity; diminishing neurogenesis and neuroprotection.

- Disconnective mechanisms by altering axonal transport caused by seemingly minor cranial traumas, and frequently ignored.

- Blood-brain barrier disruption and functional disruption of neurovascular unit precipitates the potential invasion of viral or bacterial infectious factors (herpes zoster, pneumococcus, cytomegalovirus).

- Alteration of functional support in neural structures in cognitive circuits by decreasing: brain blood flow, circulating blood quality (anemia or anemic syndromes of various etiologies), glucose metabolic support (hypoglycemia) or cell oxygenation (hypoxia).

- Cerebrovascular components with particular pathogenic mechanisms represented by cerebral amyloid angiopathy, small vessel disease or cortical and subcortical micro-bleeds.

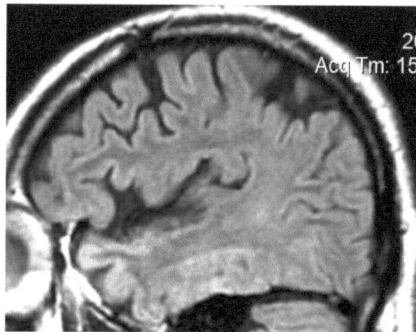

Fig. (40). Frontal and temporoparietal atrophy accentuated by the identification of a parietal pole of an area with old ischemic vascular accident and secondary cortical siderosis.

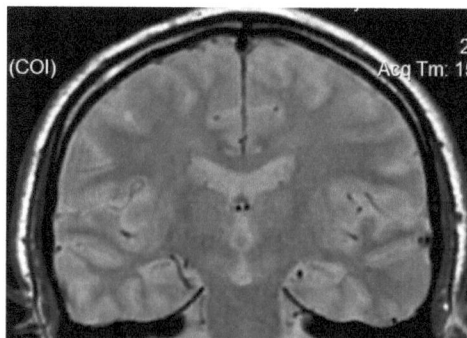

Fig. (41). Multiple cerebral microbleedds.

An important aspect in the therapy of cerebrovascular disorders in AD is the antiplatelet medication. Even if no consensus has been reached, it is considered that antiplatelet medication is associated with increased risk of cerebral micro-bleeds [147, 148]. On the other hand, in the hope of improving the quality of cerebral vascular perfusion in the small vessel disease conditions, the use of anti-platelet medications can block the development of thrombus in terms of CAA, stops the progression of βA vascular infiltration and interfere with platelet inflammatory pathways. This mechanism involving platelets can initiate new research on possible benefits of this therapy in Alzheimer's disease [149, 150].

- Imbalance of neurogenesis following the structural alteration of the emerging areas, namely the subventricular granular area (ventriculomegaly) and the granular area of the dentate gyrus (hippocampal atrophy).

- Biological and neuroimaging markers are of major importance in the diagnostic evaluation of cognitive deficits in AD and can help to develop theoretical models of dysconnectivity of cognitive neuronal circuits. A major goal of prophylactic strategies will be to prevent the development of cerebrovascular dysfunctions that have multiple pathogenic components (Fig. **42**).

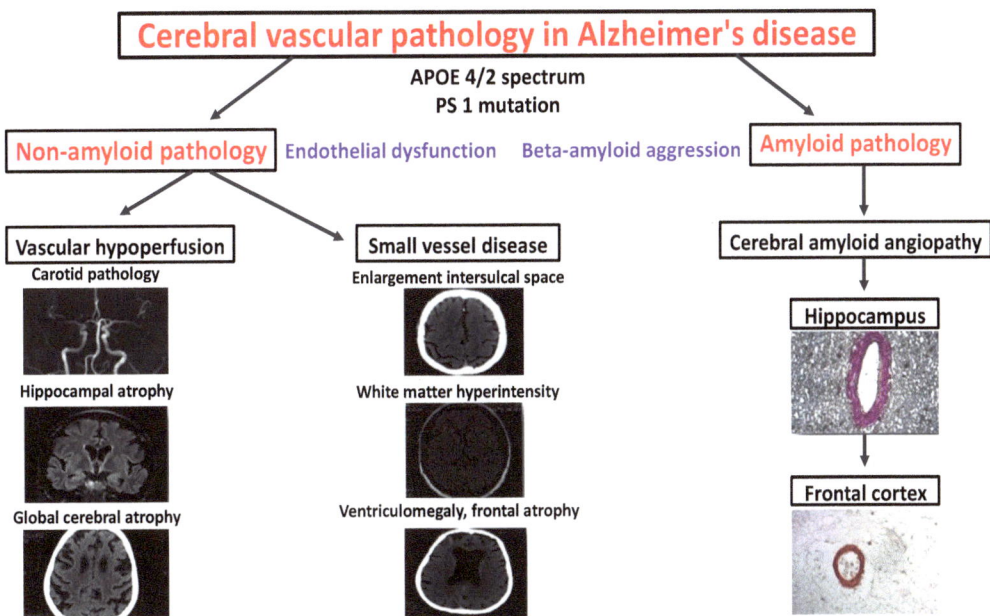

Fig. (42). Neuroimaging identification of cerebrovascular vulnerabilities; post mortem histopathological arguments [77]. Image used with permission from the main author and the editors of the RMJE.

All these mechanisms can be interdependent, can evolve individually, or can be correlated with somatic comorbidities and with the socio-cultural or cognitive level of each individual. These evolving possibilities call for a personalized diagnostic and therapeutic approach.

We consider that early identification of potential factors that may involve disconnective pathogenic mechanisms and their correct therapeutic approach can delay cognitive decline and improve the prognosis and quality of life for the elderly patient with AD.

CONCLUSIONS

Alzheimer Disease has become a major health problem at global level due to the increase in the number of patients, as well as the lack of a pharmacological therapy with spectacular results. On the other hand, the care for an AD patient comes with high costs, depending on the severity, requiring complementary therapies: customized care at home or in specialized centers, cognitive stimulation therapy with qualified medical and social personnel, behavioral cognitive therapy, psychologic therapy, therapy for the associated comorbidities, including the psychiatric ones.

Although major progress was recorded in deciphering the pathogenic mechanisms in AD (both amyloidosis and non- amyloidosis type), the current specific therapies led to slowing or stopping the cognitive decline on a small scale. Recognizing the risk factors correlated with rapid cognitive decline and with negative impact on the patient's life quality may be the solution for an early diagnosis, by involving all medical specialties consulting and treating elderly patients. We consider important the change in the approach of a patient with potential AD risk by focusing on the prevention of rapid cognitive decline dependent on risk factors that can be influenced by the medical decision (cardiovascular, infectious, related to dental area). An interesting correlation with cognitive decline is provided by the dental pathology in elderly patients.

The rapid cognitive decline is the main responsible factor for the patient's loss of independence by impacting daily activities, self-care, which leads to turning towards nursing home type of medical care, with major consequences on the quality of life of both patients and their families, but also on the costs of care. In our opinion, the specific management of some risk factors, generally minimized in the daily medical practice, could bring an essential contribution in the prevention activity. Among these factors, the following can be listed: dental risk factors (obliteration of the internal carotid artery in panoramic dental radiography, periodontitis and other chronic inflammatory processes), ophthalmological (retinal angiopathy or angiosclerosis), Doppler evaluation, neuroimaging or biological examinations.

There are major differences between the two clinical forms of AD (with early and late onset), the importance of identifying risk factors for cognitive decline being higher in the late onset form. The multidisciplinary pathogenic mechanisms brought forth in this paper allow for future elaboration of prevention strategies to improve patient's life quality by delaying cognitive decline also to maintain care within the family. Considering the hypothesis listed in this chapter, drawing a multidisciplinary, complete plan for the prophylaxis of rapid cognitive decline becomes a major challenge for the stakeholders involved in the management of AD patients.

These objectives should be a permanent concern for any medical discipline that provides a consultation to an elderly person, the multifactorial pathogenic model supported by us by offering each specialist the possibility of anticipating the risk of developing or evolving AD and of early guidance of the patient for an assessment and specific care.

We are aware of the limits of the theoretical hypotheses presented but we believe in their validity based on a long clinical experience in the treatment and care of AD patients and animal model translational research that opens new theoretical perspectives of neurobiological models that can be confirmed or refuted by subsequent research.

CONSENT FOR PUBLICATION

Not applicable.

CONFLICT OF INTEREST

The authors confirm that they have no conflict of interest to declare for this publication.

ACKNOWLEDGMENTS

We are grateful to the following colleagues for their support and for the images offered in the writing of this paper:

Daniel Pirici, Professor, Department of Pathology, University of Medicine and Pharmacy of Craiova, Craiova 200349, Romania. (Figs. **2** & **34**).

Adrian Camen, Associate Professor, Faculty of Dentistry, University of Medicine and Pharmacy of Craiova, Craiova 200349, Romania. (Figs. **9** & **10**).

Dragoş Camen, General Ultrasound Expert, Camen Medical Center, Craiova 200349, Romania. (Figs. **4**, **5**, **6**, **7** & **8**).

Andreea Tanasie, Assistant Lecturer, Department of Physiology, University of Medicine and Pharmacy of Craiova, Craiova 200349, Romania. (Fig. **3**).

REFERENCES

[1]   Evans DA, Funkenstein HH, Albert MS, *et al.* Prevalence of Alzheimer's disease in a community population of older persons. Higher than previously reported. JAMA 1989; 262(18): 2551-6.
        [http://dx.doi.org/10.1001/jama.1989.03430180093036] [PMID: 2810583]

[2]   World Alzheimer Report 2015, The Global Impact of Dementia, An analysis of prevalence, incidence, cost and trends

[3]   Mayeux R, Stern Y. Epidemiology of alzheimer disease. Cold spring harb perspect med 2012; 2(8): a006239.
        [http://dx.doi.org/10.1101/cshperspect.a006239] [PMID: 22908189]

[4]   Dragos Marinescu Didactic Collection

[5]   Forsyth E, Ritzline PD. An overview of the etiology, diagnosis, and treatment of Alzheimer disease. Phys Ther 1998; 78(12): 1325-1331, 1.
        [http://dx.doi.org/10.1093/ptj/78.12.1325] [PMID: 9859951]

[6]   Umphred Darcy Ann, Lazaro Rolando. Neurological Rehabilitation. 6th Edition., Mosby, 2012,. eBook ISBN: 9780323075879

[7]   Kumar A, Kumar P, Prasad M, Misra S, Kishor Pandit A, Chakravarty K. Association between Apolipoprotein ε4 Gene Polymorphism and Risk of Ischemic Stroke: A Meta-Analysis. Ann Neurosci 2016; 23(2): 113-21.
        [http://dx.doi.org/10.1159/000443568] [PMID: 27647962]

[8]   Sudlow C, Martínez González NA, Kim J, Clark C. Does apolipoprotein E genotype influence the risk of ischemic stroke, intracerebral hemorrhage, or subarachnoid hemorrhage? Systematic review and meta-analyses of 31 studies among 5961 cases and 17,965 controls. Stroke 2006; 37(2): 364-70.
        [http://dx.doi.org/10.1161/01.STR.0000199065.12908.62] [PMID: 16385096]

[9]   Liu C-C, Liu CC, Kanekiyo T, Xu H, Bu G. Apolipoprotein E and Alzheimer disease: risk, mechanisms and therapy. Nat Rev Neurol 2013; 9(2): 106-18.
        [http://dx.doi.org/10.1038/nrneurol.2012.263] [PMID: 23296339]

[10]   Head E, Powell D, Gold BT, Schmitt FA, Eur J. Alzheimer's Disease in Down Syndrome. Eur J Neurodegener Dis 2012; 1(3): 353-64.
        [PMID: 25285303]

[11]   Huntington Potter, Antoneta Granic, Julbert Caneus, role of trisomy 21 mosaicism in sporadic and familial alzheimer's disease, Curr Alzheimer Res. 2016; 13(1): 7–17

[12]   Shinya Yamamoto, Karen L. Schulze, Hugo J. Bellen. Introduction to Notch Signaling, Article in Methods in molecular biology, vol. 1187 (Clifton, N.J.), 2014, Springer Science Business Media New York 2014.
        [http://dx.doi.org/10.1007/978-1-4939-1139-4_1]

[13]   Reference GH. 2018. Your Guide to Understanding Genetic Conditions, PSEN1 gene,https://ghr.nlm.nih.gov/gene/PSEN1

[14]   Lendon CL, Ashall F, Goate AM. Exploring the etiology of Alzheimer disease using molecular genetics. JAMA 1997; 277(10): 825-31.
        [http://dx.doi.org/10.1001/jama.1997.03540340059034] [PMID: 9052714]

[15]   Vernooij MW, van der Lugt A, Ikram MA, *et al.* Prevalence and risk factors of cerebral microbleeds: the Rotterdam Scan Study. Neurology 2008; 70(14): 1208-14.
[http://dx.doi.org/10.1212/01.wnl.0000307750.41970.d9] [PMID: 18378884]

[16]   Poels MM, Vernooij MW, Ikram MA, *et al.* Prevalence and risk factors of cerebral microbleeds: an update of the Rotterdam scan study. Stroke 2010; 41(10) (Suppl.): S103-6.
[http://dx.doi.org/10.1161/STROKEAHA.110.595181] [PMID: 20876479]

[17]   Altmann-Schneider I, Trompet S, de Craen AJ, *et al.* Cerebral microbleeds are predictive of mortality in the elderly. Stroke 2011; 42(3): 638-44.
[http://dx.doi.org/10.1161/STROKEAHA.110.595611] [PMID: 21233474]

[18]   Pettersen JA, Sathiyamoorthy G, Gao FQ, *et al.* Microbleed topography, leukoaraiosis, and cognition in probable Alzheimer disease from the Sunnybrook dementia study. Arch Neurol 2008; 65(6): 790-5.
[http://dx.doi.org/10.1001/archneur.65.6.790] [PMID: 18541799]

[19]   Charidimou Andreas, Peeters Andre Philippe, Jäger Rolf, *et al.* Cortical superficial siderosis and intracerebral hemorrhage risk in cerebral amyloid angiopathy, Neurology 81 November 5, 2013, p 1666-1673, 2013 American Academy of Neurology

[20]   Charidimou A, Linn J, Vernooij MW, *et al.* Cortical superficial siderosis: detection and clinical significance in cerebral amyloid angiopathy and related conditions. Brain 2015; 138(Pt 8): 2126-39.
[http://dx.doi.org/10.1093/brain/awv162] [PMID: 26115675]

[21]   van der Flier WM. Clinical aspects of microbleeds in Alzheimer's disease. J Neurol Sci 2012; 322(1-2): 56-8.
[http://dx.doi.org/10.1016/j.jns.2012.07.009] [PMID: 22836015]

[22]   Kapasi A, Schneider JA. Vascular contributions to cognitive impairment, clinical Alzheimer's disease, and dementia in older persons. Biochim Biophys Acta 2016; 1862(5): 878-86.
[http://dx.doi.org/10.1016/j.bbadis.2015.12.023] [PMID: 26769363]

[23]   Nelson AR, Sweeney MD, Sagare AP, Zlokovic BV. Neurovascular dysfunction and neurodegeneration in dementia and Alzheimer's disease. Biochim Biophys Acta 2016; 1862(5): 887-900.
[http://dx.doi.org/10.1016/j.bbadis.2015.12.016] [PMID: 26705676]

[24]   Keable A, Fenna K, Yuen HM, *et al.* Deposition of amyloid β in the walls of human leptomeningeal arteries in relation to perivascular drainage pathways in cerebral amyloid angiopathy. Biochim Biophys Acta 2016; 1862(5): 1037-46.
[http://dx.doi.org/10.1016/j.bbadis.2015.08.024] [PMID: 26327684]

[25]   Hainsworth AH, Yeo NE, Weekman EM, Wilcock DM. Homocysteine, hyperhomocysteinemia and vascular contributions to cognitive impairment and dementia (VCID). Biochim Biophys Acta 2016; 1862(5): 1008-17.
[http://dx.doi.org/10.1016/j.bbadis.2015.11.015] [PMID: 26689889]

[26]   Zenaro E, Piacentino G, Constantin G. The blood-brain barrier in Alzheimer's disease. Neurobiol Dis 2017; 107: 41-56.
[http://dx.doi.org/10.1016/j.nbd.2016.07.007] [PMID: 27425887]

[27]   Zhao Zhen, Amy R. Nelson, Christer Betsholtz, Berislav V. Zlokovic, Establishment and Dysfunction of the Blood-Brain Barrier, Cell 163, November 19, 2015 Elsevier Inc

[28]   Liao H, Zhu Z, Peng Y. Potential utility of retinal imaging for alzheimer's disease: A review. Front Aging Neurosci 2018; 10: 188.www.frontiersin.org
[http://dx.doi.org/10.3389/fnagi.2018.00188] [PMID: 29988470]

[29]   Golzan SM, Goozee K, Georgevsky D, *et al.* Retinal vascular and structural changes are associated with amyloid burden in the elderly: ophthalmic biomarkers of preclinical Alzheimer's disease. Alzheimers Res Ther 2017; 9(1): 13.
[http://dx.doi.org/10.1186/s13195-017-0239-9] [PMID: 28253913]

[30]   Rivera-Rivera LA, Turski P, Johnson KM, *et al.* 4D flow MRI for intracranial hemodynamics assessment in Alzheimer's disease. J Cereb Blood Flow Metab 2016; 36(10): 1718-30.
[http://dx.doi.org/10.1177/0271678X15617171] [PMID: 26661239]

[31]   Lee KM, MacLean AG. New advances on glial activation in health and disease. World J Virol 2015; 4(2): 42-55.
[http://dx.doi.org/10.5501/wjv.v4.i2.42] [PMID: 25964871]

[32]   Guimarães Henriques JC, Kreich EM, Helena Baldani M, Luciano M, Cezar de Melo Castilho J, Cesar de Moraes L. Panoramic radiography in the diagnosis of carotid artery atheromas and the associated risk factors. Open Dent J 2011; 5: 79-83.
[http://dx.doi.org/10.2174/1874210601105010079] [PMID: 21760860]

[33]   Gil-Montoya JA, Sanchez-Lara I, Carnero-Pardo C, *et al.* Is periodontitis a risk factor for cognitive impairment and dementia? A case-control study. J Periodontol 2015; 86(2): 244-53.
[http://dx.doi.org/10.1902/jop.2014.140340] [PMID: 25345338]

[34]   Teixeira FB, Saito MT, Matheus FC, *et al.* Periodontitis and Alzheimer's Disease: A Possible Comorbidity between Oral Chronic Inflammatory Condition and Neuroinflammation. Front Aging Neurosci 2017; 9: 327.www.frontiersin.org
[http://dx.doi.org/10.3389/fnagi.2017.00327] [PMID: 29085294]

[35]   Bretz WA, Weyant RJ, Corby PM, *et al.* Systemic inflammatory markers, periodontal diseases, and periodontal infections in an elderly population. J Am Geriatr Soc 2005; 53(9): 1532-7.
[http://dx.doi.org/10.1111/j.1532-5415.2005.53468.x] [PMID: 16137283]

[36]   Poggesi A, Pasi M, Pescini F, Pantoni L, Inzitari D. Circulating biologic markers of endothelial dysfunction in cerebral small vessel disease: A review. J Cereb Blood Flow Metab 2016; 36(1): 72-94.
[http://dx.doi.org/10.1038/jcbfm.2015.116] [PMID: 26058695]

[37]   Kisler K, Nelson AR, Montagne A, Zlokovic BV. Cerebral blood flow regulation and neurovascular dysfunction in Alzheimer disease. Nat Rev Neurosci 2017; 18(7): 419-34.
[http://dx.doi.org/10.1038/nrn.2017.48] [PMID: 28515434]

[38]   Howcroft TK, Campisi J, Louis GB, *et al.* The role of inflammation in age-related disease. Aging (Albany NY) 2013; 5(1): 84-93.
[http://dx.doi.org/10.18632/aging.100531] [PMID: 23474627]

[39]   Bettcher BM, Wilheim R, Rigby T, *et al.* C-reactive protein is related to memory and medial temporal brain volume in older adults. Brain Behav Immun 2012; 26(1): 103-8.
[http://dx.doi.org/10.1016/j.bbi.2011.07.240] [PMID: 21843630]

[40]   Terry AV Jr, Buccafusco JJ. The cholinergic hypothesis of age and Alzheimer's disease-related cognitive deficits: recent challenges and their implications for novel drug development. J Pharmacol Exp Ther 2003; 306(3): 821-7.
[http://dx.doi.org/10.1124/jpet.102.041616] [PMID: 12805474]

[41]   Michael D. Kopelman, the cholinergic neurotransmitter system in human memory and dementia: A review. Q J Exp Psychol 1986; 38A: 535-73.

[42]   Nissen MJ, Knopman DS, Schacter DL. Neurochemical dissociation of memory systems. Neurology 1987; 37(5): 789-94.
[http://dx.doi.org/10.1212/WNL.37.5.789] [PMID: 3574678]

[43]   Francis PT, Palmer AM, Snape M, Wilcock GK. The cholinergic hypothesis of Alzheimer's disease: a review of progress. J Neurol Neurosurg Psychiatry 1999; 66(2): 137-47.
[http://dx.doi.org/10.1136/jnnp.66.2.137] [PMID: 10071091]

[44]   Kopelman MD, Corn TH. Cholinergic 'blockade' as a model for cholinergic depletion. A comparison of the memory deficits with those of Alzheimer-type dementia and the alcoholic Korsakoff syndrome. Brain 1988; 111(Pt 5): 1079-110.
[http://dx.doi.org/10.1093/brain/111.5.1079] [PMID: 3179685]

[45]   Contestabile A. The history of the cholinergic hypothesis. Behav Brain Res 2010.
[http://dx.doi.org/10.1016/j.bbr.2009.12.044] [PMID: 20060018]

[46]   Benson DF, Davis RJ, Snyder BD. Posterior cortical atrophy. Arch Neurol 1988; 45(7): 789-93.
[http://dx.doi.org/10.1001/archneur.1988.00520310107024] [PMID: 3390033]

[47]   Grover S, Amitava AK, Kumari N. One glasses too many: A case report of Benson's syndrome. Indian J Ophthalmol 2015; 63(3): 277-9.
[http://dx.doi.org/10.4103/0301-4738.156938] [PMID: 25971180]

[48]   Petersen RC, Smith GE, Waring SC, Ivnik RJ, Tangalos EG, Kokmen E. Mild cognitive impairment: clinical characterization and outcome. Arch Neurol 1999; 56(3): 303-8.
[http://dx.doi.org/10.1001/archneur.56.3.303] [PMID: 10190820]

[49]   Petersen RC, Roberts RO, Knopman DS, *et al.* Mild cognitive impairment: ten years later. Arch Neurol 2009; 66(12): 1447-55.
[http://dx.doi.org/10.1001/archneurol.2009.266] [PMID: 20008648]

[50]   Tarawneh R, Holtzman DM. The clinical problem of symptomatic Alzheimer disease and mild cognitive impairment. Cold Spring Harb Perspect Med 2012; 2(5): a006148.
[http://dx.doi.org/10.1101/cshperspect.a006148] [PMID: 22553492]

[51]   Petersen RC, Caracciolo B, Brayne C, Gauthier S, Jelic V, Fratiglioni L. Mild cognitive impairment: a concept in evolution. J Intern Med 2014; 275(3): 214-28.
[http://dx.doi.org/10.1111/joim.12190] [PMID: 24605806]

[52]   Tettamanti M, Garri M, Riva E, *et al.* Association of sensory deficits with dementia in the very old: the Monzino 80-plus study. Alzheimers Dement 2009; 5(4) (Suppl.): P285.
[http://dx.doi.org/10.1016/j.jalz.2009.04.389]

[53]   Jefferis JM, Mosimann UP, Clarke MP. Cataract and cognitive impairment: a review of the literature. Br J Ophthalmol 2011; 95(1): 17-23.
[http://dx.doi.org/10.1136/bjo.2009.165902] [PMID: 20807709]

[54]   Hicks K. www.alzheimers.net/the-connection-between-alzheimers-and-hearing-loss

[55]   Chen W-W, Zhang X, Huang W-J. Role of physical exercise in Alzheimer's disease. Biomed Rep 2016; 4(4): 403-7.
[http://dx.doi.org/10.3892/br.2016.607] [PMID: 27073621]

[56]   Upadhya SC, Hegde AN. Role of the ubiquitin proteasome system in Alzheimer's disease. BMC Biochem 2007; 8 (Suppl. 1): S12.
[http://dx.doi.org/10.1186/1471-2091-8-S1-S12] [PMID: 18047736]

[57]   Nithiananantharajah J, Hannan AJ. The neurobiology of brain and cognitive reserve: mental and physical activity as modulators of brain disorders. Prog Neurobiol 2009; 89(4): 369-82.
[http://dx.doi.org/10.1016/j.pneurobio.2009.10.001] [PMID: 19819293]

[58]   Pepys MB. Pathogenesis, diagnosis and treatment of systemic amyloidosis. Philos Trans R Soc Lond B Biol Sci 2001; 356(1406): 203-10.
[http://dx.doi.org/10.1098/rstb.2000.0766] [PMID: 11260801]

[59]   Schröder R, Linke RP. Cerebrovascular involvement in systemic AA and AL amyloidosis: a clear haematogenic pattern. Virchows Arch 1999; 434(6): 551-60.
[http://dx.doi.org/10.1007/s004280050383] [PMID: 10394892]

[60]   Petkova AT, Ishii Y, Balbach JJ. A structural model for Alzheimer's β-amyloid fibrils based on experimental constraints from solid state NMR. Proc Natl Acad Sci USA 2002; 99(26): 16742-7.

[61]   Walsh DM, Lomakin A, Benedek GB, Condron MM, Teplow DB. Amyloid β-protein fibrillogenesis. Detection of a protofibrillar intermediate. J Biol Chem 1997; 272(35): 22364-72.
[http://dx.doi.org/10.1074/jbc.272.35.22364] [PMID: 9268388]

[62]   Karran E, Mercken M, De Strooper B. The amyloid cascade hypothesis for Alzheimer's disease: an appraisal for the development of therapeutics. Nat Rev Drug Discov 2011; 10(9): 698-712.
[http://dx.doi.org/10.1038/nrd3505] [PMID: 21852788]

[63]   Hassan W, Al-Sergani H, Mourad W, Tabbaa R. Amyloid heart disease. New frontiers and insights in pathophysiology, diagnosis, and management. Tex Heart Inst J 2005; 32(2): 178-84.
[PMID: 16107109]

[64]   Marvanova M. Drug-induced cognitive impairment: Effect of cardiovascular agents. Ment Health Clin 2016; 6(4): 201-6.
[http://dx.doi.org/10.9740/mhc.2016.07.201] [PMID: 29955471]

[65]   Davies P. Challenging the cholinergic hypothesis in Alzheimer disease. JAMA 1999; 281(15): 1433-4.
[http://dx.doi.org/10.1001/jama.281.15.1433] [PMID: 10217061]

[66]   Wong TP, Debeir T, Duff K, Cuello AC. Reorganization of cholinergic terminals in the cerebral cortex and hippocampus in transgenic mice carrying mutated presenilin-1 and amyloid precursor protein transgenes. J Neurosci 1999; 19(7): 2706-16.
[http://dx.doi.org/10.1523/JNEUROSCI.19-07-02706.1999] [PMID: 10087083]

[67]   Mintzer J, Burns A. Anticholinergic side-effects of drugs in elderly people. J R Soc Med 2000; 93(9): 457-62.
[http://dx.doi.org/10.1177/014107680009300903] [PMID: 11089480]

[68]   Tune LE. Anticholinergic effects of medication in elderly patients. J Clin Psychiatry 2001; 62 (Suppl. 21): 11-4.
[PMID: 11584981]

[69]   Leclercq PD, Murray LS, Smith C, Graham DI, Nicoll JA, Gentleman SM. Cerebral amyloid angiopathy in traumatic brain injury: association with apolipoprotein E genotype. J Neurol Neurosurg Psychiatry 2005; 76(2): 229-33.
[http://dx.doi.org/10.1136/jnnp.2003.025528] [PMID: 15654038]

[70]   Mannix RC, Whalen MJ. Traumatic brain injury, microglia, and Beta amyloid. Int J Alzheimers Dis 2012; 2012: 608732.
[http://dx.doi.org/10.1155/2012/608732] [PMID: 22666622]

[71]   Patricia M. Washington, Sonia Villapol, Mark P. Burns, Polypathology and dementia after brain trauma: Does brain injury trigger distinct neurodegenerative diseases, or should it be classified together as traumatic encephalopathy? Exp Neurol. 2016 January ; 275(0 3): 381–388

[72]   Ding K, Puneet K, Diaz-Arrastia Ramon. Epilepsy after Traumatic Brain Injury, Chapter 14, Translational Research in Traumatic Brain Injury, Frontiers in NeuroscienceInformation Boca Raton FL: CRC Press/Taylor and Francis Group 2016. ISBN-13: 978-1-4665-8491-4

[73]   McKee AC, Cantu RC, Nowinski CJ, *et al.* Chronic traumatic encephalopathy in athletes: progressive tauopathy after repetitive head injury. J Neuropathol Exp Neurol 2009; 68(7): 709-35.
[http://dx.doi.org/10.1097/NEN.0b013e3181a9d503] [PMID: 19535999]

[74]   Foley DJ, Monjan A, Simonsick EM, Wallace RB, Blazer DG. Incidence and remission of insomnia among elderly adults: an epidemiologic study of 6,800 persons over three years. Sleep 1999; 22(S2) (Suppl. 2): S366-72.
[PMID: 10394609]

[75]   Kamel NS, Gammack JK. Insomnia in the elderly: cause, approach, and treatment. Am J Med 2006; 119(6): 463-9.
[http://dx.doi.org/10.1016/j.amjmed.2005.10.051] [PMID: 16750956]

[76]   Popoli M, Yan Z, McEwen BS, Sanacora G. The stressed synapse: the impact of stress and glucocorticoids on glutamate transmission. Nat Rev Neurosci 2011; 13(1): 22-37.
[http://dx.doi.org/10.1038/nrn3138] [PMID: 22127301]

[77]   Marinescu I, Enătescu VR, Ghelase ŞM, Marinescu D. Neurobiological arguments for a pathogenic multifactorial disconnective model of cognitive disorders from Alzheimer's disease in elderly people. Rom J Morphol Embryol 2017; 58(4): 1165-73.
        [PMID: 29556605]

[78]   Hogan DB, Maxwell CJ, Fung TS, Ebly EM. Prevalence and potential consequences of benzodiazepine use in senior citizens: results from the Canadian Study of Health and Aging. Can J Clin Pharmacol 2003; 10(2): 72-7.
        [PMID: 12879145]

[79]   Tannenbaum C, Martin P, Tamblyn R, Benedetti A, Ahmed S. Reduction of inappropriate benzodiazepine prescriptions among older adults through direct patient education: the EMPOWER cluster randomized trial. JAMA Intern Med 2014; 174(6): 890-8.
        [http://dx.doi.org/10.1001/jamainternmed.2014.949] [PMID: 24733354]

[80]   Takada M, Fujimoto M, Hosomi K. Association between Benzodiazepine Use and Dementia: Data Mining of Different Medical Databases. Int J Med Sci 2016; 13(11): 825-34.
        [http://dx.doi.org/10.7150/ijms.16185] [PMID: 27877074]

[81]   Yaffe K, Boustani M. Benzodiazepines and risk of Alzheimer's disease. BMJ 2014; 349: g5312.
        [http://dx.doi.org/10.1136/bmj.g5312] [PMID: 25205606]

[82]   Pétursson H. The benzodiazepine withdrawal syndrome. Addiction 1994; 89(11): 1455-9.
        [http://dx.doi.org/10.1111/j.1360-0443.1994.tb03743.x] [PMID: 7841856]

[83]   Högl B. REM sleep behavior disorder: Diagnostic criteria, EMG based accurate quantitative diagnostics, value and limitations of questionnaires for diagnosis and differential diagnosis, Teaching Course 1 3rd Congress of the European Academy of Neurology. Amsterdam, The Netherlands. June 24 – 27, 2017;

[84]   Ferini-Strambi L, Marelli S, Galbiati A, Rinaldi F, Giora E, Sleep REM. REM Sleep Behavior Disorder (RBD) as a marker of neurodegenerative disorders. Arch Ital Biol 2014; 152(2-3): 129-46.
        [http://dx.doi.org/10.12871/000298292014238] [PMID: 25828685]

[85]   Siepel FJ, Rongve A, Buter TC, *et al.* (123I)FP-CIT SPECT in suspected dementia with Lewy bodies: a longitudinal case study. BMJ Open 2013; 3(4): e002642.
        [http://dx.doi.org/10.1136/bmjopen-2013-002642] [PMID: 23572198]

[86]   McCarter SJ, St Louis EK, Boeve BF. REM sleep behavior disorder and REM sleep without atonia as an early manifestation of degenerative neurological disease. Curr Neurol Neurosci Rep 2012; 12(2): 182-92.
        [http://dx.doi.org/10.1007/s11910-012-0253-z] [PMID: 22328094]

[87]   Rodrigues Brazète J, Gagnon JF, Postuma RB, Bertrand JA, Petit D, Montplaisir J. Electroencephalogram slowing predicts neurodegeneration in rapid eye movement sleep behavior disorder. Neurobiol Aging 2016; 37: 74-81.
        [http://dx.doi.org/10.1016/j.neurobiolaging.2015.10.007] [PMID: 26545633]

[88]   Schenck CH, Mahowald MW. Long-term, nightly benzodiazepine treatment of injurious parasomnias and other disorders of disrupted nocturnal sleep in 170 adults. Am J Med 1996; 100(3): 333-7.
        [http://dx.doi.org/10.1016/S0002-9343(97)89493-4] [PMID: 8629680]

[89]   Schenck CH, Mahowald MW. REM sleep behavior disorder: clinical, developmental, and neuroscience perspectives 16 years after its formal identification in SLEEP. Sleep 2002;25(2):120Y138
        [http://dx.doi.org/10.1093/sleep/25.2.120]

[90]   β-Amyloid accumulation in the human brain after one night of sleep deprivation, 2018 www.pnas.org/cgi

[91]   Brown BM, Rainey-Smith SR, Villemagne VL, *et al.* The relationship between sleep quality and brain amyloid burden. Sleep (Basel) 2016; 39(5): 1063-8.

[http://dx.doi.org/10.5665/sleep.5756] [PMID: 27091528]

[92]   Ju Y-ES, Lucey BP, Holtzman DM. Sleep and Alzheimer disease pathology--a bidirectional relationship. Nat Rev Neurol 2014; 10(2): 115-9.
[http://dx.doi.org/10.1038/nrneurol.2013.269] [PMID: 24366271]

[93]   Spira AP, Gottesman RF. Sleep disturbance: an emerging opportunity for Alzheimer's disease prevention? Int Psychogeriatr 2017; 29(4): 529-31.
[http://dx.doi.org/10.1017/S1041610216002131] [PMID: 27938445]

[94]   Padayachey U, Ramlall S, Chipps J. Depression in older adults: prevalence and risk factors in a primary health care sample. S Afr Fam Pract 2017; 59(2): 61-6.
[http://dx.doi.org/10.1080/20786190.2016.1272250]

[95]   Steptoe A. Depression and Physical Illness, Cambridge University Press 2007 www.cambridge.org

[96]   Mavrides N, Nemeroff C. Treatment of depression in cardiovascular disease. Depress Anxiety 2013; 30(4): 328-41.
[http://dx.doi.org/10.1002/da.22051] [PMID: 23293051]

[97]   Laghrissi-Thode F, Wagner WR, Pollock BG, Johnson PC, Finkel MS. Elevated platelet factor 4 and beta-thromboglobulin plasma levels in depressed patients with ischemic heart disease. Biol Psychiatry 1997; 42(4): 290-5.
[http://dx.doi.org/10.1016/S0006-3223(96)00345-9] [PMID: 9270907]

[98]   Serebruany VL, Glassman AH, Malinin AI, *et al.* Enhanced platelet/endothelial activation in depressed patients with acute coronary syndromes: evidence from recent clinical trials. Blood Coagul Fibrinolysis 2003; 14(6): 563-7.
[http://dx.doi.org/10.1097/00001721-200309000-00008] [PMID: 12960610]

[99]   Serebruany VL, Glassman AH, Malinin AI, *et al.* Platelet/endothelial biomarkers in depressed patients treated with the selective serotonin reuptake inhibitor sertraline after acute coronary events: the Sertraline AntiDepressant Heart Attack Randomized Trial (SADHART) Platelet Substudy. Circulation 2003; 108(8): 939-44.
[http://dx.doi.org/10.1161/01.CIR.0000085163.21752.0A] [PMID: 12912814]

[100]   Jack C. de la Torre, Cardiovascular Risk Factors Promote Brain Hypoperfusion Leading to Cognitive Decline and Dementia, Cardiovascular Psychiatry and Neurology, Volume 2012

[101]   Bejar D, Colombo PC, Latif F, Yuzefpolskaya M. Infiltrative Cardiomyopathies. Clin Med Insights Cardiol 2015; 9 (Suppl. 2): 29-38.
[http://dx.doi.org/10.4137/CMC.S19706] [PMID: 26244036]

[102]   Biessels GJ, Kappelle LJ. Increased risk of Alzheimer's disease in Type II diabetes: insulin resistance of the brain or insulin-induced amyloid pathology? Biochem Soc Trans 2005; 33(Pt 5): 1041-4.
[http://dx.doi.org/10.1042/BST0331041] [PMID: 16246041]

[103]   Seigers R, Schagen SB, Coppens CM, *et al.* Methotrexate decreases hippocampal cell proliferation and induces memory deficits in rats. Behav Brain Res 2009; 201(2): 279-84.
[http://dx.doi.org/10.1016/j.bbr.2009.02.025] [PMID: 19428645]

[104]   Licastro F, Porcellini E. Persistent infections, immune-senescence and Alzheimer's disease. Oncoscience 2016; 3(5-6): 135-42.
[http://dx.doi.org/10.18632/oncoscience.309] [PMID: 27489858]

[105]   Kopstein M, Mohlman DJ. HIV-1 Encephalopathy and Aids Dementia Complex.StatPearls. Treasure Island, FL: StatPearls Publishing 2018.https://www.ncbi.nlm.nih.gov/books/NBK507700 Updated 2018 Jun 9 Internet

[106]   Valcour VG. HIV, aging, and cognition: emerging issues. Top Antivir Med 2013; 21(3): 119-23.
[PMID: 23981600]

[107]   Crossley KB, Peterson PK. Infections in the elderly. Clin Infect Dis 1996; 22(2): 209-15.

[http://dx.doi.org/10.1093/clinids/22.2.209] [PMID: 8838174]

[108] Pisa D, Alonso R, Rábano A, Rodal I, Carrasco L. Different brain regions are infected with fungi in alzheimer's disease. Sci Rep 2015; 5: 15015.
[http://dx.doi.org/10.1038/srep15015] [PMID: 26468932]

[109] de Pauw BE, What AFI. What are fungal infections? Mediterr J Hematol Infect Dis 2011; 3(1): e2011001.
[http://dx.doi.org/10.4084/mjhid.2011.001] [PMID: 21625304]

[110] Tate Judith A. Infection Hospitalization Increases Risk of Dementia in the Elderly, Crit Care Med. 2014 May; 42(5): 1037–1046

[111] Pepys MB. Amyloidosis. Annu Rev Med 2006; 57(1): 223-41.
[http://dx.doi.org/10.1146/annurev.med.57.121304.131243] [PMID: 16409147]

[112] Perry VH, Newman TA, Cunningham C. The impact of systemic infection on the progression of neurodegenerative disease. Nat Rev Neurosci 2003; 4(2): 103-12.
[http://dx.doi.org/10.1038/nrn1032] [PMID: 12563281]

[113] Chen Y-M, Chen LK, Lan JL, Chen DY. Geriatric syndromes in elderly patients with rheumatoid arthritis. Rheumatology (Oxford) 2009; 48(10): 1261-4.
[http://dx.doi.org/10.1093/rheumatology/kep195] [PMID: 19651885]

[114] Tutuncu Z, Kavanaugh A. Treatment of elderly rheumatoid arthritis. Future Rheumatol 2007; 2(3): 313.
[http://dx.doi.org/10.2217/17460816.2.3.313]

[115] Aiello F, Dueñas EP, Musso CG, Nephropathy S. Senescent nephropathy: The new renal syndrome. Healthcare (Basel) 2017; 5(4): 81.
[http://dx.doi.org/10.3390/healthcare5040081] [PMID: 29143769]

[116] Abrass Christine K. In: online geriatric nephrology curriculum. chicago: American society of nephrology, 2009. Available form: www.asn-online

[117] Galešić K, Ljubanović D, Sabljar-Matovinović M, Prkačin I, Horvatić I, Račić I. Nephrotic Syndrome in the Elderly, Acta clinica Croatica, Vol.42 No.4 Prosinac 2003

[118] Mathews PM, Cataldo AM, Kao BH, *et al.* Brain expression of presenilins in sporadic and early-onset, familial Alzheimer's disease. Mol Med 2000; 6(10): 878-91.
[http://dx.doi.org/10.1007/BF03401825] [PMID: 11126202]

[119] Joy T, Wang J, Hahn A, Hegele RA. APOA1 related amyloidosis: a case report and literature review. Clin Biochem 2003; 36(8): 641-5.
[http://dx.doi.org/10.1016/S0009-9120(03)00110-3] [PMID: 14636880]

[120] Assman G, Schmitz G, Funke H, von Eckardstein A. Apolipoprotein A-1 and HDL deficiency. Curr Opin Lipidol 1990; 1: 110-5.
[http://dx.doi.org/10.1097/00041433-199004000-00005]

[121] Gillmore JD, Lovat LB, Persey MR, Pepys MB, Hawkins PN. Amyloid load and clinical outcome in AA amyloidosis in relation to circulating concentration of serum amyloid A protein. Lancet 2001; 358(9275): 24-9.
[http://dx.doi.org/10.1016/S0140-6736(00)05252-1] [PMID: 11454373]

[122] Kovacs GG. Molecular pathological classification of neurodegenerative diseases: Turning towards precision medicine. Int J Mol Sci 2016; 17(2): E189.
[http://dx.doi.org/10.3390/ijms17020189] [PMID: 26848654]

[123] Fang YS, Tsai KJ, Chang YJ, *et al.* Full-length TDP-43 forms toxic amyloid oligomers that are present in frontotemporal lobar dementia-TDP patients. Nat Commun 2014; 5: 4824.
[http://dx.doi.org/10.1038/ncomms5824] [PMID: 25215604]

[124] Amador-Ortiz C, Lin WL, Ahmed Z, *et al.* TDP-43 immunoreactivity in hippocampal sclerosis and

Alzheimer's disease. Ann Neurol 2007; 61(5): 435-45.
[http://dx.doi.org/10.1002/ana.21154] [PMID: 17469117]

[125] Larner AJ. Epileptic seizures in neurodegenerative dementia syndromes, Journal of neurology and neuroscience 2010, Vol.1, No. 1:3

[126] Penke B, Bogár F, Fülöp L. β-Amyloid and the Pathomechanisms of Alzheimer's Disease: A Comprehensive View. Molecules 2017; 22(10): 1692.
[http://dx.doi.org/10.3390/molecules22101692] [PMID: 28994715]

[127] van der Kant R, Goldstein LS. Cellular functions of the amyloid precursor protein from development to dementia. Dev Cell 2015; 32(4): 502-15.
[http://dx.doi.org/10.1016/j.devcel.2015.01.022] [PMID: 25710536]

[128] Thinakaran G, Koo EH. Amyloid precursor protein trafficking, processing, and function. J Biol Chem 2008; 283(44): 29615-9.
[http://dx.doi.org/10.1074/jbc.R800019200] [PMID: 18650430]

[129] Banks WA. Physiology and pathology of the blood-brain barrier: implications for microbial pathogenesis, drug delivery and neurodegenerative disorders. J Neurovirol 1999; 5(6): 538-55.
[http://dx.doi.org/10.3109/13550289909021284] [PMID: 10602396]

[130] Sheng JG, Boop FA, Mrak RE, Griffin WS. Increased neuronal beta-amyloid precursor protein expression in human temporal lobe epilepsy: association with interleukin-1 alpha immunoreactivity. J Neurochem 1994; 63(5): 1872-9.
[http://dx.doi.org/10.1046/j.1471-4159.1994.63051872.x] [PMID: 7931344]

[131] Jeffrey L. Noebels, a perfect storm: Converging paths of epilepsy and alzheimer's dementia intersect in the hippocampal formation. Epilepsia 2011; 52 (Suppl. 1): 39-46.
[http://dx.doi.org/10.1111/j.1528-1167.2010.02909.x]

[132] Nag S, Yu L, Capuano AW, et al. Hippocampal sclerosis and TDP-43 pathology in aging and Alzheimer disease. Ann Neurol 2015; 77(6): 942-52.
[http://dx.doi.org/10.1002/ana.24388] [PMID: 25707479]

[133] Rowczenio DM, Noor I, Gillmore JD, et al. Online registry for mutations in hereditary amyloidosis including nomenclature recommendations. Hum Mutat 2014; 35(9): E2403-12.
[http://dx.doi.org/10.1002/humu.22619] [PMID: 25044787]

[134] Kang HG, Bybee A, Ha IS, et al. Hereditary amyloidosis in early childhood associated with a novel insertion-deletion (indel) in the fibrinogen Aalpha chain gene. Kidney Int 2005; 68(5): 1994-8.
[http://dx.doi.org/10.1111/j.1523-1755.2005.00653.x] [PMID: 16221199]

[135] Morie A. Gertz, John P. Garton, Robert A. Kyle, Primary amyloidosis (AL) in families, American Journal of Hematology - Wiley Online Library, First published: June 1986

[136] Lavatelli F, Albertini R, Di Fonzo A, Palladini G, Merlini G. Biochemical markers in early diagnosis and management of systemic amyloidoses. Clin Chem Lab Med 2014; 52(11): 1517-31.
[http://dx.doi.org/10.1515/cclm-2014-0235] [PMID: 24870609]

[137] Morie A, Gertz, Ray Comenzo, et al. Definition of organ involvement and treatment response in immunoglobulin light chain amyloidosis (AL): A consensus opinion from the 10th international symposium on amyloid and amyloidosis. Am J Hematol 2005; 79: 319-28.
[PMID: 16044444]

[138] Wechalekar AD. Biomarkers in AL amyloidosis: is the summit in sight? Blood 2018; 131(14): 1502-3.
[http://dx.doi.org/10.1182/blood-2018-02-832113] [PMID: 29622535]

[139] Fuchs T, Trollor JN, Crawford J, et al. Macrophage inhibitory cytokine-1 is associated with cognitive impairment and predicts cognitive decline - the Sydney Memory and Aging Study. Aging Cell 2013; 12(5): 882-9.
[http://dx.doi.org/10.1111/acel.12116] [PMID: 23758647]

[140] Wechalekar AD, Gillmore JD, Wassef N, Lachmann HJ, Whelan C, Hawkins PN. Abnormal N-terminal fragment of brain natriuretic peptide in patients with light chain amyloidosis without cardiac involvement at presentation is a risk factor for development of cardiac amyloidosis. Haematologica 2011; 96(7): 1079-80.
[http://dx.doi.org/10.3324/haematol.2011.040493] [PMID: 21606171]

[141] Buerger K, Uspenskaya O, Hartmann O, *et al*. Prediction of Alzheimer's disease using midregional proadrenomedullin and midregional proatrial natriuretic peptide: a retrospective analysis of 134 patients with mild cognitive impairment. J Clin Psychiatry 2011; 72(4): 556-63.
[http://dx.doi.org/10.4088/JCP.09m05872oli] [PMID: 21208578]

[142] Marinescu D, Mogoanta L, Sfera A, Gradini R, Osorio C, Preda A. An experimental model of ischemia-induced changes into the glymphatic system of the brain. Rom J Psychopharma 2013; 13(4): 243-51.

[143] Kapural M, Krizanac-Bengez Lj, Barnett G, *et al*. Serum S-100β as a possible marker of blood-brain barrier disruption. Brain Res 2002; 940(1-2): 102-4.
[http://dx.doi.org/10.1016/S0006-8993(02)02586-6] [PMID: 12020881]

[144] van Engelen BG, Lamers KJ, Gabreels FJ, Wevers RA, van Geel WJ, Borm GF. Age-related changes of neuron-specific enolase, S-100 protein, and myelin basic protein concentrations in cerebrospinal fluid. Clin Chem 1992; 38(6): 813-6.
[PMID: 1375875]

[145] Marinescu Ileana, Stovicek PO, Marinescu D. Blood-brain barrier disruption - a risk factor in the evolution of alzheimer's disease, national alzheimer conference, bucharest 2018, 8th edition, latest dementia news, scientific abstract book, vol VIII, 2018, p.36

[146] Verheggen ICM, Van Boxtel MPJ, Verhey FRJ, Jansen JFA, Backes WH. Interaction between blood-brain barrier and glymphatic system in solute clearance. Neurosci Biobehav Rev 2018; 90: 26-33.
[http://dx.doi.org/10.1016/j.neubiorev.2018.03.028] [PMID: 29608988]

[147] Qiu J, Ye H, Wang J, Yan J, Wang J, Wang Y. Antiplatelet Therapy, Cerebral Microbleeds, and Intracerebral Hemorrhage: A Meta-Analysis. Stroke 2018; 49(7): 1751-4.
[http://dx.doi.org/10.1161/STROKEAHA.118.021789] [PMID: 29798835]

[148] Liu S, Li C. Antiplatelet drug use and cerebral microbleeds: A meta-analysis of published studies. J Stroke Cerebrovasc Dis 2015; 24(10): 2236-44.
[http://dx.doi.org/10.1016/j.jstrokecerebrovasdis.2015.05.022] [PMID: 26272868]

[149] Gowert NS, Donner L, Chatterjee M, *et al*. Blood platelets in the progression of Alzheimer's disease. PLoS One 2014; 9(2): e90523.
[http://dx.doi.org/10.1371/journal.pone.0090523] [PMID: 24587388]

[150] Zhang W, Huang W, Jing F. Contribution of blood platelets to vascular pathology in Alzheimer's disease. J Blood Med 2013; 4: 141-7. [doi].
[http://dx.doi.org/10.2147/JBM.S45071] [PMID: 24235853]

SUBJECT INDEX

A

Acetylcholine 154, 155, 156, 157, 159, 160, 161, 162, 165, 187, 191, 192, 206, 207
  deficiency 154, 156, 207
  esterases 160, 165
  hydrolysis of 160, 161
  levels 154, 157, 207
Acetylcholinesterase Inhibitors 34
Acetylcoenzymes 160
AChE 42, 49, 163, 168, 171, 173
  and monoamine oxidase inhibitors 173
  by binding to active gorge of enzyme 168
  dual binding and β-amyloid aggregation inhibition 173
  compounds acts 163
  drugs in MTDL approach 42
  inhibition 49, 171
AChE-induced 176, 180
  Aβ anti-aggregation activity 180
  -induced BChE 176
AChE inhibitors (AChEi) 35, 38, 41, 42, 154, 165, 171, 173, 178
  active 171
  rivastigmine 165
AChE inhibitory 43, 163
  activity 163
  drugs 43
Acid 4, 58, 83, 87, 89, 90, 96, 97, 160, 180
  amino 4, 83, 87, 89, 90, 96, 97
  lipoic 180
  valproic 58
Actions 53, 138, 139, 140, 141
  anti-inflammatory 53, 139, 140, 141
  anti-oxidative 138, 141
Activity 60, 84, 136, 138, 192, 212, 213, 214, 228
  functional 60, 84
  locomotor 136, 138
  physical 212, 213, 214, 228
  prophylactic 192, 212
AD-associated 86, 102, 103

autophagy 103
  neuroinflammation 86, 102, 103
Agents, antidiabetic 51
Aggression 132, 134, 136
  significant improvements in 132
Aggressiveness 211, 213, 218, 225, 240
Aging, progressive 81, 82
Agitation 132, 133, 134, 135, 136, 226
Alzheimer 44, 81, 82, 96, 107, 165
Alzheimer's disease 36, 37, 169
  assessment scale-cognitive 169
  drug development pipeline 36, 37
Alzheimer's disease international 193
Ameliorate disease pathology 106
Amphotericin 81, 82, 108
Amplification 202, 219, 220, 225, 234, 240
Amyloid angiopathy, cerebral 192, 197, 247, 248
Amyloidogenesis 49
Amyloidosis 216, 217, 233, 237, 242, 250
  processes 216, 217
Amyloid plaques 35, 48, 91, 106, 107, 135, 139, 216
Angiotensin-converting enzyme (ACE) 84
Annonacin 9, 16
Anti-Aβ Aggregation 45, 46
Anti-AD Drugs 39, 42, 165
Antidepressants 218, 225, 226, 230
Antioxidant 33, 38, 50, 51, 57, 61, 138, 173, 176, 179, 180, 181
  activity 33, 50, 173, 180, 181
  enzyme activity 179
Antipsychotic medications 131, 241, 242
Antiviral medication 235
Apoptosis 8, 84, 90, 141, 244, 245
Approach 39, 191, 192
  multidisciplinary 191, 192
  multi-targeting 39
APP transgenic mice 103, 135
Areas 200, 212, 249
  cortical projection 200, 212
  granular 212, 249

binding site 163, 164
Cholinergic 35, 156, 157, 159, 162, 163, 177,
    207, 227
  abnormalities 157
  neurons 35, 157, 159, 162, 163
  system 156, 162, 177
  transmission 162, 163, 207, 227
Cholinesterases 33, 40, 41, 160, 162, 165, 169,
    174, 176, 179
  inhibitors, irreversible 165, 169
Choroid plexus 215
Chromosome 2, 18, 83, 162, 163, 196
Circuits 212, 247, 249
  cognitive neuronal 247, 249
  functional cognitive brain 212
Circulating 205
  endothelial cells (CEC) 205
  progenitors cells (CPC) 205
Citalopram 135
Clinical-biological parameters 191, 192
Cognitive decline 3, 10, 34, 41, 57, 58, 61, 94,
    95, 191, 192, 193, 194, 198, 200, 201,
    206, 212, 217, 218, 219, 220, 227, 229,
    231, 235, 237, 238, 243, 249
  rapid progression of 191, 192, 194, 201, 229,
    235
  deficits 3, 61, 193, 198, 200, 206, 212, 217,
    219, 227, 231, 237, 243, 249
  dysfunction 206, 217, 218, 220, 231, 238
  function 10, 34, 41, 57, 58, 94, 95
Cognitive impairment 98, 157, 158, 180, 192,
    193, 194, 199, 204, 229, 238, 241, 247
  mild 98, 157, 158, 180
  rapid 193, 199, 204, 238, 241
  rapid progression of 192, 194, 229, 247
Comorbidities 192, 214, 216, 235
Complex IV, mitochondrial 14, 15
Corpus striatum (CS) 16, 18
C-reactive protein (CRP) 94, 205, 238
Cytochrome 12, 13, 15, 17, 45, 90
Cytokines 85, 86, 102, 139, 229
  proinflammatory 86, 102, 139, 229

D

Danger-associated molecular patterns
    (DAMPs) 85
Deficiencies, immune 237
Degeneration 7, 8, 53, 84, 221, 222, 223, 224
  axonal 84, 221, 222
  neuronal 7, 8, 53, 223, 224
Derivatives, selenium-containing clioquinol
    107
Dexamethasone 225, 234
Diffuse intraparenchymal hemorrhage 222
Dihydrolipoic acid 180
Dilated cardiomyopathy 217, 231, 232, 233
Discontinuation syndrome 226, 227
Disease 13, 18, 52, 172, 191, 204, 207, 209,
    231, 248, 250
  alzheimer 18, 250
  cardiovascular 191, 192, 231
  mitochondrial muscle 13
  multifactorial 52, 172
  small vessel 204, 207, 209, 248
Disease 6, 9, 33, 34, 36, 38, 49, 61, 171
  -associated mutant proteins 6
  -modifying agents (DMAs) 34, 36, 38, 49,
    61
  -modifying therapeutic (DMTs) 33, 171
  progression, age-related 9
Disorders 154, 173, 207, 226, 227, 228, 229,
    230, 247
  behavioral 154, 173, 227
  depressive 207, 226, 228, 229, 230, 247
  somatic 213, 216, 218, 224, 242
Dizziness 218
DNA, nuclear 13
Donepezil 34, 35, 38, 41, 42, 49, 90, 163, 164,
    165, 167, 169, 171, 173, 174, 176, 188
Dopamine 104, 142, 191, 206, 207
Doxycycline 81, 91, 93, 94, 95
Drugs 34, 45, 61, 91, 177
  anti-leprosy 91
  multi-functional 34, 45

www.ingramcontent.com/pod-product-compliance
Lightning Source LLC
Chambersburg PA
CBHW050817220326
41598CB00006D/234